Introduction to ENVIRONMENTAL TOXICOLOGY

Introduction to ENVIRONMENTAL TOXICOLOGY

Volume–2

Patricia Shaw
Diana Hutchinson

Kruger Brentt
Publishers
2024

Kruger Brentt Publishers UK. LTD.
Company Number 9728962

Regd. Office: 68 St Margarets Road, Edgware, Middlesex HA8 9UU

ISBN: 978-1-78715-204-5 (Vol. 2)

For information on all our publications visit our website at http://krugerbrentt.com/

PREFACE

Environmental toxicology is the study of the effects of chemicals and other substances on living organisms and their environment. It is a multidisciplinary field that combines elements of toxicology, environmental science, and ecology. The impact of toxic substances on the environment has become a growing concern in recent years, and the field of environmental toxicology plays an important role in assessing and mitigating these effects. This essay will explore the field of environmental toxicology, its importance, and some of the challenges it faces. Environmental toxicology plays a critical role in protecting human health and the environment. Toxic substances can contaminate air, water, and soil, and can have serious consequences for both human and animal health. By studying the effects of these substances, environmental toxicologists can identify potential hazards and develop strategies to mitigate their effects. Environmental toxicology is also important for regulatory agencies that set standards for exposure to toxic substances. These standards are designed to protect human health and the environment, and are often based on the results of environmental toxicology studies.

One of the major challenges of environmental toxicology is the complexity of the environment. The effects of toxic substances on living organisms can be influenced by a wide range of factors, including temperature, humidity, and other environmental conditions. This complexity makes it difficult to predict the effects of toxic substances in different environments. Another challenge is the lack of information about the effects of many chemicals on the environment. There are thousands of chemicals in use today, and only a fraction of these have been thoroughly tested for their effects on the environment. This lack of information makes it difficult to assess the risks associated with many chemicals. The future of environmental toxicology will be shaped by advances in technology and a growing awareness of the importance of protecting the environment. New tools and techniques, such as computer modeling and high-throughput screening, will allow researchers to study the effects of toxic substances more efficiently and effectively. These advances will also allow researchers to better predict the effects of toxic substances in different environments. Environmental toxicology is a critical field that plays an important role in protecting human health and the environment. While the field faces many challenges, advances in technology and growing awareness of the importance of environmental protection offer opportunities for progress. By continuing to study the effects of toxic substances on the environment, environmental toxicologists can help ensure a safer and healthier world for future generations.

The present book provides a good general introduction to all the major areas of environmental toxicology,

Fundamental aspects of environmental health, occupational health, detection of pollutants, and risk assessment are discussed. This book provides a thorough, systematic introduction to environmental toxicology and addresses many of the effects of pollutants on humans, animals, and the environment. The book will be of interest to biologists, environmental engineers, chemists, environmental scientists, microbiologists, medical doctors and all academics, professionals, policy makers and practitioners involved in the environmental toxicology and hazardous waste management.

We are grateful to all those persons as well as various books, manuals, periodicals, magazines, journals etc. that helped in the preparation of this book. In spite of the best efforts, it is possible that some errors may have occurred into the compilation and editing of the book. Further queries, constructive suggestions and criticisms for the improvement of the book are always welcomed and shall be thankfully acknowledged.

Patricia Shaw

Diana Hutchinson

CONTENTS

4 Chapter TOXICOLOGY

If you want to re-use this chapter, for e.g. in your electronic learning environment, feel free to copy this url: maken.wikiwijs.nl/120176/4__Toxicology

4.1. TOXICOKINETICS

Toxicology usually distinguishes between toxicokinetics and toxicodynamics. Toxicokinetics involves all processes related to uptake, internal transport and the accumulation inside an organism, while toxicodynamics deals with the interaction of a compound with a receptor, induction of defence mechanisms, damage repair and toxic effects. Of course the two sets of processes may interact, for instance defence may feed-back to uptake and damage may change the internal transport. However, often toxicokinetics analysis just focuses on tracking of the chemical itself and ignores possible toxic effects. This holds up to a critical threshold, the critical body concentration, above which effects become obvious and the normal toxicokinetics analysis is no longer valid. The assumption that toxicokinetic rate parameters are independent of the internal concentration is due the limited amount of information that can be obtained from animals in the environment. However, in so called physiology-based pharmacokinetic and pharmacodynamic models (PBPK models) kinetics and dynamics are analyzed as an integrated whole. The use of such models is however mostly limited to mammals and humans.

It must be emphasized that toxicokinetics considers fluxes and rates, i.e. mg of a substance moving per time unit from one compartment to another. Fluxes may lead to a dynamic equilibrium, i.e. an equilibrium that is due to inflow being equal to outflow; when only the equilibrium conditions are considered, this is called partitioning.

In this Chapter 4.1 we will explore the various approaches in toxicokinetics, including the fluxes of toxicants through individual organisms and through trophic levels as well as the biological processes that determine such fluxes. We start by comparing the concentrations of toxicants between organisms and their environment (section 4.1.1), and between organisms of different trophic levels (section 4.1.6). This leads to the famous concept of bioaccumulation, one of the properties of a substance often leading to environmental problems. While in the past dilution was sometimes seen as a solution to pollution, this is not correct for bioaccumulating substances, since they may turn up elsewhere in the next level food-chain and reach an even higher concentration. The bioaccumulation factor is one of the best-investigated properties characterizing the environmental behaviour of a substance . It may be predicted from properties of the substance such as the octanol-water partitioning coefficient.

In section 4.1.2 we discuss the classical theory of uptake-elimination kinetics using the one-compartment linear model. This theory is a crucial part of toxicological analysis. One of the first things you want to know about a substance is how quickly it enters an organism and how quickly it is removed. Since toxicity is basically a time-dependent process, the turnover rate of the internal concentration and the build-up of a residue depend upon the exposure time. An understanding of toxicokinetics is therefore critical to any interpretation of a toxicity experiment. Rate parameters may partly be predicted from substance properties, but properties of the organism play a much greater role here. One of these is simply the body mass; prediction of elimination rate constants from body mass is done by allometric scaling relationships, explored in section 4.1.5.

In two sections, 4.1.3 and 4.1.4, we present the biological processes that underlie the turnover of toxicants in an organism. These are very different for metals than for organic substances, hence, two separate sections are devoted to this topic, one on tissue accumulation of metals and one on defence mechanisms for organic xenobiotics.

Finally, if we understand all toxicokinetic processes we will also be able to understand whether the concentration inside a target organ will stay below or just passes the threshold that can be tolerated. The critical body concentration, explored in section 4.1.7 is an important concept linking toxicokinetics to toxicity.

4.1.1. Bioaccumulation

Introduction: Terminology for Bioaccumulation

The term bioaccumulation describes the transfer and accumulation of a chemical from the environment into an organism." For a chemical like hexachlorobenzene, the concentration in fish is more than 10,000 times higher than in water, which is a clear illustration of "bioaccumulation". A chemical like hexachlorobenzene is hydrophobic, so has a very low aqueous solubility. It therefore prefers to escape the aqueous phase to enter (or partition into) a more lipophilic phase such as the lipid phase in biota.

Uptake may take place from different sources. Fish mostly take up chemicals from the aqueous phase, organisms living at the sediment-water interphase are exposed via the overlying water and sediment particles, organisms living in soil or sediment via pore water and by ingesting soil or sediment, while predators will be exposed via their food. In many cases, uptake is related to more than one source. The different uptake routes are also reflected in the parameters and terminology used in bioaccumulation studies. The different parameters include the bioconcentration factor (BCF), bioaccumulation factor (BAF), biomagnification factor (BMF) and biota-to-sediment or biota-to-soil accumulation factor (BSAF). Figure 1 summarizes the definition of these parameters. Bioconcentration refers to uptake from the aqueous phase, bioaccumulation to uptake via both the aqueous phase and the ingestion of sediment or soil particles, while biomagnification expresses the accumulation of contaminants from food.

Source		Organism	Uptake	Parameter		
Water C_a	→	Fish C_{org}	Uptake from water only	Bioconcentration factor	BCF (L/kg)	C_{org} / C_{aq}
Sediment C_s / Pore water C_a	→ →	Sediment organism C_{org}	Uptake from water and sediment or soil	Bioaccumulation factor	BAF (L/kg)	C_{org} / C_{aq}
Sediment C_s / Pore water C_a	→ →	Sediment organism C_{org}	Uptake from water and sediment or soil	Biota-to-sediment or Biota-to-soil accumulation factor	BSAF (kg soil / kg org.)	C_{org} / C_{sed} or C_{org} / C_{soil}
Food C_{food}	→	Fish or birds C_{org}	Uptake from food	Biomagnification factor	BMF (kg food/ kg org.)	C_{org} / C_{food}

Figure 1. Parameters used to describe the bioaccumulation of chemicals.

C_{aq} concentration in water (aqueous phase)

C_{org} concentration in organism

C_f concentration in food

C_s concentration in sediment or soil

Please note that the bioaccumulation factor (BAF) is defined in a similar way as the bioconcentration Factor (BCF), but that uptake can be both from the aqueous phase and the sediment or soil and that the exposure concentration usually is expressed per kg dry sediment or soil. Other definitions of the BAF are possible, but we have followed the one from Mackay et al. (2013). "The bioaccumulation factor (BAF) is defined here in a similar fashion as the BCF; in other words, BAF is CF/CW at steady state, except that in this case the fish is exposed to both water and food; thus, an additional input of chemical from dietary assimilation takes place".

All bioaccumulation factors are steady state constants: the concentration in the organism constant and the organisms is in equilibrium with its surrounding phase. It will take time before such an steady state is reached. Steady state is reached when uptake rate (for example from an aqueous phase) equals the elimination rate. Models that include the factor time in describing the uptake are called kinetic models; see section on Bioaccumulation kinetics.

Effect of Biological Properties on Accumulation

Uptake of chemicals is determined by properties of both the organism and the chemical. For xenobiotic lipophilic chemicals in water, organism-specific factors usually play a minor role and concentrations in organisms can pretty well be predicted from chemical properties (see section on Structure-property relationships).

For metals, on the contrary, uptake is to a large extent determined by properties of the organism, and a direct consequence of its mineral requirements. A chemical with low bioavailability (low uptake compared to concentration in exposure medium) may nevertheless accumulate to high levels when the organism is not capable of excreting or metabolising the chemical.

Factors related to the organism are:

- **Fat content:** Because lipophilic chemicals mainly accumulate in the fat of organisms, it is reasonable to assume that lipid rich organisms will have higher concentrations of lipophilic chemicals. See Figure 2 for an example of bioconcentration factors of 1,2,4-trichlorobenzene in a number of organisms with varying lipid content. This was one of the explanations for the high PCB levels in eel (high lipid content) in Dutch rivers and seals from the Wadden Sea. Nevertheless, large differences are still found between species when concentrations are expressed on a lipid basis. This may e.g. be explained by the fact that lipids are not all identical: PCBs seem to dissolve better in lipids from anchovies than in lipids from algae (see data in Table 1). Also more recent research has shown that not all lipids are the same and that differences in bioaccumulation between species may be due to differences in lipid composition (Van der Heijden and Jonker, 2011). Related to this is the development of BCF models that are based on multiple compartments and that make a separation between storage and membrane lipids and also include a protein fraction as additional sink (Armitage et al.,

2013). In this model, the overall distribution coefficient DBW (or BCF) is estimated via equation 1. The equation is using distribution coefficients because the model also "accounts for the presence of neutral and charged chemical species" (Armitage et al., 2013).

$$BCF \quad or \quad D_{BW} = f_{SL}D_{SL-W} + f_{ML}D_{ML-W} + f_{NLOM}D_{NLOM-W} + f_W$$

DBW	overall organism-water distribution coefficient (or surrogate BCF) at a given pH
DSL-W	storage lipid-water distribution ratio
DML-W	membrane lipid-water distribution ratio
DNLOM-W	sorption coefficient to NLOM (non-lipid organic matter, for example proteins)
fSL	fraction of storage lipids
fML	fraction of membrane lipids
fNLOM	fraction of non-lipid organic matter (e.g. proteins, carbohydrates)
fW	fraction of water

- **Sex:** Chemicals (such as DDT, PCB) accumulated in milk fat may be transferred to juveniles upon lactation. This was found in marine mammals. In this way, females have an additional excretion mechanism. A typical example is shown in Figure 3, taken from a study of Abarnou et al. (1986) on the levels of organochlorinated compounds in the Antarctic dolphin Cephalorhyncus commersonii. In males, concentrations increase with increasing age, but concentrations in mature females decrease with increasing age.

- **Weight**: (body mass) of the organism relative to the surface area across which exchange with the water phase takes place. Smaller organisms have a larger surface-to-volume ratio and the exchange with the surrounding aqueous phase is faster. Therefore, although lipid normalized concentrations will be the same at equilibrium, this equilibrium is reached earlier in smaller than in larger organisms.

- **Difference in uptake route.** The relative importance of uptake through the skin (in fish e.g. the gills) and (oral) uptake through the digestive system. It is generally accepted that for most free-living organism's direct uptake from the water dominates over uptake after digestion in the digestive tract.

- **Metabolic activity**. Even at the same weight or the same age, the balance between uptake and excretion may change due to an increased metabolic activity, e.g. in times of fast growth or high reproductive activity.

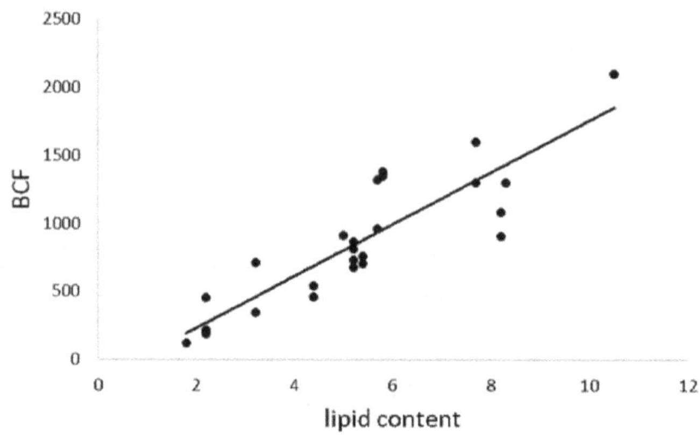

Figure 2. The influence of lipid content on the bioconcentration of 1,2,4-trichlorobenzene in different fish species (reproduced using data from Geyer et al., 1985).

Table 1. Mean PCB concentrations in algae (Dunaliella spec.), rotifers (Brachionus plicatilis) and anchovies larvae (Angraulis mordax), expressed on a dry-weight basis and on a lipid basis. From Moriarty (1983).

Organism	Lipid content (%)	PCB-concentration based on dry weight (μg g-1)	PCB-concentration based on lipid weight (μg g-1)	BCF based on concentration in the lipid phase
algae	6.4	0.25	3.91	0.48 x 106
rotifer	15.0	0.42	2.80	0.34 x 106
fish (anchovies) larvae	7.5	2.06	27.46	13.70 x 106

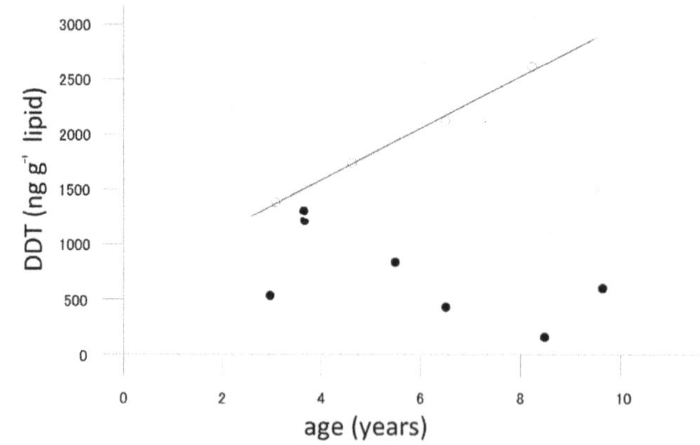

Figure 3. Concentrations of DDT in dolphins of different age and the difference between male and female dolphins Redrawn from Abernou et al. (1986) by Wilma IJzerman.

Cited References

Abarnou, A., Robineau, D., Michel, P. (1986). Organochlorine contamination of commersons dolphin from the Kerguelen islands. Oceanologica Acta 9, 19-29.

Armitage, J.M., Arnot, J.A., Wania, F., Mackay, D. (2013). Development and evaluation of a mechanistic bioconcentration model for ionogenic organic chemicals in fish. Environmental Toxicology and Chemistry 32, 115-128.

Geyer, H., Scheunert, I., Korte, F. (1985). Relationship between the lipid-content of fish and their bioconcentration potential of 1,2,4-trichlorobenzene. Chemosphere 14, 545-555.

Mackay, D., Arnot, J.A., Gobas, F., Powell, D.E. (2013). Mathematical relationships between metrics of chemical bioaccumulation in fish. Environmental Toxicology and Chemistry 32, 1459-1466.

Moriarty, F. (1983). Ecotoxicology: The Study of Pollutants in Ecosystems. Publisher: Academic Press, London.

Van der Heijden, S.A., Jonker, M.T.O. (2011). Intra- and interspecies variation in bioconcentration potential of polychlorinated biphenyls: are all lipids equal? Environmental Science and Technology 45, 10408-10414.

Suggested Reading

Mackay, D., Fraser, A. (2000). Bioaccumulation of persistent organic chemicals: Mechanisms and models. Environmental Pollution 110, 375-391.

Van Leeuwen, C.J., Vermeire, T.G. (Eds.) (2007). Risk Assessment of Chemicals: An Introduction. Springer, Dordrecht, The Netherlands. Chapter 3.

4.1.2. Toxicokinetics

In the section "Bioaccumulation", the process of bioaccumulation is presented as a steady state process. Differences in the bioaccumulation between chemicals are expressed via, for example, the bioconcentration factor BCF. The BCF represents the ratio of the chemical concentration in, for instance, a fish versus the aqueous concentration at a situation where the concentrations in water and fish do not change in time.

$$BCF = \frac{Corg}{Caq} \qquad (1)$$

where:

C_{aq} concentration in water (aqueous phase) (mg/L)

C_{org} concentration in organism (mg/kg)

The unit of BCF is L/kg.

Kinetic Models

Steady state can be established in a simple laboratory set-up where fish are exposed to a chemical at a constant concentration in the aqueous phase. From the start of the exposure (time 0, or t=0), it will take time for the chemical concentration in the fish to reach steady state and in some cases, this will not be established within the exposure period. In the environment, exposure concentrations may fluctuate and, in such scenarios, constant concentrations in the organism will often not be established. Steady state is reached when the uptake rate (for example from an aqueous phase) equals the elimination rate. Models that include the factor time in describing the uptake of chemicals in organisms are called kinetic models.

Toxicokinetic models for the uptake of chemicals into fish are based on a number of processes for uptake and elimination. An overview of these processes is presented in Figure 1. In the case of fish, the major process of uptake is by diffusion from the surrounding water compartment via the gill to the blood. Elimination can be via different processes: diffusion via the gill from blood to the surrounding water compartment, via transfer to offspring or eggs by reproduction, by growth (dilution) and by internal degradation of the chemical (biotransformation).

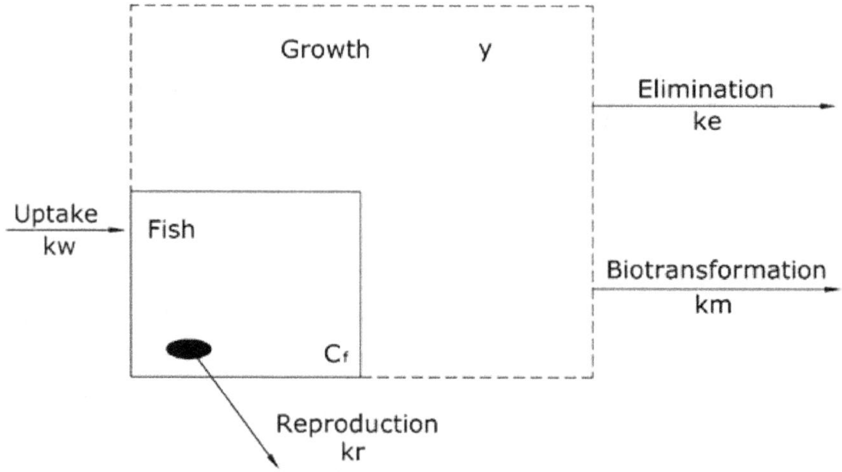

Figure 1. Uptake and elimination processes in fish and the rate constants (k) for each process. Reproduced from Van Leeuwen and Vermeire (2007) by Wilma IJzerman.

Kinetic models to describe uptake of chemicals into organisms are relatively simple with the following assumptions:

First order kinetics:

Rates of exchange are proportional to the concentration. The change in concentration with time (dC/dt) is related to the concentration and a rate constant (k):

$$\frac{dC}{dt} = kC \tag{2}$$

One compartment:

It is often assumed that an organism consists of only one single compartment and that the chemical is homogeneously distributed within the organism. For "simple" small organisms this assumption is intuitively valid, but for large fish this assumption looks unrealistic. But still, this simple model seems to work well also for fish. To describe the internal distribution of a chemical within fish, more sophisticated kinetic models are needed, similar to the ones applied in mammalian studies. These more complex models are the "physiologically based toxicokinetic" (PBTK) models (Clewell, 1995; Nichols et al., 2004)

Equations for the kinetics of accumulation process

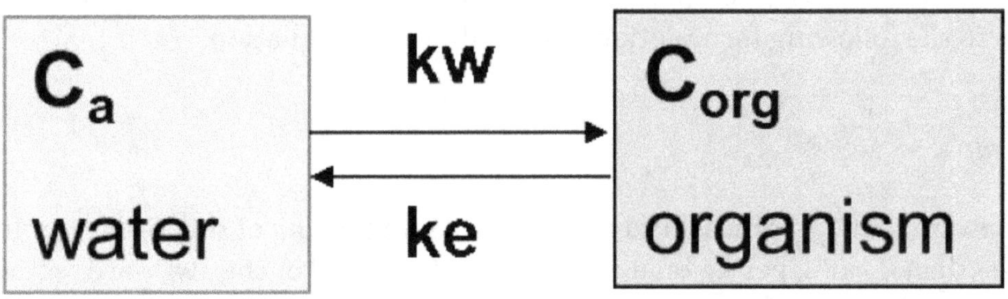

The accumulation process can be described as the sum of rates for uptake and elimination.

$$\frac{dC_{org}}{dt} = k_w C_{aq} - k_e C_{org} \tag{3}$$

Integration of this differential equation leads to equation 4.

$$C_{org}(t) = \frac{k_w}{k_e} C_{aq} (1 - e^{k_e t}) \tag{4}$$

C_{org} concentration in organisms (mg/kg)

C_{aq} concentration in aqueous phase (mg/L)

k_w uptake rate constant (L/kg·day)

k_e elimination rate constant (1/day)

t time (day)

(dimensions used are: amount of chemical: mg; volume of water: L; weight of organism: kg; time: day); see box.

Box: The units of toxicokinetic rate constants

The differential equation underlying toxicokinetic analysis is basically a mass balance equation, specifying conservation of mass. A mass balance implies that the amount of chemical is expressed in absolute units such as mg. If Q is the amount in the animal and F the amount in the environmental compartment the mass balance reads:

$$\frac{dQ}{dt} = k_1^{'} F(t) - k_2 Q(t)$$

where is the uptake rate constant and k_2 the elimination rate constant, both with dimension time-1. However, it is often more practical to work with the concentration in the animal (e.g. expressed in mg/kg). This can be achieved by dividing the left and right sides of the equation by w, the body-weight of the animal and defining Cint = Q/w. In addition, we define the external concentration as Cenv = F/V, where V is the volume (L or kg) of the environmental compartment. This leads to the following formulation of the differential equation:

$$\frac{dC_{int}}{dt} = k_1^{'} \frac{V}{w} C_{env} - k_2 C_{int}$$

Beware that C_{env} is measured in other units (mg per kg of soil, or mg per litre of water) than C_{int} (mg per kg of animal tissue). To get rid of the awkward factor V/w it is convenient to define a new rate constant, k_1:

$$k_1 = \frac{V}{w} k_1^{'}$$

This is the uptake rate constant usually reported in scientific papers. Note that it has other units than : it is expressed as kg of soil per kg of animal tissue per time unit (kg kg-1 h-1), and in the case of water exposure as L kg-1 h-1. The dimension of k2 remains the same whether mass or concentrations are used (time-1). We also learn from this analysis that when dealing with concentrations, the body-weight of the animal must remain constant.

References

Moriarty, F. (1984) Persistent contaminants, compartmental models and concentration along food-chains. Ecological Bulletins 36: 35-45.

Skip, B., A.J. Bednarska, & R. Laskowski (2014) Toxicokinetics of metals in terrestrial invertebrates: making things straight with the one-compartment principle. PLoS ONE 9(9): e108740.

Equation 4 describes the whole process with the corresponding graphical representation of the uptake graph (Figure 2).

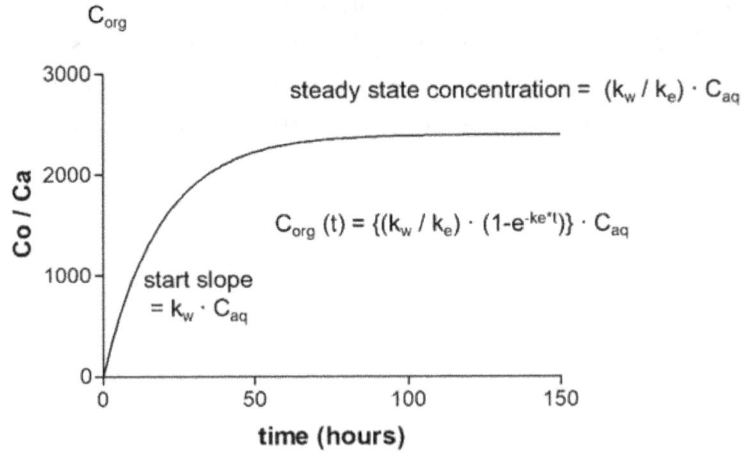

Figure 2. The basic equation for uptake of a chemical from the aqueous phase to a fish: one compartment model with first order kinetics.

The concentration in the organism is the result of the net process of uptake and elimination. At the initial phase of the accumulation process, elimination is negligible and the ratio of the concentration in the organism is given by:

$$\frac{dC_{org}}{dt} = k_w C_{aq} \tag{5}$$

$$C_{org}(t) = k_w C_{aq} t \tag{6}$$

Steady State

After longer exposure time, elimination becomes more substantial and the uptake curve starts to level off. At some point, the uptake rate equals the elimination rate and the ratio C_{org}/C_{aq} becomes constant. This is the steady state situation. The constant Corg/Caq at steady state is called the bioconcentration factor BCF. Mathematically, the BCF can also be calculated from k_w/k_e. This follows directly from equation 4: after long exposure time (t), eket becomes 0 leading to

$$C_{org}(t) = \frac{k_w}{k_e} C_{aq}$$

$$\frac{C_{org}}{C_{aq}}(at \quad steady \quad state) = BCF = \frac{k_w}{k_e} \tag{7}$$

Elimination

Elimination is often measured following an uptake experiment. After the organism has reached a certain concentration, fish are transferred to a clean environment and concentration in the organism will decrease in time. Because this is also a first order kinetic process, the elimination rate will depend on the concentration in the organism (C_{org}) and the elimination rate constant (k_e) (see equation 8). Concentration will decrease exponentially in time (equation 9) as shown in Figure 3A. Concentrations are often transformed to the natural logarithmic values (ln Corg) because this results in a linear relationship with slope -ke.(equation 10 and figure 3B).

$$\frac{dC_{org}}{dt} = -k_e C_{org} \tag{8}$$

$$C_{org}(t) = C_{org}(t=0)e^{-k_e t} \tag{9}$$

$$lnC_{org}(t) = lnC_{org}(t=0) - k_e t \tag{10}$$

Where Corg(t=0) is the concentration in the organism when the elimination phase starts.

The half-life ($T_{1/2}$ or DT_{50}) is the time needed to eliminate half the amount of chemical from the compartment. The relationship between ke and T1/2 is: T1/2 = (In 2) / ke. The half-life increases when ke decreases.

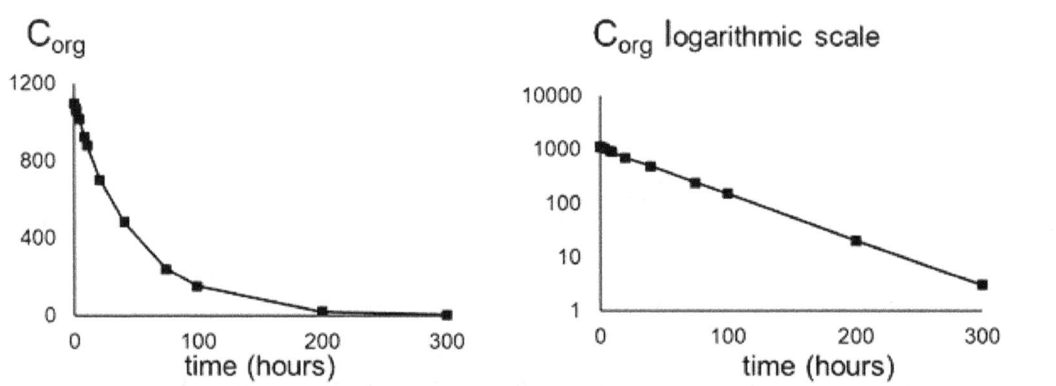

Figure 3. Elimination of a chemical from fish to water: one compartment model. Left: concentrations given on a normal scale. Right: concentration expressed as the natural logarithm to enable linear regression against time to yield the rate constant as the slope.

Multicompartment Models

Very often, organisms cannot be considered as one compartment, but as two or even more compartments (Figure 4A). Deviations from the one-compartment system usually are seen when elimination does not follow an exponential pattern as expected: no linear relationship is obtained after logarithmic transformation.

Figure 4B shows the typical trend of the elimination in a two-compartment system. The decrease in concentration (on a logarithmic scale) shows two phases: phase I with a relatively fast decrease and phase II with a relatively slow decrease. According to the linear compartment theory, elimination may be described as the sum of two (or more) exponential terms, like:

$$C_{org}(t) = C_{org}(t=0)F(I)e^{-k_e(I)*t} + F(II)e^{-k_e(II)*t} \quad (11)$$

where

k_e(I) and ke(II) represent elimination rate constants for compartment I and II,

F(I) and F(II) are the size of the compartments (as fractions)

Typical examples of two compartment systems are:

- Blood (I) and liver (II)
- Liver tissue (I) and fat tissue (II)

Elimination from fat tissue is often slower than from, for example, the liver. The liver is a well perfused organ while the exchange between lipid tissue and blood is much less. That explains the faster elimination from the liver.

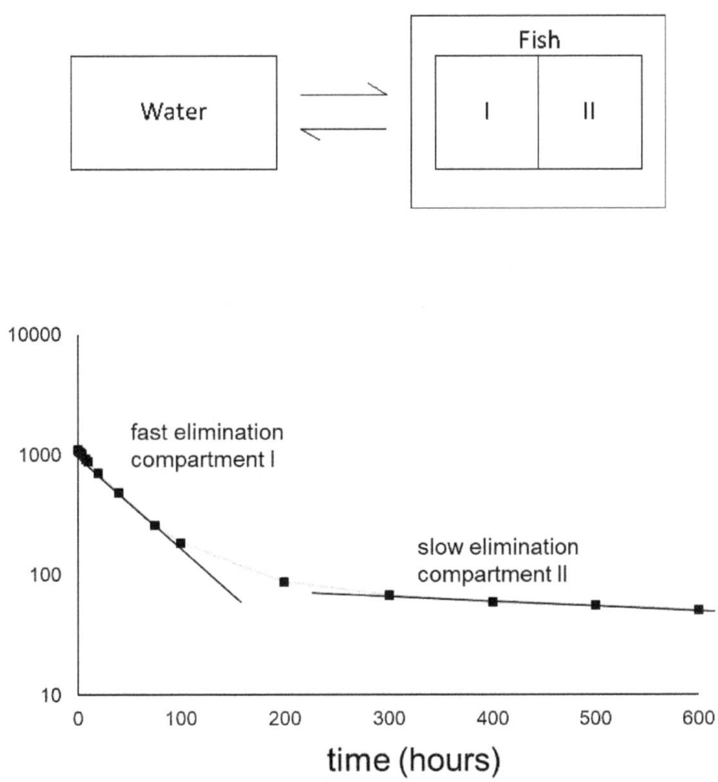

Figure 4. Elimination of chemical from fish to water (top) in a two-compartment model (bottom).

Examples of Uptake Curves for Different Chemicals and Organisms

Figure 5 gives uptake curves for two different chemicals and the corresponding kinetic parameters. Chemical 2 has a BCF of 1000, chemical 1 a BCF of 10,000. Uptake rates (k_w) are the same, which is often the case for organic chemicals. Half-lives (time to reach 50 % of the steady state level) are 14 and 140 hours. This makes sense because it will take a longer time to reach steady state for a chemical with a higher BCF. Elimination rate constants also differ a factor of 10.

In figure 6, uptake curves are presented for one chemical, but in two organisms of different size/weight. Organism 1 is much smaller than organism 2 and reaches steady state much earlier. $T_{1/2}$ values for the chemical in organisms 1 and 2 are 14 and 140 hours, respectively. The small size explains this fast equilibration. Rates of uptake depend on the surface-to-volume ratio (S/V) of an organism, which is much higher for a small organism. Therefore, kinetics in small organisms is faster resulting in shorter equilibration times. The effect of size on kinetics is discussed in more detail in Hendriks et al. (2001) and in the Section on Allometric Relationships.

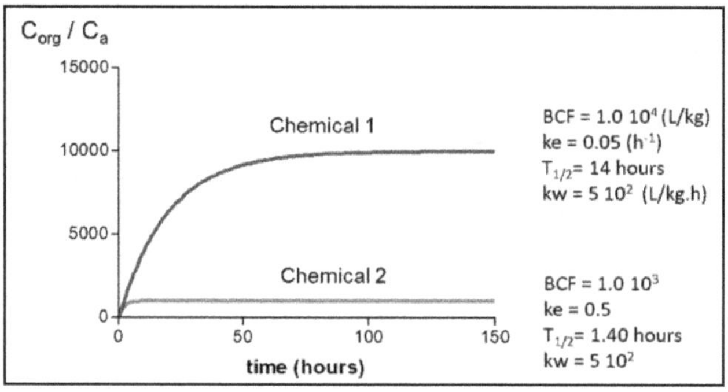

Figure 5. Uptake curves for two chemicals, having different properties, in the same organism.

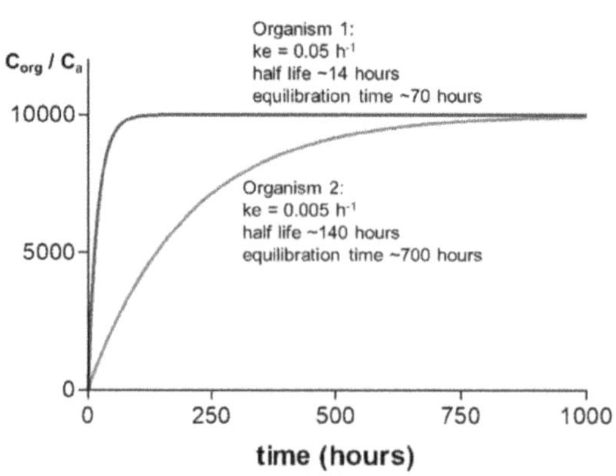

Figure 6. Uptake curves for the same chemical in two organisms having different sizes; organism 2 is much bigger than organism 1.

Bioaccumulation Involving Biotransformation and Different Routes of Uptake

In equation 2, elimination only includes gill elimination. If other processes such as biotransformation and growth are considered, the equation can be extended to include these additional processes (see equation 12).

$$\frac{dC_{org}}{dt} = k_w C_a - (k_e + k_m + k_r) C_{org} \qquad (12)$$

For organisms living in soil or sediment, different routes of uptake may be of importance: dermal (across the skin), or oral (by ingestion of food and/or soil or sediment particles). Mathematically, the uptake in an organism in sediment can be described as in equation 13.

$$C_{org}(t) = \frac{(k_w C_{aq} + k_s C_s)}{k_e}(1 - e^{-k_e t}) \qquad (13)$$

C_{org}	concentration in organisms (mg/kg)
C_{aq}	concentration in aqueous phase (mg/L)
C_s	concentration soil or sediment (mg/kg)
k_w	uptake rate constant from water (L/kg/day)
k_s	uptake rate constant from soil or sediment (kgsoil/kgorganism/day)
k_e	elimination rate constant (1/day)
t	time (day)

(dimensions used are: amount of chemical: mg; volume of water: L; weight of organism: kg; time: day)

In this equation, kw and ks are the uptake rate constants from water and sediment, ke is the elimination rate constant and Caq and Cs are the concentrations in water and sediment or soil. For soil organisms, such as earthworms, oral uptake appears to become more important with increasing hydrophobicity of the chemical (Jager et al., 2003). This is because the concentration in soil (Cs) will become higher than the porewater concentration Ca for the more hydrophobic chemicals (see section on Sorption).

References

Clewell, H.J., 3rd (1995). The application of physiologically based pharmacokinetic modeling in human health risk assessment of hazardous substances. Toxicology Letters 79, 207-217.

Hendriks, A.J., van der Linde, A., Cornelissen, G., Sijm, D. (2001). The power of size. 1. Rate constants and equilibrium ratios for accumulation of organic substances related to octanol-water partition ratio and species weight. Environmental Toxicology and Chemistry 20, 1399-1420.

Jager, T., Fleuren, R., Hogendoorn, E.A., De Korte, G. (2003). Elucidating the routes of exposure for organic chemicals in the earthworm, Eisenia andrei (Oligochaeta). Environmental Science and Technology 37, 3399-3404.

Nichols, J.W., Fitzsimmons, P.N., Whiteman, F.W. (2004). A physiologically based toxicokinetic model for dietary uptake of hydrophobic organic compounds by fish - II. Simulation of chronic exposure scenarios. Toxicological Sciences 77, 219-229.

Van Leeuwen, C.J., Vermeire, T.G. (Eds.) (2007). Risk Assessment of Chemicals: An Introduction. Springer, Dordrecht, The Netherlands.

4.1.3. Tissue Accumulation of Metals

Synopsis

The issue of metal speciation, which is crucially important to understand metal fate in the environment, is equally important for internal distribution in organisms and toxicity inside the cell. Many metals tend to accumulate in specific organs, for example the hepatopancreas of crustaceans, the chloragogen tissue of annelids, and the kidney of mammals. In addition, there is often a specific organ or tissue where dysfunction or toxicity is first observed, e.g., in the human body, the primary effects of chronic exposure to mercury are seen in the brain, for lead in bone marrow and for cadmium in the kidney. This module is aimed at increasing the insight into the different mechanisms by which metals accumulate in biological tissues.

Introduction

Metals will be present in biological tissues in a large variety of chemical forms: free metal ion, various inorganic species with widely varying solubility, such as chlorides or carbonates, plus all kind of metal species bound to low-molecular and high-molecular weight biotic ligands. The free metal ion is considered the most relevant species relating to toxicity.

To explain the affinities of metals with specific targets, a system has been proposed based on the physical properties of the ion; according to this system, metals are divided into "oxygen-seeking metals" (class A, e.g. lithium, beryllium, calcium and lanthanum), and "sulfur-seeking metals" (class B, e.g. silver, mercury and lead) (See section on Metals and metalloids). However, most of the metals of environmental relevance fall in an intermediate class, called "borderline" (chromium, cadmium, copper, zinc, etc.). This classification is to some extent predictive of the binding of metals to specific cellular targets, such as SH-groups in proteins, nitrogen in histidine or carbonates in bone tissue.

Not only do metals differ enormously in their physicochemical properties, also the organisms themselves differ widely in the way they deal with metals. The type of ligand to which a metal is bound, and how this ligand is transported or stored in the body, determines to a great extent where the metal will accumulate and cause toxicity. Sensitive targets or critical biochemical processes differ between species and this may also lead to differential toxicity.

Inorganic Metal Binding

Many tissues contain "mineral concretions", that is, granules with a specific mineral composition that, due to the nature of the mineral, attract different metals. Especially the gut epithelium of invertebrates, and their digestive glands (hepatopancreas, midgut gland, chloragogen tissue) may be full of such concretions. Four classes of granules are distinguished (Figure 1):

- Calcium-pyrophosphate granules with magnesium, manganese, often also zinc, cadmium, lead and iron
- Sulfur granules with copper, sometimes also cadmium
- Iron granules, containing exclusively iron
- Calcium carbonate granules, containing mostly calcium

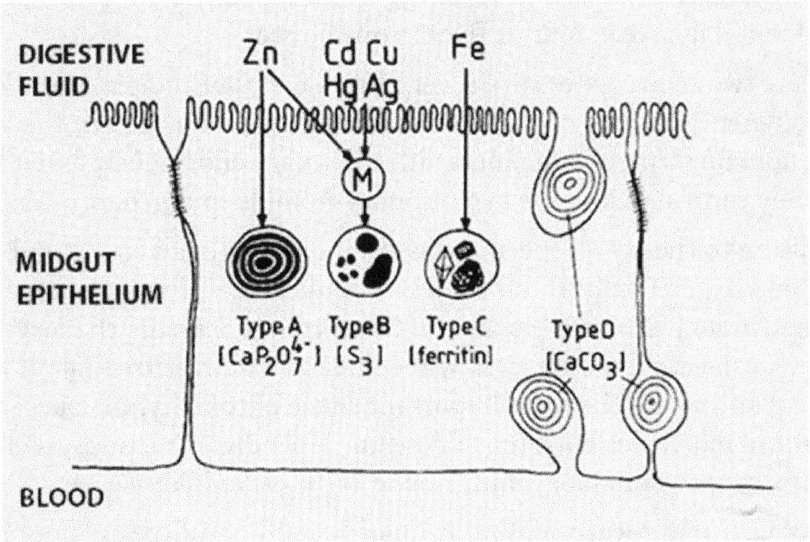

Figure 1. Schematic diagram of the gut epithelium of an invertebrate animal, showing four different types of granules and the metals they usually contain. Redrawn from Hopkin (1989) by Wilma IJzerman.

The type B granules are assumed to be lysosomal vesicles that have absorbed metal-loaded peptides such as metallothionein or phytochelatin, and have developed into inorganic granules by degrading almost all organic material; the high sulfur content derives from the cysteine residues in the peptides.

Tissues or cells that specialize in the synthesis of intracellular granules are also the places where metals tend to accumulate. Well-known are the "S cells" in the hepatopancreas of isopods. These cells (small cells, B-type cells sensu Hopkin 1989) contain very large amounts of copper. Most likely the large stores of copper in woodlice and other crustaceans relate to their use of hemocyanin, a copper-dependent protein, as an oxygen-transporting molecule. Similar tissues with high loadings of mineral concretions have been described for earthworms, snails, collembolans and insects.

Organic Metal Binding

The second class of metal-binding ligands is of organic nature. Many plants but also several animals synthesize a peptide called phytochelatin (PC). This is an oligomer derived from glutathione with the three amino acids, γ-glumatic acid, cysteine and glycine, arranged in the following way: (γ-glu-cys)n-gly, where n can vary from 2 to 11. The thiol groups of several cysteine residues are involved in metal binding.

The other main organic ligand for metals is metallothionein (MT). This is a low-molecular weight protein with hydrophilic properties and an unusually large number of cysteine residues. Several cysteines (usually nine or ten) can bind a number of metal ions (e.g. four or five) in one cluster.

There are two such clusters in the vertebrate metallothionein. Metallothioneins occur throughout the tree of life, from bacteria to mammals, but the amino acid sequence, domain structure and metal affinities vary enormously and it is doubtful whether they represent a single evolutionary-homologous group.

In addition to these two specific classes of organic ligands, MT and PC, metals will also bind aspecifically to all kind of cellular constituents, such as cell wall components, albumen in the blood, etc. Often this represents the largest store of metals; such aspecific binding sites will constantly deliver free metal ions to the cellular pool and so are the most important cause of toxicity. Of course metals are also present in molecules with specific metal-dependent functions, such as iron in hemoglobin, copper in hemocyanin, zinc in carbonic anhydrase, etc.

The distinction between inorganic ligands and organic ones is not as strict as it may seem. After binding to metallothionein or phytochelatin metals may be transferred to a more permanent storage compartment, such as the intracellular granules mentioned above, or they may be excreted.

Regulation of Metal Binding

Free metal ions are strong inducers of stress response pathways. This can be due to the metal ion itself but more often the stress response is triggered by a metal-induced disturbance of the redox state, i.e. an induction of oxidative stress. The stress response often involves the synthesis of metal-binding ligands such as phytochelatin and metallothionein. Because this removes metal ions from the active pool it is also called metal scavenging.

The binding capacity of phytochelatin is enhanced by activation of the enzyme phytochelatin synthase (PC synthase). According to one model of its action, the C-terminus of the enzyme has a "metal sensor" consisting of a number of cysteines with free SH-groups. Any metal ions reacting with this nucleophile center (and cadmium is a strong reactant) will activate the enzyme which then catalyzes the reaction from (γ-glu-cys)n-gly to (γ-glu-cys)n+1-gly, thus increasing the binding capacity of cellular phytochelatin (Figure 2). This reaction of course relies on the presence of sufficient glutathione in the cell. In plants the PC-metal complex is transported into the central vacuole, where it can be stabilized through incorporation of acid-labile sulfur (S2). The PC moiety is degraded, resulting in the formation of inorganic metal-sulfide crystallites. Alternatively, complexes of metals with organic acids may be formed (e.g. citrates or oxalates). The fate of metal-loaded PC in animal cells is not known, but it might be absorbed in the lysosomal system to form B-type granules (see above).

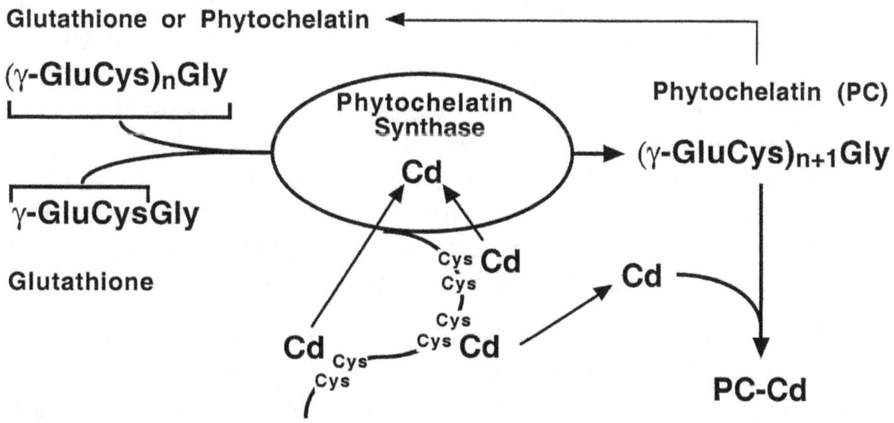

Figure 2. Model for the regulation of phytochelatin synthase, an enzyme catalyzing the extension of phytochelatin and increasing the binding capacity for metals. From Cobbet (2000). Source: http://www.plantphysiol.org/content/123/3/825

The upregulation of metallothionein (MT) occurs in a quite different manner, since it depends on de novo synthesis of the apoprotein. It is a classic example of gene regulation contributing to protection of the cell. In a wide variety of animals, including vertebrates and invertebrates, metallothionein genes (Mt) are activated by a transcription factor called metal-responsive transcription factor 1 (MTF-1). MTF -1 binds to so-called metal-responsive elements (MREs) in the promoter of Mt. MREs are short motives with a characteristic base-pair sequence that form the core of a transcription factor binding site in the DNA. Under normal physiological conditions MTF-1 is inactive and unable to induce Mt. However, it may be activated by Zn2+ ions, which are released, from unspecified ligands, by metals such as cadmium that can replace zinc (Figure 3).

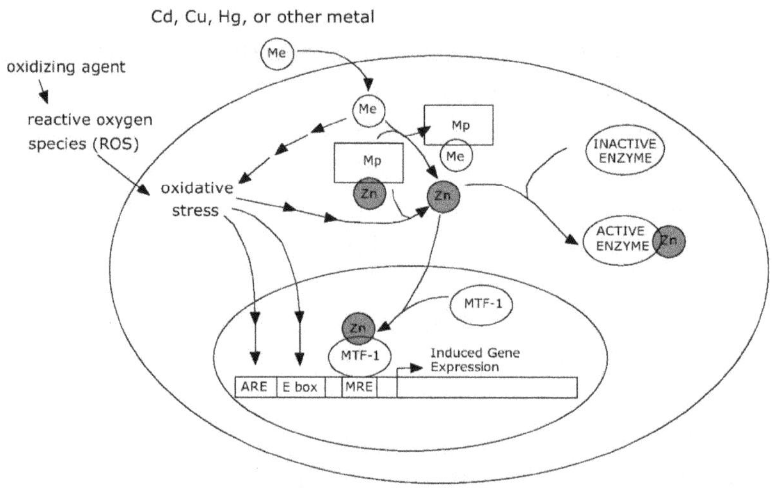

Figure 3. Model for metallothionein induction by cadmium. Cadmium ions (Me) displace zinc ions from ligands (MP). Free zinc ions (Zn) then activate Metal Transcription Factor 1 (MTF-1) in the nucleus to bind to transcription factor binding sites containing Metal-Responsive Elements (MRE). This induces expression of Mt and other genes. These genes may also be induced by oxidative stress, acting upon antioxidant-responsive elements (ARE) and Enhancer Box (E box). Oxidative stress may come from external oxidizing agents or from the metal itself. Finally, zinc ions may also directly activate antioxidant enzymes. Adapted from Haq et al. (2003) by Wilma IJzerman.

It must be emphasized that the model discussed above is inspired by work on vertebrates. Arthropods (Drosophila, Orchesella, Daphnia) could have a similar mechanism since they also have an MTF-1 homolog that activates Mt, however, the situation for other invertebrates such as annelids and gastropods is unclear; their Mt genes seem to lack MREs, despite being inducible by cadmium. In addition, the variability of metallothioneins in invertebrates is extremely large and not all metal-binding proteins may be orthologs of the vertebrate metallothionein. In snails, a cadmium-binding, cadmium-induced MT functions alongside a copper-binding MT while the two MTs have different tissue distributions and are also regulated quite differently.

While both phytochelatin and metallothionein will sequester essential as well as non-essential metals (e.g. Cd) and so contribute to detoxification, the widespread presence of these systems throughout the tree of life suggests that they did not evolve primarily to deal with anthropogenic metal pollution. The very strong inducibility of these systems by non-essential elements like cadmium may be considered a side-effect of a different primary function, for example regulation of the cellular redox state or binding of essential metals.

Target Organs

Any tissue-specific accumulation of metals can be explained by turnover of metal-binding ligands. For example, accumulation of cadmium in mammalian kidney is due to the fact that metallothionein loaded with cadmium cannot be excreted. High concentrations of metals in the hind segments of earthworms are due to the presence of "residual bodies" which are fully packed with intracellular granules. Accumulation of cadmium in the human prostrate is due to the high concentration of zinc citrate in this organ, which serves to protect contractile proteins in sperm tails from oxidation; cadmium assumedly enters the prostrate through zinc transporters.

It is often stated that essential metals are subject to regulatory mechanisms, which would imply that their body burden, over a large range of external exposures, is constant. However, not all "essential" metals are regulated to the extent that the whole-body concentration is kept constant. Many invertebrates have body compartments associated with the gut (midgut gland, hepatopancreas, Malpighian tubules) in which metals, often in the form of mineral concretions, are inactivated and stored permanently or exchanged very slowly with the active pool. Since these compartments are outside the reach of regulatory mechanisms but usually not separated in whole-body metal analysis, the body burden as a whole is not constant. Some invertebrates even carry a "backpack" of metals accumulating over life. This holds, e.g., for zinc in barnacles, copper in isopods and zinc in earthworms.

Accumulation of metals in target organs may lead to toxicity when the critical binding or excretion capacity is exhausted and metal ions start binding aspecifically to cellular constituents. The organ in which this happens is often called the target organ. The total metal concentration at which toxicity starts to become apparent is called the critical body concentration (CBC) or critical tissue concentration. For example, the critical concentration for cadmium in kidney, above which renal damage is observed to occur, is estimated to be 50 μg/g. A list of critical organs for metals in the human body is given in Table 1.

The concept of CBC assumes that the complete metal load in an organ is in equilibrium with the active fraction causing toxicity and that there is no permanent storage pool. In the case of storage detoxification the body burden at which toxicity appears will depend on the accumulation history.

Table 1. Critical organs for chronic toxicity of metals in the human body

Metal or Metalloid	Critical Organ	Symptoms
Al	Brain	Alzheimer's disease
As	Lung, liver, heart gut	Multisystem energy disturbance
Cd	Kidney, liver	Kidney damage
Cr	Skin, lung, gut	Respiratory system damage
Cu	Liver	Liver damage
Hg	Brain, liver	Mental illness
Ni	Skin, kidney	Allergic reaction, kidney damage
Pb	Bone marrow, blood, brain	Anemia, mental retardation

References

Cobbett, C., Goldsbrough, P. (2002). Phytochelatins and metallothioneins: roles in heavy metal detoxification and homeostasis. Annual Review of Plant Biology 53, 159-182.

Dallinger, R., Berger, B., Hunziker, P., Kägi, J.H.R. (1997). Metallothionein in snail Cd and Cu metabolism. Nature 388, 237-238.

Dallinger, R., Höckner, M. (2013). Evolutionary concepts in ecotoxicology: tracing the genetic background of differential cadmium sensitivities in invertebrate lineages. Ecotoxicology 22, 767-778.

Haq, F., Mahoney, M., Koropatnick, J. (2003) Signaling events for metallothionein induction. Mutation Research 533, 211-226.

Hopkin, S.P. (1989) Ecophysiology of Metals in Terrestrial Invertebrates. London, Elsevier Applied Science.

Nieboer, E., Richardson, D.H.S. (1980) The replacement of the nondescript term "heavy metals" by a biologically and chemically significant classification of metal ions. Environmental Pollution Series B 1, 3-26.

Rainbow, P.S. (2002) Trace metal concentrations in aquatic invertebrates: why and so what? Environmental Pollution 120, 497-507.

4.1.4. Xenobiotic Defence and Metabolism

Synopsis

All organisms are equipped with metabolic defence mechanisms to deal with foreign compounds. The reactions involved, jointly called biotransformation, can be divided into three phases, and usually aim to increase water solubility and excretion. The first step (phase I) is catalyzed by cytochrome P450, which is followed by a variety of conjugation reactions (phase II) and excretion (phase III). The enzymes and transporters involved are often highly inducible, i.e. the amount of protein is greatly enhanced by the xenobiotic compounds themselves. The induction involves binding of the compound to cytoplasmic receptor proteins, such as the arylhydrocarbon receptor (AhR), or the constitutive androstane receptor (CAR). In some cases the intermediate metabolites, produced in phase I are extremely reactive and a main cause of toxicity, a well-known example being the metabolic activation of polycyclic aromatic hydrocarbons such as benzo(a) pyrene, which readily forms DNA adducts and causes cancer. In addition, some compounds greatly induce metabolizing enzymes but are hardly degraded by them and cause chronic cellular stress. The various biotransformation reactions are a crucial aspect of both toxicokinetics and toxicodynamics of xenobiotics.

Introduction

The term "xenobiotic" ("foreign to biology") is generally used to indicate a chemical compound that does not normally have a metabolic function. We will use the term extensively in this module, despite the fact that it is somewhat problematic (e.g., can a compound be considered "foreign" if it circulates in the body, is metabolized or degraded in the body?, and: what is "foreign" for one species is not necessarily "foreign" for another species).

The body has an extensive defence system to deal with xenobiotic compounds, loosely designated as biotransformation. The ultimate result of this system is excretion the compound in some form or another. However, many xenobiotics are quite lipophilic, tend to accumulate and are not easily excreted due to low water solubility. Molecular modifications are usually required before such compounds can be removed from the body, as the main circulatory and excretory systems (blood, urine) are water-based. By introducing hydrophilic groups in the molecule (-OH, =O, -COOH) and by conjugating it to an endogenous compound with good water-solubility, excretion is usually accomplished. However, as we will see below, intermediate metabolites may have enhanced reactivity and it often happens that a compound becomes more toxic while being metabolized. In the case of pesticides, deliberate use is made of such responses, to increase the toxicity of an insecticide once it is in the target organism.

The study of xenobiotic metabolism is a classical subject not only in toxicology but also in pharmacology. The mode of action of a drug often depends critically on the rate and mode of metabolism. Also, many drugs show toxic side-effects as a consequence of metabolism. Finally, xenobiotic metabolism is also studied extensively in entomology, as both toxicity and resistance of pesticides are often mediated by metabolism.

The most problematic xenobiotics are those with a high octanol-water partition coefficient (Kow) that are strongly lipophilic and very hydrophobic. They tend to accumulate, in proportion to their Log Kow, in tissues with a high lipid content such as the subcutis of vertebrates, and may cause tissue damage due to disturbance of membrane functions. This mode of action is called "minimum toxicity". Well-known are low-molecular weight aliphatic petroleum compounds and chlorinated alkanes such as chloroform. These compounds cause their primary damage to cell membranes; especially neurons are sensitive to this effect, hence minimum toxicity is also called narcotic toxicity. Lipophilic chemicals with high Log Kow do not reach concentrations high enough to cause minimum toxicity because they induce biotransformation at lower concentrations. The toxicity is then usually due to a reactive metabolite.

Xenobiotic metabolism involves three subsequent phases (Figure 1):

- Activation (usually oxidation) of the compound by an enzyme known as cytochrome P450, which acts in cooperation with NADPH cytochrome P450 reductase and other factors.
- Conjugation of the activated product of phase I to an endogenous compound. A host of different enzymes is available for this task, depending on the compound, the tissue and the species. There are also (slightly polar) compounds that enter phase II directly, without being activated in phase I.
- Excretion of the compound into circulation, urine, or other media, usually by means of membrane-spanning transporters belonging to the class of ATP-binding cassette (ABC) transporters, including the infamous multidrug resistance proteins. Hydrophilic compounds may pass on directly to phase III, without being activated or conjugated.

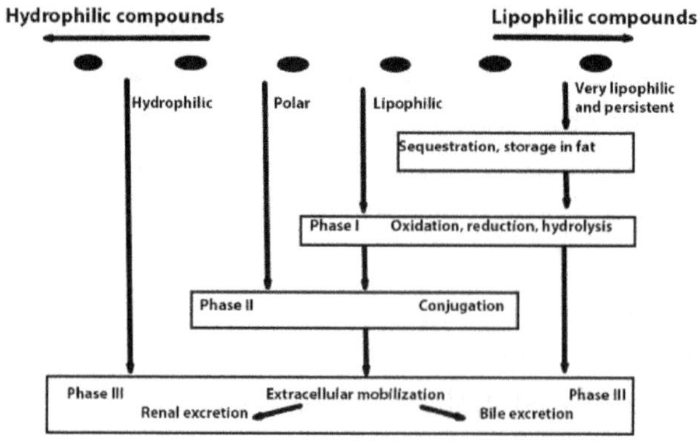

Figure 1. The various phases of xenobiotic metabolism. Redrawn by Wilma Ijzerman.

Phase I Reactions

Cytochrome P450 is a membrane-bound enzyme, associated with the smooth endoplasmic reticulum. It carries a porphyrin ring containing an Fe atom, which is the active center of the molecule. The designation P450 is derived from the fact that it shows an absorption maximum at 450 nm when inhibited by carbon monoxide, a now outdated method to demonstrate its presence. Other (outdated) namings are MFO (mixed function oxygenase) and drug metabolizing enzyme complex. Cytochrome P450 is encoded by a gene called CYP, of which there are many paralogs in the genome, all slightly differing from each other in terms of inducibility and substrate specificity. Three classes of CYP genes are involved with biotransformation, designated CYP1, CYP2 and CYP3 in vertebrates. Each class has several isoforms; the human genome has 57 different CYP genes in total. The CYP complement of invertebrates and plants often involves even more genes; many evolutionary lineages have their own set, arising from extensive gene duplications within that lineage. In humans, the genetic complement of a person's CYP genes is highly relevant as to its drug metabolizing profile (see the section on Genetic variation in toxicant metabolism).

Cytochrome P450 operates in conjunction with an enzyme called NADPH cytochrome P450 reductase, which consists of two flavoproteins, one containing Flavin adenine dinucleotide (FAD), the other flavin mononucleotide (FMN). The reduced Fe^{2+} atom in cytochrome P450 binds molecular oxygen, and is oxidized to Fe^{3+} while splitting O2; one O atom is introduced in the substrate, the other reacts with hydrogen to form water. Then the enzyme is reduced by accepting an electron from cytochrome P450 reductase. The overall reaction can be written as:

$$RH + O_2 + NADPH + H^+ \rightarrow ROH + H_2O + NADP^+$$

where R is an arbitrary substrate.

Cytochrome P450 is expressed to a great extent in hepatocytes (liver cells), the liver being the main organ for xenobiotic metabolism in vertebrates (Figure 2), but it is also present in epithelia of the lung and the intestine. In insects the activity is particularly high in the Malpighian tubules in addition to the gut and the fat body. In mollusks and crustaceans the main metabolic organ is the hepatopancreas.

Phase II Reactions

After activation by cytochrome P450 the oxidized substrate is ready to be conjugated to an endogenous compound, e.g. a sulphate, glucose, glucuronic acid or glutathione group. These reactions are conducted by a variety of different enzymes, of which some reside in the sER like P450, while others are located in the cytoplasm of the cell (Figure 2). Most of them transfer a hydrophilic group, available from intermediate metabolism, to the substrate, hence the enzymes are called transferases. Usually the compound becomes more polar in phase II, however, not all phase II reactions increase water solubility; for example, methylation (by methyl transferase) decreases reactivity but increases apolarity. Other phase II reactions are conjugation with glutathione, conducted by glutathione-S-transferase (GST) and with glucuronic acid, conducted by UDP-glucuronyl transferase. In invertebrates other conjugations may dominate, e.g. in arthropods and plants, conjugation with malonyl glucose is a common reaction, which is not seen in vertebrates.

Conjugation with glutathione in the human body is often followed by splitting off glutamic acid and glycine, leaving only the cysteine residue on the substrate. Cysteine is subsequently acetylated, thus forming a so-called mercapturic acid. This is the most common type of metabolite for many xenobiotics excreted in urine by humans.

Like cytochrome P450, the phase II enzymes consist in various isoforms, encoded by different paralogs in the genome. Especially the GST family is quite extensive and polymorphisms in these genes contribute significantly to the personal metabolic profile (see the section on Genetic variation in toxicant metabolism).

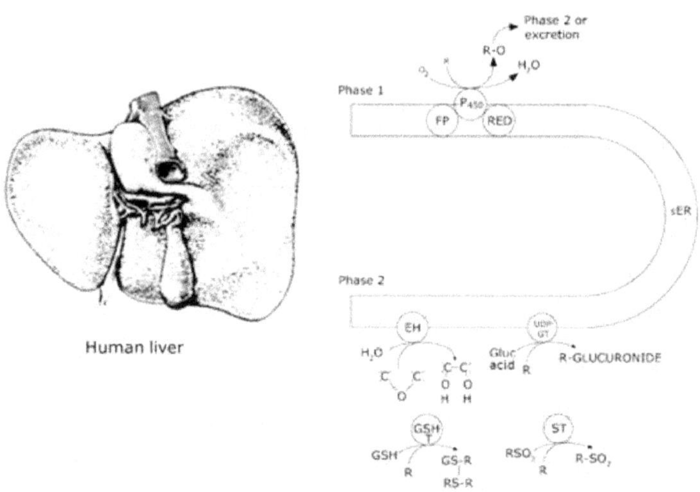

Figure 2. Schematic view of xenobiotic metabolism (phase I and phase II) in human liver. P450 = cytochrome P450, FP = flavoprotein, RED = cytochrome P450 reductase, EH = epoxide hydrolase, UDP-GT = uridyldiphospho-glucuronyl transferase, GSH-T = glutathione-S-transferase, ST = sulphotransferase. Reproduced from Vermeulen & Van den Broek (1984) by Wilma Ijzerman.

Phase III Reactions

In the human body, there are two main pathways for excretion, one from the liver into the bile (and further into the gut and the faeces), the other through the kidney and urine. These two pathways are used by different classes of xenobiotics:

very hydrophobic compounds such as high-molecular weight polycyclic aromatic hydrocarbons are still not readily soluble in water even after metabolism but can be emulsified by bile salt and excreted in this way. It sometimes happens that such compounds, once arriving in the gut, are assimilated again, transported to the liver by the portal vein and metabolized again. This is called "entero-hepatic circulation". Lower molecular weight compounds and hydrophilic compounds are excreted through urine. Volatile compounds can leave the body through the skin and exhaled air.

Excretion of activated and conjugated compounds from tissues out of the cell usually requires active transport, which is mediated by ABC (ATP binding cassette) transporters, a very large and diverse family of membrane proteins which have in common a binding cassette for ATP. Different subgroups of ATP transporters transport different types of chemicals, e.g. positively charged hydrophobic molecules, neutral molecules and water-soluble anionic compounds. One well-known group consists of multidrug resistance proteins or P-glycoproteins. These transporters export drugs aiming to attack tumor cells. Because their activity is highly inducible, these proteins can enhance excretion enormously, making the cell effectively resistant and thus case major problems for cancer treatment.

Introduction

All enzymes of xenobiotic metabolism are highly inducible: their activity is normally on a low level but is greatly enhanced in the presence of xenobiotics. This is achieved through a classic case of transcriptional regulation upon CYP and other genes, leading to de novo synthesis of protein. In addition, extensive proliferation of the endoplasmic reticulum may occur, and in extreme cases even swelling of the liver (hepatomegaly).

The best investigated pathway for transcriptional activation of CYP genes is due to the arylhydrocarbon receptor (AhR). Under normal conditions, this peptide is stabilized in the cytoplasm by heat-shock proteins, however, when a xenobiotic compound binds to AhR, it is activated and can join with another protein called Ah receptor nuclear translocator (ARNT) to translocate to the nucleus and bind to DNA elements present in the promotor of CYP and other genes. It thus acts as a transcriptional activator or transcription factor on these genes (Figure 3). The DNA motifs to which AhR binds are called xenobiotic responsive elements (XRE) or dioxin-responsive elements (DRE). The compounds acting in this manner are called 3-MC-type inducers, after the (highly carcinogenic) model compound 3-methylcholanthrene. The inducing capacity of a compound is related to its binding affinity to the AhR, which in itself is determined by the spatial structure of the molecule. The lock-and key fitting between AhR and xenobiotics explains why induction of biotransformation by xenobiotics shows a very strong stereospecificity.

For example, among the chlorinated biphenyls and chlorinated dibenzodioxins, some compounds are extremely strong inducers of CYP1 genes, while others, even with the same number of chlorine atoms, are no inducers at all. The precise position of the chlorine atoms determines the molecular "fit" in the Ah receptor (see Section on Receptor interaction).

In addition to 3MC-type of induction there are other modes in which biotransformation enzymes are induced, but these are less well-known. A common class is PB-type induction (named after another model compound, phenobarbital). PB-type induction is not AhR-dependent, but acts through activation of another nuclear receptor, called constitutive androstane receptor (CAR). This receptor activates CYP2 genes and some CYP3 genes.

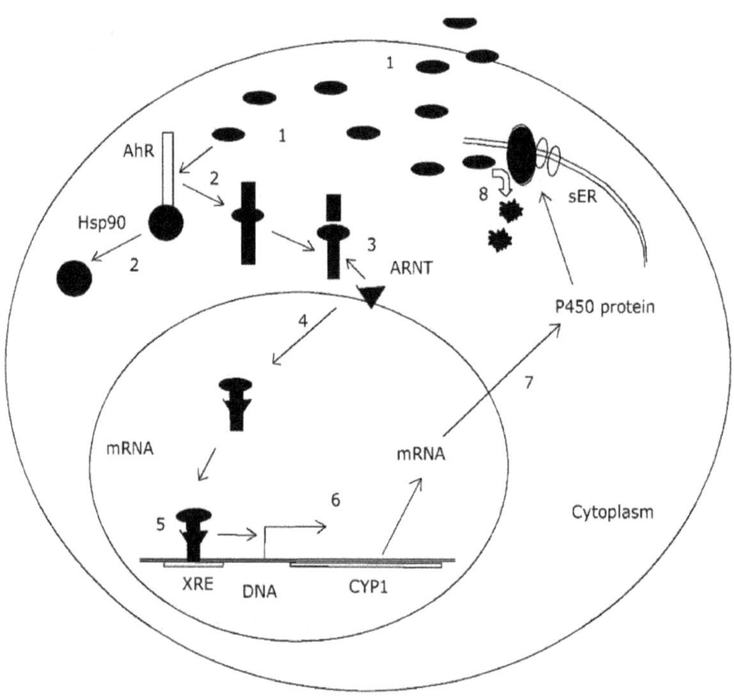

Figure 3. Scheme illustrating how AhR-reactive compounds induce biotransformation activity. 1 xenobiotic compounds enter the cell through diffusion, 2 compound binds to AhR, releasing it from Hsp90, 3 AhR complex is activated by ARNT, 4 translocation to nucleus, 5 binding to XRE, 6 enhanced transcription, 7 translocation of mRNA to cytoplasm and translation, 8 incorporation of P450 enzyme in sER and biotransformation of compounds. AhR = arylhydrocarbon receptor, Hsp90 = heat shock protein 90, ARNT = aryl hydrocarbon nuclear translocator, XRE = xenobiotic responsive element, CYP1 = gene encoding cytochrome P450 1, sER = smooth endoplasmic reticulum. Drawn by Wilma Ijzerman.

The high inducibility of biotransformation can be exploited in a reverse manner: if biotransformation is seen to be highly upregulated in a species living in the environment, this indicates that that species is being exposed to xenobiotic compounds. Assays addressing cytochrome P450 activity can therefore be exploited in bioindication and biomonitoring systems. The EROD (ethoxyresorufin-O-deethylase) assay is often used for this purpose, although it is not 100% specific to the isoform of P450. Another approach is to address CYP expression directly, e.g. through reverse transcription-quantitative PCR, a method to quantify the amount of CYP mRNA.

Secondary Effects of Biotransformation

Although the main aim of xenobiotic metabolism is to detoxify and excrete foreign compounds, some pathways of biotransformation actually enhance toxicity. This is mostly due to the first step, activation by cytochrome P450. The activation may lead to intermediate metabolites which are highly reactive and the actual cause of toxicity. The best investigated examples are due to bioactivation of polycyclic aromatic hydrocarbons (PAHs), a group of chemicals present in diesel, soot, cigarette smoke and charred food products. Many of these compounds, e.g. benzo(a) pyrene, benz(a)anthracene and 3-methylcholanthrene, are not reactive or toxic as such but are activated by cytochrome P450 to extremely reactive molecules. Benzo(a)pyrene, for instance, is activated to a diol-epoxide, which readily binds to DNA, especially to the free amino-group of guanine (Figure 4). The complex is called a DNA adduct, the double helix is locally disrupted and this results in a mutation. If this happens in an oncogene, a tumor may develop (see the Section on Carcinogenesis and genotoxicity).

Not all PAHs are carcinogenic. Their activity critically depends on the spatial structure of the molecule, which again determines its “fit” in the Ah receptor. PAHs with a “notch” (often called bay-region) in the molecule tend to be stronger carcinogens than compounds with a symmetric (round or linear) molecular structure.

Another mechanism for biotransformation-induced toxicity is due to some very recalcitrant organochlorine compounds such as polychlorinated dibenzodioxins (PCDDs, or dioxins for short) and polychlorinated biphenyls (PCBs). Some of these compounds are very potent inducers of biotransformation, but they are hardly degraded themselves. The consequence is that the highly upregulated cytochrome P450 activity continues to generate a large amount of reactive oxygen (ROS), causing oxidative stress and damage to cellular constituents. It is assumed that the chronic toxicity of 2,3,7,8-tetrachlorodibenzo(para)dioxin (TCDD), one of the most toxic compounds emitted by human activity, is due to its high capacity to induce prolonged oxidative stress. On the molecular level, there is a close link between oxidative stress and biotransformation activity. Many toxicants that primarily induce oxidative stress (e.g. cadmium) also upregulate CYP enzymes. Two defence mechanisms, oxidative stress defence and biotransformation are part of the same integrated stress defence system of the cell.

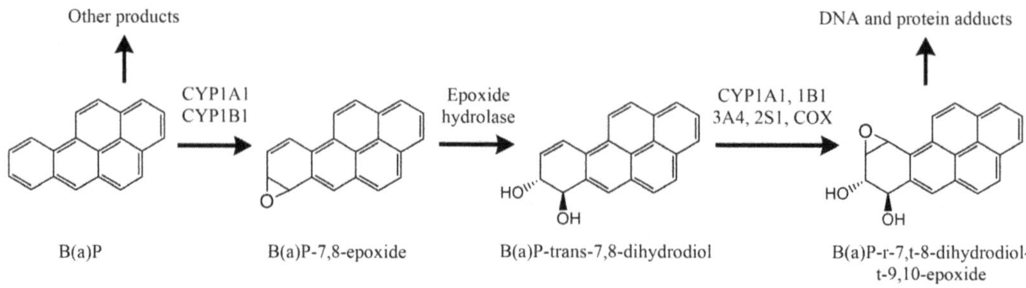

Figure 4. Metabolism of benzo(a)pyrene (B[a]P) to a mutagenic and carcinogenic metabolite, BaP-diol epoxide. Cytochrome P450 activity (CYP1A1 and CYP1B1) introduces an epoxide on the 7,8 position of the molecule, which is then hydrolyzed by epoxide hydrolase to a dihydrodiol (two OH groups next to each other, which can be in trans position, pointing in different directions relative to the plane of the molecule, or in cis, pointing in the same direction). The trans metabolite is preferably formed. Subsequently another epoxide is introduced on the 9,10 position. The diolepoxide is highly reactive, and may bind proteins and to guanine in DNA, causing a DNA adduct. Modified from Bui et al. (2009) by Steven Droge.

References

Bui, P.H., Hsu, E.L., Hankinson, O. (2009), Fatty acid hydroperoxides support cytochrome P450 2S1-mediated bioactivation of benzo[a]pyrene-7-8-dihydrodiol. Molecular Pharmacology 76, 1044-1052.

Stroomberg, G.J., Zappey, H., Steen, R.J.C.A., Van Gestel, C.A.M., Ariese, F., Velthorst, N.H., Van Straalen, N.M. (2004). PAH biotransformation in terrestrial invertebrates – a new phase II metabolite in isopods and springtails. Comparative Biochemistry and Physiology Part C 138, 129-137.

Timbrell, J.A. (1982). Principles of Biochemical Toxicology. Taylor & Francis Ltd, London.

Van Straalen, N.M., Roelofs, D. (2012). An Introduction to Ecological Genomics, 2nd Ed. Oxford University Press, Oxford.

Vermeulen, N.P.E., Van den Broek, J.M. (1984). Opname en verwerking van chemicaliën in de mens. Chemisch Magazine. Maart: 167-171.

4.1.5. Allometric Relationships

Introduction

Globally more than 100,000,000 chemicals have been registered. In the European Union more than 100,000 compounds are awaiting risk assessment to protect ecosystem and human health, while 1,500,000 contaminated sites potentially require clean-up. Likewise, 8,000,000 species, of which 10,000 are endangered, need protection worldwide, with one lost per hour (Hendriks, 2013). Because of financial, practical and ethical (animal welfare) constraints, empirical studies alone cannot cover so many substances and species, let alone their combinations. Consequently, the traditional approach of ecotoxicological testing is gradually supplemented or replaced by modelling approaches. Environmental chemists and toxicologists for long have developed relationships allowing extrapolation across chemicals. Nowadays, so-called Quantitative Structure Activity Relationships (QSARs) provide accumulation and toxicity estimates for compounds based on their physical-chemical properties. For instance bioaccumulation factors and median lethal concentrations have been related to molecular size and octanol-water partitioning, characteristic properties of a chemical that are usually available from its industrial production process.

In analogy with the QSAR approach in environmental chemistry, the question may be asked whether it is possible to predict toxicological, physiological and ecological characteristics of species from biological traits, especially traits that are easily measured, such as body size. This approach has gone under the name "Quantitative Species Sensitivity Relationships" (QSSR) (Notenboom, 1995).

Among the various traits available, body-size is of particular interest. It is easily measured and a large part of the variability between organisms can be explained from body size, with r2 > 0.5. Not surprisingly, body size also plays an important role in toxicology and pharmacology. For instance, toxic endpoints, such as LC50s, are often expressed per kg body weight. Recommended daily intake values assume a "standard" body weight, often 60 kg. Yet, adult humans can differ in body weight by a factor of 3 and the difference between mouse and human is even larger. Here it will be explored how body-size relationships, which have been studied in comparative biology for a long time, affect the extrapolation in toxicology and can be used to extrapolate between species.

Fundamentals of Scaling in Biology

Do you expect a 104 kg elephant to eat 104 times more than a 1 kg rabbit per day? Or less, or more? On being asked, most people intuitively come up with the right answer. Indeed, daily consumption by the proboscid is less than 104 times that of the rodent. Consequently, the amount of food or water used per kilogram of body weight of the elephant is less than that of the rabbit. Yet, how much less exactly? And why should sustaining 1 kg of rabbit tissue require more energy than 1 kg of elephant flesh in the first place?

A century of research (Peters, 1983) has demonstrated that many biological characteristics Y scale to size X according to a power function:

$$Y = a\ X^b$$

where the independent variable X represents body mass, and the dependent variable Y can virtually be any characteristic of interest ranging, e.g., from gill area of fish to density of insects in a community.

Plotted in a graph, the equation produces a curved line, increasing super-linearly if $b > 1$ and sub-linearly if $b < 1$. If $b=1$, Y and X are directly proportional and the relationship is called isometric. As curved lines are difficult to interpret, the equation is often simplified by taking the logarithm of the left and right parts. The formula then becomes:

$$\log Y = \log a + b \log X$$

When log Y is plotted against log X, a straight line results with slope b and intercept log a. If data are plotted in this way, the slope parameter b may be estimated by simple linear regression.

Across wide size ranges, slope b often turns out to be a multitude of ¼ or, occasionally, ⅓. Rates [kg·d-1] of consumption, growth, reproduction, survival and what not, increase with mass to the power ¾ , while rate constants, sometimes called specific rates [kg·kg-1·d-1], decrease with mass to the power –¼. So, while the elephant is 104 kg heavier than the 1 kg rabbit, it eats only (104)¾ = 103 times more each day. Vice versa, 1 kg of proboscid apparently requires a consumption of (104)-¼ kg·kg-1·d-1, i.e., 10 times less. Variables with a time dimension [d] like lifespan or predator-prey oscillation periods scale inversely to rate constants and thus change with body mass to the power ¼. So, an elephant becomes (104)¼ = 10 times older than a rabbit. Abundance, i.e., the number of individuals per surface area [m-2] decreases with body mass to the power –¾. Areas, such as gill surface or home range, scale inversely to abundance, typically as body mass to the power ¾.

Now, why would sustaining 1 kg of elephant require 10 times less food than 1 kg of rabbit? Biologists, pharmacologists and toxicologists first attributed this difference to area-volume relationships. If objects of the same shape but different size are compared, the volume increases with length to the power 3 and the surface increases with length to the power 2. For a sphere with radius r, for example, area A and volume V increase as A ~ r2 and V ~ r3, so area scales to volume as A ~ r2 ~ $(V^{1/3})2$ ~ $V^{2/3}$. So, larger animals have relatively smaller surfaces, as long as the shape of the organism remains the same. Since many biological processes, such as oxygen and food uptake or heat loss deal with surfaces, metabolism was, for long, thought to slow down like geometric structures, i.e., with multitudes of ⅓. Yet, empirical regressions, e.g. the "mouse-elephant curve" developed by Max Kleiber in the early 1930s show a universal slope of ¼ (Peters, 1983. This became known as the "Kleiber's law". While the data leave little doubt that this is the case, it is not at all clear why it should be ¼ and not ⅓. Several explanations for the ¼ slope have been proposed but the debate on the exact value as well as the underlying mechanism continues.

Application of Scaling in Toxicology

Since chemical substances are carried by flows of air and water, and inside the organism by sap and blood, toxicokinetics and toxicodynamics are also expected to scale to size. Indeed, data confirm that uptake and elimination rate constants decrease with size, with an exponent of about –¼ (Figure 1). Slopes vary around this value, the more so for regressions that cover small size ranges and physiologically different organisms. The intercept is determined by resistances in unstirred water layers and membranes through which the substances pass, as well as by delays in the flows by which they are carried. The resistances mainly depend on the affinity and molecular size of the chemicals, reflected by, e.g., the octanol-water partition coefficient Kow for organic chemicals or atomic mass for metals. The upper boundary of the intercept is set by the delays imposed by consumption and, subsequently, egestion and excretion. The lower end is determined by growth dilution. Both uptake and elimination scale to mass with the same exponent so that their ratio, reflecting the bioconcentration or biomagnification factor in equilibrium, is independent of body-size.

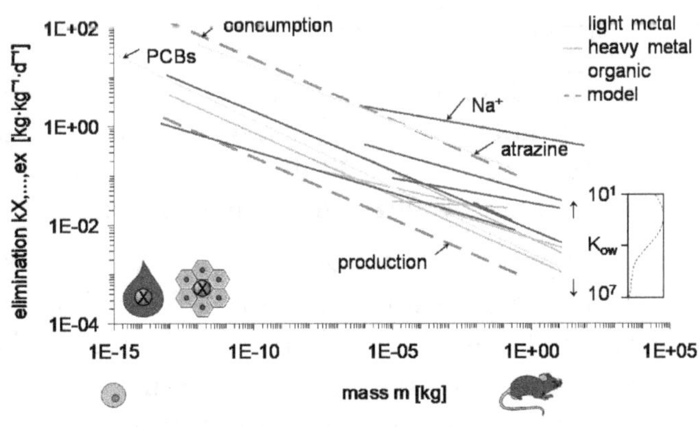

Figure 1. Regressions of elimination rate constants [kg·kg-1·d-1 = d-1] as a function of organism mass m [kg], ranging from algae to mammals, and organic chemicals' octanol-water partition ratio Kow and metal mass within the limits set by consumption and production (redrawn by author, based on Hendriks et al. 2001).

Scaling of rate constants for uptake and elimination, such as in Figure 1, implies that small organisms reach a given internal concentration faster than large ones. Vice versa, lethal concentrations in water or food needed to reach the same internal level after equal (short-term) exposure duration are lower in smaller compared to larger organisms. Thus, the apparent "sensitivity" of daphnids can, at least partially, be attributed to their small body-size. This emphasizes the need to understand simple scaling relationships before developing to more elaborate explanations.

Using Figure 1, one can, within strict conditions not elaborated here, theoretically relate median lethal concentrations LC50 [μg L-1] to the Kow of the chemical and the size of the organism, with $r^2 > 0.8$ (Hendriks, 1995; Table 1). Complicated responses like susceptibility to toxicants can be predicted only from Kow and body size, which illustrates the generality and power of allometric scaling. Of course, the regressions describe the general trends and in individual cases the deviations can be large. Still, considering the challenges of risk assessment as outlined above, and in the absence of specific data, the predictions in Table 1 can be considered as a reasonable first approximation.

Table 1. Lethal concentrations and doses as a function of test animal body-mass

Species	Endpoint	Unit	b (95% CI)	r2	nc	ns	Source
Guppy	LC50	mg·L-1	0.66 (0.51-0.80)	0.98	6	1	1
Mammals	LD10≈MTD	mg·animal-1	0.73	0.69-0.77	27	5	2
Birds	Oral LD50	mg·animal-1	1.19 (0.67-0.82)	0.76	194	3...37	3
Mammals	Oral LD50	mg·animal-1	0.94 (1.18-1.20)	0.89	167	3...16	4
Mammals	Oral LD50	mg·animal-1	1.01 (1.00-1.01)		>5000	2...8	5

MTD = maximum threshold dose, repeated dosing. LD50 single dose, b = slope of regression line, nc = number of chemicals, ns = number of species. Sources: 1 Anderson & Weber (1975), 2 Travis & White (1987), 4 Sample & Arenal (1999), 5 Burzala-Kowalczyk & Jongbloed (2011).

Allometry is also important when dealing with other levels of biological organisation. Leaf or gill area, the number of eggs in ovaries, the number of cell types and many other cellular and organ characteristics scale to body-size as well. Likewise, intrinsic rates of increase (r) of populations and the production-biomass ratios (P/B) of communities can also be obtained from the (average) species mass. Even the area needed by animals in laboratory assays scales to size, i.e., by $m^{3/4}$, approximately the same slope noted for home ranges of individuals in the field.

Future Perspectives

Since almost any physiological and ecological process in toxicokinetics and toxicodynamics depends on species size, allometric models are gaining interest. Such an approach allows one to quantitatively attribute outliers (like apparently "sensitive" daphnids) to simple biological traits, rather than detailed chemical-toxicological mechanisms.

Scaling has been used in risk assessment at the molecular level for a long time. The molecular size of a compound is often a descriptor in QSARs for accumulation and toxicity. If not immediately evident as molecular mass, volume or area often pops up as an indicator of steric properties. Scaling does not only apply to bioaccumulation and toxicity from molecular to community levels, size dependence is also observed in other sections of the environmental cause-effect chain. Emissions of substances, e.g., scale non-linearly to the size of engines and cities. Concentrations of chemicals in rivers depend on water discharge, which in itself is an allometric function of catchment size. Hence, understanding the principles of cross-disciplinary scaling is likely to pay off in protecting many species against many chemicals.

References

Anderson, P.D., Weber, L.J. (1975). Toxic response as a quantitative function of body size. Toxicology and Applied Pharmacology 33, 471-483.

Burzala-Kowalczyk, L., Jongbloed, G. (2011). Allometric scaling: Analysis of LD50 data. Risk Analysis 31, 523-532.

Hendriks, A.J. (1995). Modelling response of species to microcontaminants: Comparative ecotoxicology by (sub)lethal body burdens as a function of species size and octanol-water partitioning of chemicals. Ecotoxicology and Environmental Safety 32, 103-130.

Hendriks, A.J. (2013). How to deal with 100,000+ substances, sites, and species: Overarching principles in environmental risk assessment. Environmental Science and Technology 47, 3546–3547.

Hendriks, A.J., Van der Linde, A., Cornelissen, G., Sijm, D.T.H.M. (2001). The power of size: 1. Rate constants and equilibrium ratios for accumulation of organic substances. Environmental Toxicology and Chemistry 20, 1399-1420.

Notenboom, J., Vaal, M.A., Hoekstra, J.A. (1995). Using comparative ecotoxicology to develop quantitative species sensitivity relationships (QSSR). Environmental Science and Pollution Research 2, 242-243.

Peters, R.H. (1983). The Ecological Implications of Body Size. Cambridge University Press, Cambridge.

Sample, B.E., Arenal, C.A. (1999) Allometric models for interspecies extrapolation of wildlife toxicity data. Bulletin of Environmental Contamination and Toxicology 62, 653-66.

Travis, C.C., White, R.K. (1988) Interspecific scaling of toxicity data. Risk Analysis 8, 119-125.

4.1.6. Food Chain Transfer

Accumulation of Chemicals Across Different Trophic Levels.

Chemicals may be transferred from one organism to another. Grazers will ingest chemicals that are in the vegetation they eat. Similarly, predators are exposed to chemicals in their prey items. This so-called food web accumulation is governed by properties of the chemical, but also by some traits of the receiving organism (e.g. grazer or predator).

Chemical Properties Driving Food Web Accumulation

Some chemicals are known to accumulate in food webs, reaching the highest concentrations in top-predators. Examples of such chemicals are organochlorine pesticides like DDT and brominated flame retardants (e.g. PBDEs; see section on POPs). Such accumulating chemicals have a few properties in common: they need to be persistent and they need to have affinity for the organismal body. Organic chemicals with a relatively high log Kow, indicating a high affinity for lipids, will enter organisms quite effectively (see section on Bioconcentration and kinetics modelling). Once in the body, these chemicals will be distributed to lipid rich tissues, and excretion is rather limited. In case of persistent chemicals that are not metabolised, concentrations will increase over time when uptake is higher than excretion. Furthermore, such chemicals are likely to be passed on to organisms at the next trophic level in case of prey-predator interactions.

Some of these chemicals may be metabolised by the organism, most often into more water soluble metabolites (see section on Xenobiotic metabolism & defence). These metabolites are more easily excreted, and in this way concentrations of metabolizable chemicals do not increase so much over time, and will therefore also transfer less to higher trophic levels. The effects of metabolism on the internal concentrations of organisms is clearly illustrated by a study on the uptake of organic chemicals by different aquatic species (Kwok et al., 2013). In that study the uptake of persistent chemicals (organochlorine pesticides; OCPs) was compared with the uptake of chemicals that may be metabolised (polycyclic aromatic hydrocarbons; PAHs). The authors compared shrimps with fish, the former having a limited capacity to metabolise PAHs while fish can. Figure 1 shows the Biota-to-Sediment Accumulation Factors (BSAFs; see section on Bioaccumulation), which is the ratio between the concentration in the organism and in the sediment. It is shown that OCPs accumulate to a high extent in both species, reflecting persistent, non-metabolizable chemicals. For PAHs the results are different per species, fish are able to metabolise and as a result the concentrations of PAHs in the fish are low, while in shrimp, with a limited metabolic capacity, the accumulation of PAHs is comparable to the OCPs. These results show that not only the properties of the chemicals are of importance, but also some traits of the organisms involved, in this case the metabolic capacity.

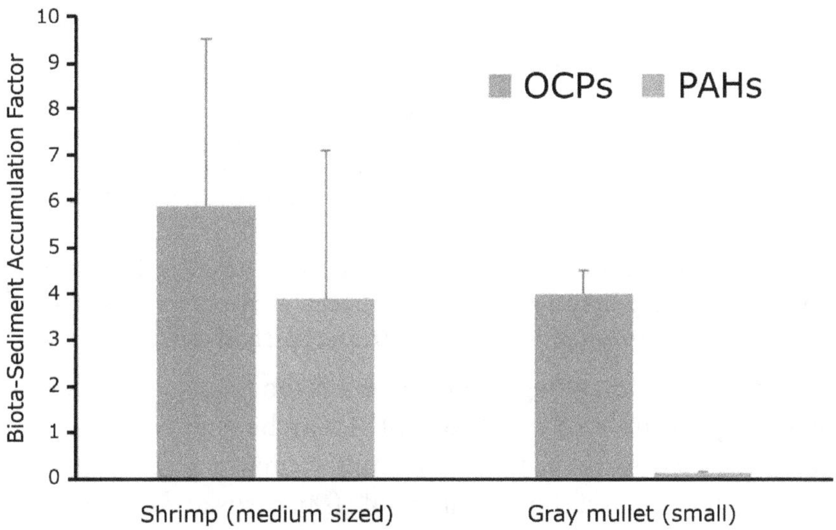

Figure 1. Biota–sediment accumulation factors organochlorine pesticides (OCPs) and polycyclic aromatic hydrocarbons (PAHs) in fish and shrimp. Redrawn from Kwok et al. (2013).

Effects of Species Traits and Food Web Structure on Food Web Accumulation

Food-web accumulation of chemicals is driven by food uptake. At lower trophic levels, most organisms will acquire relatively low concentrations from the ambient environment. First consumers, foraging on these organisms will accumulate the chemicals of all of them, and in case of persistent chemicals that enter the body easily, concentrations in the consumers will be higher than in their diet. Similarly, concentrations will increase when the chemicals are transferred to the next trophic level. This process is called biomagnification, indicating increasing concentrations of persistent and accumulative chemicals in food webs. The most iconic example of this is on the increasing concentrations of DDTs in fish eating American Osprey (Figure 2), a casus which has led to the ban of a lot of organochlorine chemicals.

DDT and Bioaccumulation(absorbed or stored in organs) and **Biomagnification** (through a food chain)

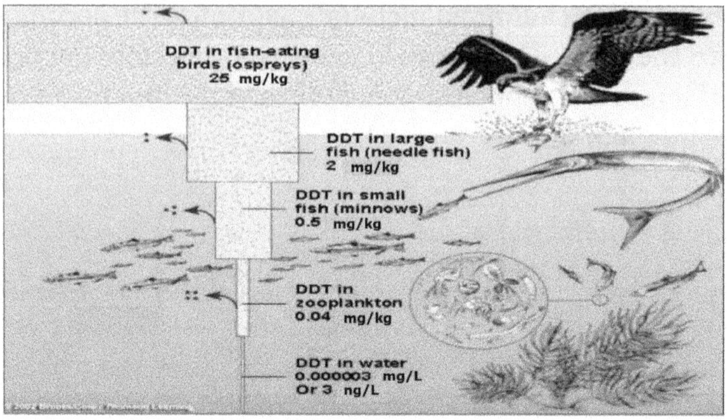

Figure 2. Example of biomagnification of DDT in an aquatic food web of ospreys. Slightly modified from: http://naturalresources.anthro-seminars.net/concepts/ecological-concepts-biomagnification/.

Since biomagnification along trophic levels is food driven, it is of importance to include diet composition into the studies. This can be explained by an example on small mammals in the Netherlands. Two similar small mammal species, the bank vole (Myodes glareolus) and the common vole (Microtus arvalis) co-occur in larger part of the Netherlands. Although the species look very similar, they are different in their diet and habitat use. The bank vole is a omnivorous species, inhabiting different types of habitat while the common vole is strictly vegetarian living in pastures. In a study on the species-specific uptake of cadmium, diet items of both species were analysed, indicating nearly 3 orders of magnitudes differences in cadmium concentrations between earthworms and berries from vegetation (Fig 3A, van den Brink et al., 2010). Stable isotopic ratios of carbon and nitrogen were used to assess the general diets of the organisms. The common vole ate mostly stinging nettle and grass, including seeds, while the bank vole showed to forage on grass herbs and earthworms. This difference in diet was reflected in increased concentrations of cadmium in the bank vole in comparison to the common vole (both inhabiting the same area). The concentrations of one bank vole appeared to be extremely low (red diamond in Figure 3b), and initially this was considered to be an artefact. However, detailed analysis of the stable isotopic ratios in this individual revealed that it had foraged on stinging nettle and grass, hence a diet more reflecting the common vole. This emphasises once more that organisms accumulate through their diet (you accumulate what you eat!)

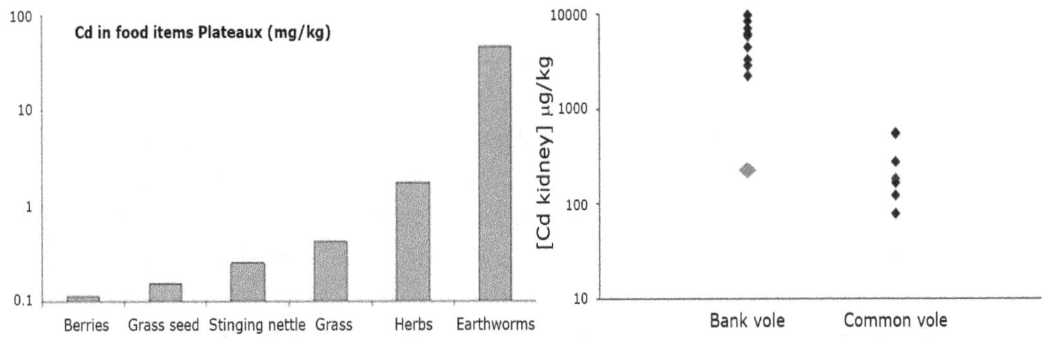

Figure 3. Left: Concentrations of cadmium in diet items of small mammals collected in the Plateaux area near Eindhoven, the Netherlands; Right: Cadmium concentrations in kidneys of bank voles and common voles collected in the Plateaux area. See text for further explanation. Redrawn from van den Brink et al. (2010).

Case Studies

Orcas or Killer whales (Orcinus orca) are large marine predatory mammals, which roam all around the oceans, from the Arctic to the deep south region of the Antarctic. Although they appear ubiquitous around the world, generally different pods of Orcas occur in different regions of the marine ecosystem. Often, each pod has developed specialised foraging behaviours targeted at specific prey species. Although Orcas are generally apex top-predators at the top of the (local) food web, the different foraging strategies would suggest that exposure to accumulating chemicals may differ considerably between pods. This was indeed shown to be the case in a very elaborate study on different pods of Orcas of the West-coast of Canada by Ross et al. (2000). In the Vancouver region there is a resident pod while the region is also often visited by two transient groups of Orcas. PCB concentrations were high in all animals, but the transient animals contained significantly higher levels. The transient whales mainly fed on marine mammals, while the resident animals mainly fed on fish, and this difference in diet was thought to be the cause of the differences in PCB levels between the groups. In that study, it was also shown that PCB levels increased with age, due to the persistence of the PCBs, while female orcas contained significant lower concentrations of PCBs. The latter is caused by the lactation of the female Orcas during which they feed their calves with lipid rich milk, containing relatively high levels of (lipophilic) PCBs. By this process, females offload large parts of their PCB body burden, however by transferring these PCBs to their developing calves (see also Figure 3 in the section on Bioaccumulation). A recent study showed that although PCBs have been banned for decades now, they still pose threats to populations of Orcas (Deforges et al., 2018). In that study, regional differences in PCB burdens were confirmed, likely due to differences in diet preferences although not specifically mentioned. It was shown that PCB levels in most of the Orca populations were still above toxic threshold levels and concerns were raised regarding the viability of these populations. This study confirms that 1) Orcas are exposed to different levels of PCBs according to their diet, which influences the biomagnification of the PCBs, 2) Orca populations are very inefficient in clearing PCBs from the individual due to little metabolism but also from the population due to the efficient maternal transfer from mother to calve, and 3) persistent, accumulating chemicals may pose threats to organisms even decades after their use. Understanding the mechanisms and processes underlying the biomagnification of persistent and toxic compounds is essential for a in depth risk assessment.

References

Desforges, J.-P., Hall, A., McConnell, B., Rosing-Asvid, A., Barber, J.L., Brownlow, A., De Guise, S., Eulaers, I., Jepson, P.D., Letcher, R.J., Levin, M., Ross, P.S., Samarra, F., Víkingson, G., Sonne, C., Dietz, R. (2018). Predicting global killer whale population collapse from PCB pollution. Science 361, 1373-1376.

Ford, J.K.B., Ellis, G.A., Matkin, D.R., Balcomb, K.C., Briggs, D., Morton, A.B. (2005). Killer whale attacks on minke whales: Prey capture and antipredator tactics. Marine Mammal Science 21, 603-618.

Guinet, C. (1992). Predation behaviour of killer whales (Orcinus orca) around Grozet Islands. Canadian Journal of Zoology 70, 1656-1667.

Kwok, C.K., Liang, Y., Leung, S.Y., Wang, H., Dong, Y.H., Young, L., Giesy, J.P., Wong, M.H. (2013). Biota–sediment accumulation factor (BSAF), bioaccumulation factor (BAF), and contaminant levels in prey fish to indicate the extent of PAHs and OCPs contamination in eggs of waterbirds. Environmental Science and Pollution Research 20, 8425-8434.

Ross, P.S., Ellis, G.M., Ikonomou, M.G., Barrett-Lennard, L.G., Addison, R.F. (2000). High PCB concentrations in free-ranging Pacific killer whales, Orcinus orca: Effects of age, sex and dietary preference. Marine Pollution Bulletin 40, 504-515.

Samarra, F.I.P., Bassoi, M., Beesau, J., Eliasdottir, M.O., Gunnarsson, K., Mrusczok, M.T., Rasmussen, M., Rempel, J.N., Thorvaldsson, B., Vikingsson, G.A. (2018). Prey of killer whales (Orcinus orca) in Iceland. Plos One 13, 20.

van den Brink, N., Lammertsma, D., Dimmers, W., Boerwinkel, M.-C., van der Hout, A. (2010). Effects of soil properties on food web accumulation of heavy metals to the wood mouse (Apodemus sylvaticus). Environmental Pollution 158, 245-251.

4.1.7. Critical Body Concentration

Introduction

One of the quests in ecotoxicology is how to link toxicity to exposure and to understand why some organisms experience toxic effects while others do not at the same level of exposure. A generally accepted approach for assessing possible adverse effects on biota, no matter what kind of species, is the Critical Body Concentration (CBC) concept (McCarty 1991). According to this concept, toxicity is determined by the amount of chemical taken up, so by the internal concentration, which has a relationship with the duration of the exposure as well as the exposure concentration.

Figure 1 shows the relationship between the development with time of the internal concentrations of a chemical in an organism and the time when mortality occurs at different exposure concentrations under constant exposure. Independent of exposure time or exposure concentration, mortality occurs at a more or less fixed internal concentration. The CBC is defined as the highest internal concentration of a substance in an organism that does cause a defined effect, e.g. 50% mortality or 50% reduction in the number of offspring produced. By comparing internal concentrations measured in exposed organisms to CBC values derived in the laboratory, a measure of risk is obtained. The CBC applies to lethality as well as to sub-lethal effects like reproduction or growth inhibition.

Relating Toxicity to Toxicokinetics

From Figure 1A, it may also become clear that chemicals that have fast uptake kinetics will reach the CBC faster than chemicals that have slow kinetics (see Section on Bioaccumulation kinetics). As a consequence, also the time to reach a constant LC50 (indicated as the ultimate LC50: LC50¥; Figure 1B) depends on kinetics. Hence, both toxic effects and chemical concentration are controlled by the same kinetics. The CBC can be derived from the LC50-time relationship and be linked to the LC50¥ using uptake and elimination rate constants (k_1 and k_2). It should be noted that the k2 in this case does not reflect the rate of chemical excretion but rather the rate of elimination of toxic effects caused by the chemical (so, note the difference here with Section on Bioaccumulation kinetics).

The time needed to reach steady state depends on the body size of the organisms, with larger organisms taking longer time to attain steady state compared to smaller organisms (McCarty 1991). The time needed to reach steady state depends also on the exposure surface area of the exposed organisms (Pawlisz and Peters 1993) as well as their metabolic activity. Organisms not capable of excreting or metabolizing a chemical will continue accumulating with time, and the LC50¥ will be zero. This is e.g. the case for cadmium in isopods (Crommentuijn et al. 1994), but kinetics are so slow that cadmium in these animals never reaches lethal concentrations in relatively clean environments as their life span is too short.

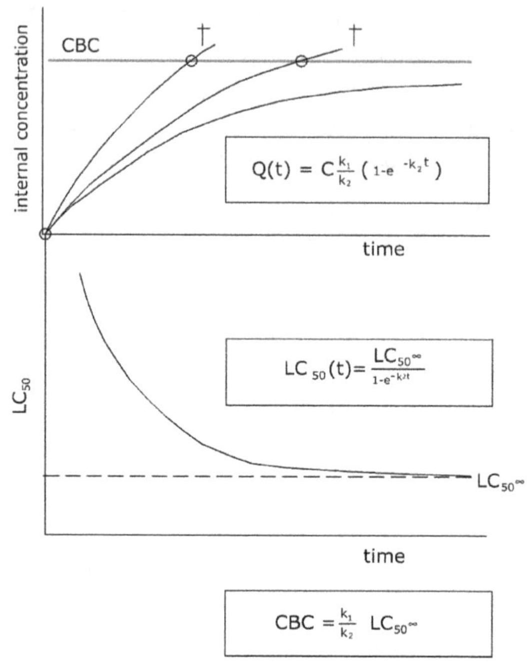

$$Q(t) = C\frac{k_1}{k_2}(1-e^{-k_2 t})$$

$$LC_{50}(t) = \frac{LC_{50\infty}}{1-e^{-k_2 t}}$$

$$CBC = \frac{k_1}{k_2}\, LC_{50\infty}$$

Figure 1: A. The relationship between uptake kinetics of a chemical in an organism and its toxicity according to the Critical Body Concentration (CBC) concept under constant exposure. The red line depicts the highest internal concentration of the chemical in the organism killing the organism (CBC); exceedance of that line results in mortality. B. The relationship between LC50 and time. LC50 will reach a constant value with time, indicated as the ultimate LC50 (LC50¥). The CBC can be calculated from the LC50¥ using uptake and elimination rate constants (k1 and k2) derived by first order kinetics as shown here. Note: the CBC approach is not limited to first order kinetics. And the same curves may be seen when focusing on sublethal effects, LC50 then reads EC50. Drawn by Wilma Ijzerman.

The CBC integrates environmentally available fractions with bioavailable concentrations and toxicity at specific receptors (McCarty and MacKay 1993). See also Section on Bioavailability. In this way, the actual exposure concentration in the environment does not need to be known for performing a risk assessment. The internal concentration of the chemical in the organism is the only concentration required for a risk assessment. Therefore many difficulties are overcome regarding bioavailability issues, e.g. it removes some of the disadvantages of the exposure concentration expressed per unit of soil, as well as of dealing with exposures that vary over time or space.

Proof of the CBC Concept

A convincing body of evidence was collected to support the CBC approach. For organic compounds with a narcotic mode of action, effects could be assessed over a wide range of organisms, test compounds and exposure media. For narcotic compounds with octanol-water partition coefficients (K_{ow}) varying from 10 to 1,000,000 (see for details Section on Relevant chemical properties), the concentration of chemical required for lethality through narcosis is approximately 1-10 mmol/kg: Figure 2 (McCarty and MacKay 1993).

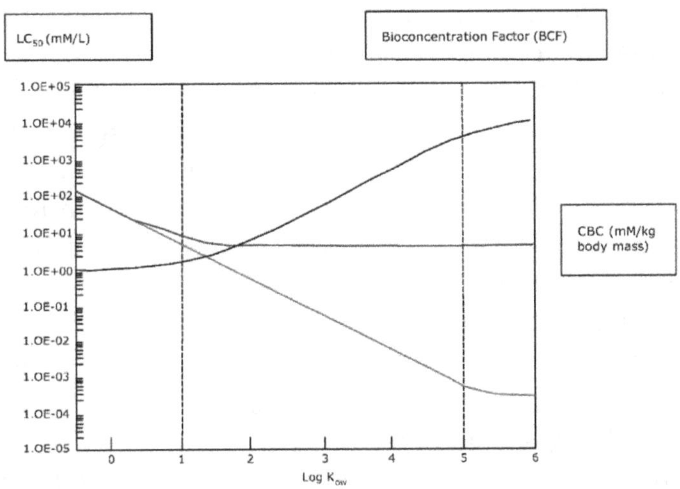

Figure 2: Theoretical plot supporting the Critical Body Concentration (CBC) concept for non-polar organic chemicals acting by narcosis. The bioconcentration factor (BCF in L/kg body mass; black line) of these chemicals increases with increasing log Kow while their acute toxicity also increases (LC50 in test solution in mM decreases; red line). The product of LC50 and BCF is the critical body concentrations (in mM/kg body mass; blue line). Adapted from McCarty and Mackay (1993) by Wilma Ijzerman.

To reduce the variation in bioconcentration factor (BCF) values for the accumulation of chemicals in organisms from water, normalization by lipid content has been suggested allowing to determine the chemical activity within an organism's body (US EPA 2003). For that reason, lipid extraction protocols are intensively described within the updated OECD Guideline for the testing of chemicals No. 305 for fish bioaccumulation tests, along with a sampling schedule of lipid measurement in fish. Correction of the BCF for differences in lipid content is also described in the same OECD guideline No. 305. If chemical and lipid analyses have been conducted on the same fish, this requires correction for the corresponding lipid content of each individual measured concentration in the fish. This should be done prior to using the data to calculate the kinetic BCF. If lipid content is not measure on all sampled fish, a mean lipid content of approx. 5% must be used to normalize the BCF. It should be noted that this correction holds only for chemicals accumulating in lipids and not for chemicals that do primarily bind to proteins (e.g. perfluorinated substances).

When does the CBC Concept not Apply?

The CBC concept also has some limitations. Crommentuijn *et al.* (1994) found that the toxicity of metals to soil invertebrates could not be explained using critical body concentrations. The way different organisms deal with accumulated metals has a large impact on the magnitude of body concentrations reached and the accompanying metal sensitivity (Rainbow 2002). Moreover, adaptation or development of metal tolerance limits the application of CBCs for metals. When the internal metal concentration does not show a monotonic relationship with the exposure concentration, it is not possible to derive CBCs. This means that whenever organisms are capable of trapping a portion of the metal in forms that are not biologically reactive, a direct relationship between body metal concentrations and toxicity may be absent or less evident (Luoma and Rainbow 2005, Vijver et al. 2004). Consequently, for metals a wide range of body concentrations with different biological significance exists. It therefore remains an open question whether the approach is applicable to modes of toxic action other than narcosis. Another important point is the question to what extent the CBC approach is applicable to assessing the effect of chemical mixtures, especially in cases the chemicals have a different mode of action.

References

Crommentuijn, T., Doodeman, C.J.A.M., Doornekamp, A., Van der Pol, J.J.C., Bedaux, J.J.M., Van Gestel, C.A.M. (1994). Lethal body concentrations and accumulation patterns determine time-dependent toxicity of cadmium in soil arthropods. Environmental Toxicology and Chemistry 13, 1781-1789.

Luoma, S.N., Rainbow, P.S. (2005). Why is metal bioaccumulation so variable? Biodynamics as a unifying concept. Environmental Science and Technology 39, 1921-1931

McCarty, L.S. (1991). Toxicant body residues: implications for aquatic bioassays with some organic chemicals. In: Mayes, M.A., Barron, M.G. (Eds.), Aquatic Toxicology and Risk Assessment: Fourteenth Volume. ASTM STP 1124. Philadelphia: American Society for Testing and Materials. pp. 183-192. DOI: 10.1520/STP23572S

McCarty, L.S., Mackay, D. (1993). Enhancing ecotoxicological modeling and assessment. Environmental Science and Technology 27, 1719-1727

Pawlisz, A.V., Peters, R.H. (1993). A test of the equipotency of internal burdens of nine narcotic chemicals using Daphnia magna. Environmental Science and Technology 27, 2801-2806

Rainbow P.S. (2002). Trace metal concentrations in aquatic invertebrates: why and so what? Environmental Pollution 120, 497-507.

U.S. EPA. (2003). In: Methodology for Deriving Ambient Water Quality Criteria for the Protection of Human Health: Technical Support Document. Volume 2. United States Environmental Protection Agency, Washington, D.C: Development of National Bioaccumulation Factors.

Vijver M.G., Van Gestel, C.A.M., Lanno, R.P., Van Straalen, N.M., Peijnenburg, W.J.G.M. (2004) Internal metal sequestration and its ecotoxicological relevance: a review. Environmental Science and Technology 18, 4705-4712.

4.2. TOXICODYNAMICS & MOLECULAR INTERACTIONS

If you want to re-use this section, for e.g. in your electronic learning environment, feel free to copy this url: https://maken.wikiwijs.nl/162477/4_2__Toxicodynamics___Molecular_Interactions

4.2.1. Protein Inactivation

Introduction

Proteins play an important role in essential biochemical processes including catalysis of metabolic reactions, DNA replication and repair, transport of messengers (e.g. hormones), or receptor responses to such messengers. Many toxic compounds exert their toxic action by binding to a protein and thereby disturbing these vital protein functions.

Inhibition of the Protein Transport Function

Binding of xenobiotic compounds to a transporter protein may hamper binding of the natural ligand of the protein, thereby inhibiting the transporter function of the protein. An example of such inhibition is the binding of halogenated phenols to transthyretin (TTR). TTR is a transport protein for thyroid hormones, present in the blood. It has two binding places for the transport of thyroid hormone, i.e. mainly thyroxine (T4) in mammals and mainly triiodothyronine (T3) in other vertebrates (Figure 1). Compounds with high structural resemblance with thyroid hormone (especially halogenated phenols, such as hydroxylated metabolites of PCBs or PBDEs), are capable to compete with thyroid hormone for TTR binding. Apart from the fact that this enhances distribution of the toxic compounds, this also causes an increase of unbound thyroid hormone in the blood, which is then freely available for uptake in the liver, metabolic conjugation, and urinary excretion. Ultimately, this may lead to decreased thyroid hormone levels in the blood.

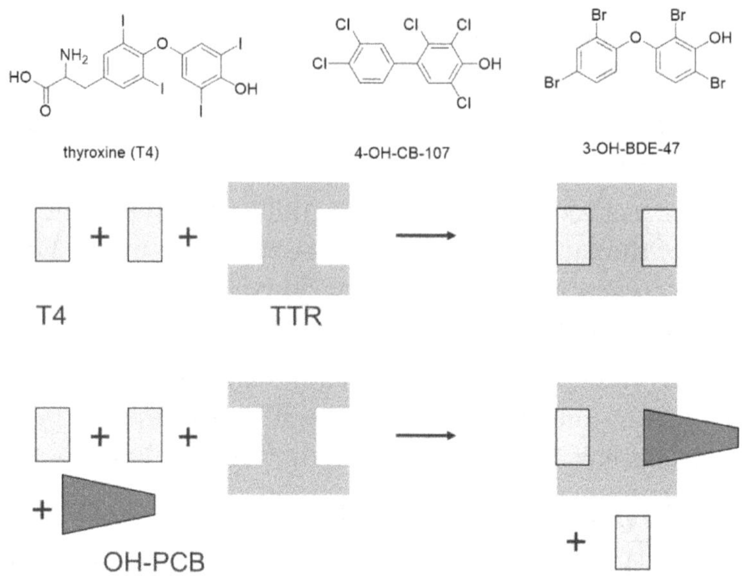

Figure 1: Structural resemblance between T4, a hydroxylated PCB metabolite (4-OH-CB-107) and a hydroxylated PBDE metabolite (3-OH-BDE-47). The lower panel illustrates how halogenated phenols (red; e.g. OH-PCB), given their structural resemblance with T_4, can compete with T4 (cyan) for TTR-binding (pink), thereby increasing the levels of unbound T4.

Inhibition of the Protein Enzymatic Activity

Proteins involved in the catalysis of a metabolic reaction are called enzymes. The general formula of such a reaction is

$$Substrate \xrightarrow{enzyme} Product$$

Binding of a toxic compound to an enzyme usually causes an inhibition of the enzyme activity, i.e. a decrease in the conversion rate of the endogenous substrate(s) of the enzyme into its/their corresponding product(s). In practice, this causes a toxic response due to a surplus of substrate and/or a deficit of product. One of the classical examples of enzyme inhibition by toxic compounds is the inhibition of the enzyme acetylcholinesterase (AChE) by organophosphate insecticides. AChE catalyzes the hydrolysis of the neurotransmitter acetylcholine (ACh), in the cholinergic synapses. During transfer of an action potential from one cell to the other, ACh is released in these synapses from the presynaptic cell into the synaptic cleft in order to stimulate the acetylcholine-receptor (AChR) on the membrane of the postsynaptic cell. AChE, which is also present in these synapses, is then responsible to break down the ACh into acetic acid and choline:

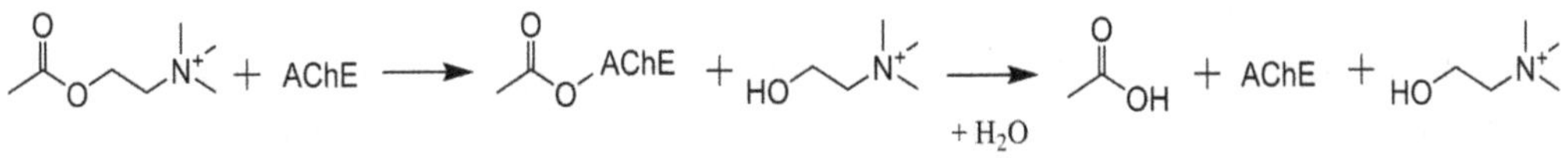

By covalent binding to serine residues in the active site of the AChE enzyme, organophosphate insecticides can inhibit this reaction causing accumulation of the ACh neurotransmitter in the synapse (Fig. 2). As a consequence, the AChR is overstimulated causing convulsions, hypertension, muscle weakness, salivation, lacrimation, gastrointestinal problems, and slow heartbeat.

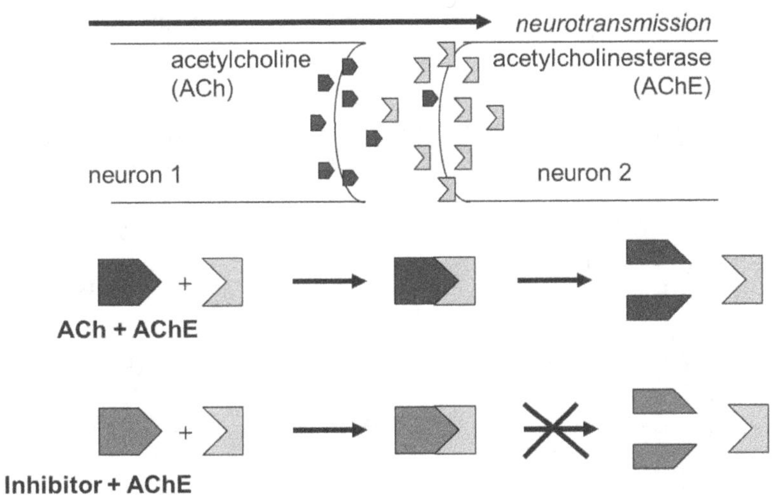

Figure 2: ACh (blue) is released from the presynaptic neuron into the synapse where it merges to and activates the AChR present on membrane of the postsynaptic cell (not shown). Meanwhile, AChE (grey) present in the synaptic cleft hydrolyses the ACh neurotransmitter to avoid overstimulation of the postsynaptic membrane. Organophosphate insecticides (red) bind to the AChE and prevent its reaction with ACh, causing accumulation of ACh.

Irreversible vs Reversible Enzyme Inhibition

Organophosphate insecticides bind covalently to the AChE enzyme thereby causing irreversible enzyme inhibition. Irreversible enzyme inhibition progressively increases in time following first-order kinetics (link to section on Bioaccumulation and kinetic modelling). Recovery of enzyme activity can only be obtained by de novo synthesis of enzymes. In contrast to AChE inhibition, inhibition of the T4 transport function of TTR is reversible because the halogenated phenols bind to TTR in a non-covalent way. Similarly, non-covalent binding of a toxic compound to an enzyme causes reversible inhibition of the enzyme activity.

In addition to covalent and non-covalent enzyme binding, irreversible enzyme inhibition may occur when toxic compounds cause an error during enzyme synthesis. For instance, ions of essential metals, which are present as cofactors in the active site of many enzymes, may be replaced by ions of other metals during enzyme synthesis, yielding inactive enzymes. A classic example of such decreased enzyme activity is the inhibition of δ-aminolevulinic acid dehydratase (δ-ALAD) by lead. In this case, lead replaces zinc in the active site of the enzyme, thereby inhibiting a catalytic step in the synthesis of a precursor of heme, a cofactor of the protein hemoglobulin (link to section on Toxicity mechanisms of metals).

With respect to reversible enzyme inhibition, three types of inhibition can be distinguished, i.e. competitive, non-competitive, and uncompetitive inhibition (Figure 3).

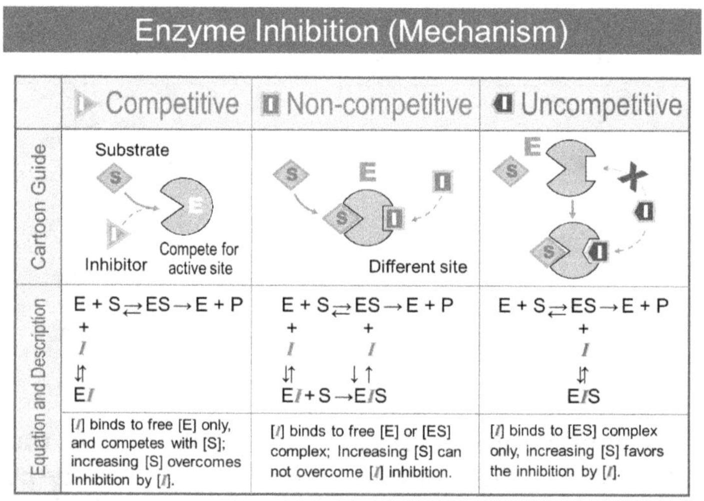

Figure 3: Three types of reversible enzyme inhibition, i.e. competitive (left), non-competitive (middle), and uncompetitive (right) binding. See text for further explanation. Source: juang.bst.ntu.edu.tw/files/Enz04%20inhibition.PPT

Competitive inhibition refers to a situation where the chemical competes ("fights") with the substrate for binding to the active site of the enzyme. Competitive inhibition is very specific, because it requires that the inhibitor resembles the substrate and fits in the same binding pocket of the active site. The TTR-binding example described above is a typical example of competitive inhibition between thyroid hormone and halogenated phenols for occupation of the TTR-binding site. A more classic example of competitive inhibition is the inhibition of beta-lactamase by penicillin. Beta-lactamase is an enzyme responsible for the hydrolysis of beta-lactam, which is the final step in bacterial cell wall synthesis. By defective cell wall synthesis, penicillin is an antibiotic causing bacterial death.

Non-competitive inhibition refers to a situation where the chemical binds to an allosteric site of the enzyme (i.e. not the active site), thereby causing a conformational change of the active site. As a consequence, the substrate cannot enter the active site, or the active site becomes inactive, or the product cannot be released from the active site. For instance, echinocandin antifungal drugs non-competitively inhibit the enzyme 1,3-beta glucan synthase, which is responsible for the synthesis of beta-glucan, a major constituent of the fungal cell wall. Lack of beta-glucan in fungal cell walls prevents fungal resistance against osmotic forces, leading to cell lysis.

Uncompetitive inhibition refers to a situation where the chemical can only bind to the enzyme if the substrate is simultaneously bound. Substrate binding leads to a conformational change of the enzyme, which leads to the formation of an allosteric binding site for the inhibitor. Uncompetitive inhibition is more common in two-substrate enzyme reactions than in one-substrate enzyme reactions. An example of uncompetitive inhibition is the inhibition by lithium of the enzyme inositol mono phosphatase (IMPase), which is involved in recycling of the second messenger inositol-3-phospate (I3P) (link to section on Receptor interaction). IMPase is involved in the final step of dephosphorylating inositol monophosphate into inositol. Since lithium is the primary treatment for bipolar disorder, this observation has led to the inositol depletion hypothesis that inhibition of inositol phosphate metabolism offers a plausible explanation for the therapeutic effects of lithium.

4.2.2. Receptor Interaction

Introduction

Receptor proteins specifically bind and respond to endogenous signalling ligands such as hormones, prostaglandins, growth factors, or neurotransmitters, by causing a typical cellular response. Receptor proteins can be located in the cell membrane, in the cytosol, and in the nucleus of a cell. Agonistic receptor ligands activate the receptor protein whereas antagonistic ligands inactivate the receptor and prevent (endogenous) agonists from activating the receptor (Figure 1). Based on the role of the receptor protein, binding by ligands may interfere with:

- ion channels
- G-protein coupled receptors
- enzyme-linked receptors
- nuclear receptors.

Xenobiotic ligands can interfere with these cellular responses by acting as agonistic or antagonistic ligands.

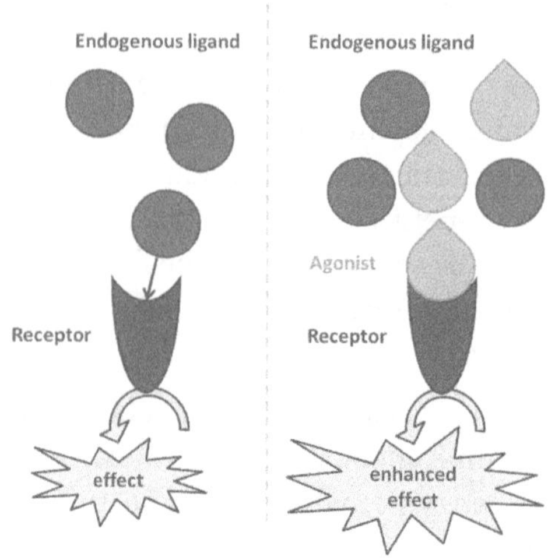

Figure 1: Activation by the endogenous ligand of a receptor leads to an effect. An agonistic compound may also activate the receptor and leads in cooperation with the endogenous ligand to an enhanced effect. An antagonistic compound also has binding affinity for the receptor, but cannot activate it. Instead, it prevents the endogenous ligand from binding, and activating the receptor, thereby preventing the effect.

Ion channels

Ion channels are transmembrane protein complexes that transport ions across a phospholipid bilayer membrane. Ion channels are especially important in neurotransmission, when stimulating neurotransmitters (e.g. acetylcholine or ACh) bind to the (so-called ionotropic) receptor part of the ion channel and open the ion channel for a very short (i.e. millisecond) period of time. As a result, ions can cross the membrane causing a change in transmembrane potential (Figure 2). On the other hand, receptor-binding by inhibiting neurotransmitters (e.g. gamma-aminobutyric acid or GABA) prevents the opening of ion channels.

Compounds interfering with sodium channels, for instance, are neurotoxic compounds (see section on Neurotoxicity). They can either block the ion channels or keep them in a prolonged or permanently open state. Many compounds known to interfere with ion channels are natural toxins. For instance, tetrodotoxin (TTX), which is produced by marine bacteria and highly accumulated in puffer fish, and saxitoxin, which is produced by dinoflagellates and is accumulated in shellfish are capable of blocking voltage-gated sodium channels in nerve cells. In contrast, ciguatoxin, which is another persistent toxin produced by dinoflagellates that accumulates in predatory fish positioned high in the food chain, causes prolongation of the opening of voltage-gated sodium channels. Some pesticides like DDT and pyrethroid insecticides also prevent closure of voltage-gated sodium channels in nerve cells. As a consequence, full repolarization of the membrane potential is not achieved. As a consequence, the nerve cells do not reach the resting potential and any new stimulus that would be too low to reach the threshold for depolarization under normal conditions, will now cause a new action potential. In other words, the nerve cells become hyperexcitable and undergo a series of action potentials (repetitive firing) causing tremors and hyperthermia.

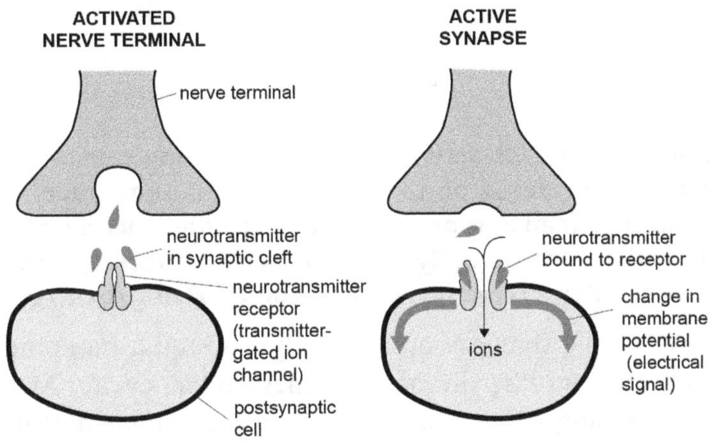

Figure 2. The acetylcholine receptor (AChR) is a sodium channel. During neurotransmission from the presynaptic to the postsynaptic cell, binding of the neurotransmitter acetylcholine (ACh) to AChR causes opening of the sodium channel allowing depolarisation of the postsynaptic membrane and propagation of the action potential. Drawn by Evelin Karsten-Meessen.

G-protein Coupled Receptors (GPCRs)

GPCRs are transmembrane receptors that transfer an extracellular signal into an activated G-protein that is connected to the receptor on the intracellular side of the membrane. G-proteins are heterotrimer proteins consisting of three subunits alpha, beta, and gamma, of which the alpha subunit – in inactivated form – contains a guanosine diphosphate (GDP) molecule. Upon binding by endogenous ligands such as hormones, prostaglandins, or neurotransmitters (i.e. the signal or “first messenger”) to the (so-called metabotropic) receptor, a conformational change in the GPCR complex leads to an exchange of the GDP for a guanosine triphosphate (GTP) molecule in the alpha monomer part of the G-protein, causing release of the activated alpha subunit from the beta/gamma dimer part. The activated alpha monomer can interact with several target enzymes causing an increase in “second messengers” starting signal transduction pathways (see point 3 Enzyme-linked receptors). The remaining beta-gamma complex may also move along the inner membrane surface and affect the activity of other proteins (Figure 3).

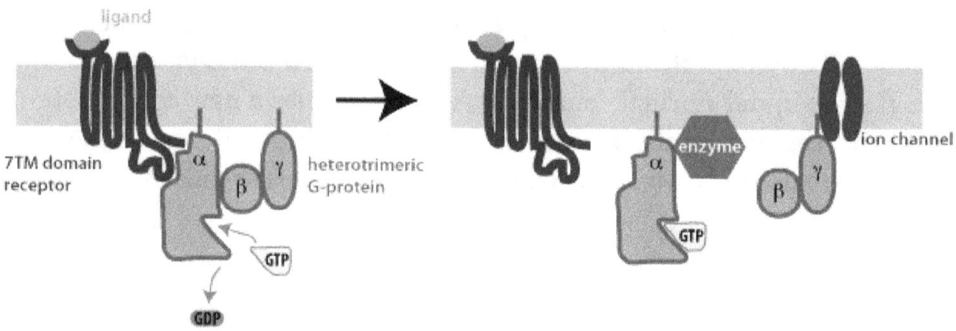

Figure 3. Mechanism of GPCR-activation: ligand binding causes a conformational change leading to the release of an activated alpha monomer, which interacts with a target enzyme (causing an increase of second messengers), and a beta-gamma dimer, which may directly affect activity of other proteins (e.g. an ion channel). Source: http://courses.washington.edu/conj/bess/gpcr/gpcr.htm

Two major enzymes that are activated by the alpha monomer are adenylyl cyclase causing an increase in second messenger cyclic AMP (cAMP) and phospholipase C causing an increase in second messenger diacylglycerol (DAG). In turn, cAMP and DAG activate protein kinases, which can phosphorylate many other enzymes. Activated phospholipase C also causes an increase in levels of the second messenger inositol-3-phosphate (I3P), which opens ion channels in the endoplasmic reticulum causing a release of calcium from the endoplasmic store, which also acts as a second messenger. On the other hand, the increase in cytosolic calcium levels is simultaneously tempered by the beta/gamma dimer, which can inhibit voltage-gated calcium channels in the cell membrane. Ultimately, the GPCR signal is extinguished by slow dephosphorylation of GTP into GDP by the activated alpha monomer, causing it to rearrange with the beta/gamma dimer into the original inactivated trimer G-protein (see also https://courses.washington.edu/conj/bess/gpcr/gpcr.htm).

The most well-known example of disruption of GPCR signalling is by cholera toxin (see text block Cholera toxin below).

Despite the recognized importance of GPRCs in medicine and pharmacology, little attention has so-far been paid in toxicology to interaction of xenobiotics with GPCRs. Although a limited number of studies have demonstrated that endocrine disrupting compounds including PAHs, dioxins, phthalates, bisphenol-A, and DDT can interact with GPCR signalling, the toxicological implications of these interactions (especially with respect to disturbed energetic metabolism) remain subject for further research (see review by Le Ferrec and Øvrevik, 2018).

Cholera Toxin

Cholera toxin is a so-called AB exotoxin by Vibrio cholerae bacteria, consisting of an "active" A-part and a "binding" B-part (see http://www.sumanasinc.com/webcontent/animations/content/diphtheria.html). Upon binding by the B-part to the intestinal epithelium membrane, the entire AB complex is internalized into the cell via endocytosis, and the active A-part is released. This A-part adds an ADP-ribose group to G-proteins making the GTP dephosphorylation of activated G-proteins impossible. As a consequence, activated G-proteins remain in a permanent active state, adenylyl cyclase is permanently activated and cAMP levels rise, which in turn cause an imbalance in ion housekeeping, i.e. an excessive secretion of chloride ions to the gut lumen and a decreased uptake of sodium ions from the gut lumen. Due to the increased osmotic pressure, water is released to the gut lumen causing dehydration and severe diarrhoea ("rice-water stool").

Enzyme-linked Receptors

Enzyme-linked receptors are transmembrane receptors that transfer an extracellular signal into an intracellular enzymatic activity. Most enzyme-linked receptors belong to the family of receptor tyrosine kinase (RTK) proteins. Upon binding by endogenous ligands such as hormones, cytokines, or growth factors (i.e. the signal or primary messenger) to the extracellular domain of the receptors, the receptor monomers dimerize and develop kinase activity, i.e. become capable of coupling of a phosphate group donated by a high-energy donor molecule to an acceptor protein. The first substrate for this phosphorylation activity is the dimerized receptor itself, which accepts a phosphate group donated by ATP on its intracellular tyrosine residues. This autophosphorylation is the first step of a signalling pathway consisting of a cascade of subsequent phosphorylation steps of other kinase proteins (i.e. signal transduction), ultimately leading to transcriptional activation of genes followed by a cellular response (Figure 4).

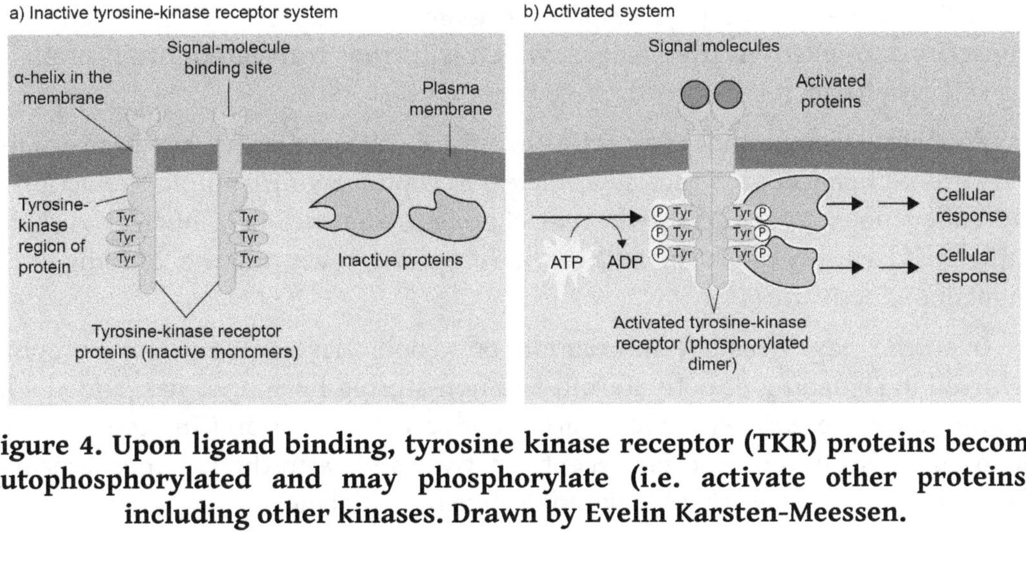

Figure 4. Upon ligand binding, tyrosine kinase receptor (TKR) proteins become autophosphorylated and may phosphorylate (i.e. activate other proteins), including other kinases. Drawn by Evelin Karsten-Meessen.

Xenobiotic compounds can interfere with these signalling pathways in many different ways. Compounds may avoid binding of the endogenous ligand, by blocking the receptor or by chelating the endogenous ligands.

Most RTK inhibitors inhibit the kinase activity directly by acting as a competitive inhibitor for ATP binding to the tyrosine residues. Many RTK inhibitors are used in cancer treatment, because RTK overactivity is typical for many types of cancer. This overactivity may for instance be caused by increased levels of receptor-activating growth factors, or to spontaneous dimerization when the receptor is overexpressed or mutated).

Nuclear Receptors

Nuclear receptors are proteins that are activated by endogenous compounds (often hormones) leading ultimately to expression of genes specifically regulated by these receptors. Apart from ligand binding, activation of most nuclear receptors requires dimerization with a coactivating transcription factor. While some nuclear receptors are located in the nucleus in inactive form (e.g. the thyroid hormone receptor), most nuclear receptors are located in the cytosol, where they are bound to co-repressor proteins (often heat-shock proteins) keeping them in an inactive state.

Upon ligand binding to the ligand binding domain (LBD) of the receptor, the co-repressor proteins are released and the receptor either forms a homodimer with a similar activated nuclear receptor or forms a heterodimer with a different nuclear receptor, which is often the retinoid-X receptor (RXR) for nuclear hormone receptors. Before or after dimerization, activated nuclear receptors are translocated to the nucleus. In the nucleus, they bind through their DNA-binding domain (DBD, or "zinc finger") to a responsive element in the DNA located in the promotor region of receptor-responsive genes. Consequently, these genes are transcribed to mRNA in the nucleus, which is further translated into proteins in the cell cytoplasm, see Figure 5).

Xenobiotic compounds may act as agonist or antagonists of nuclear receptor activation. Chemicals that act as a nuclear receptor agonist mimic the action of the endogenous activator(s), whereas chemicals that act as a nuclear receptor antagonist basically block the LBD of the receptor, preventing the binding of the endogenous activator(s).

Over the past decades, interaction of xenobiotics with nuclear receptors involved in signalling of both steroid and non-steroid hormones has gained a lot of attention of researchers investigating endocrine disruption (link to section on Endocrine Disruption). Nuclear receptor activation is also the key mechanism in dioxin-like toxicity (see text block dioxin-like toxicity below).

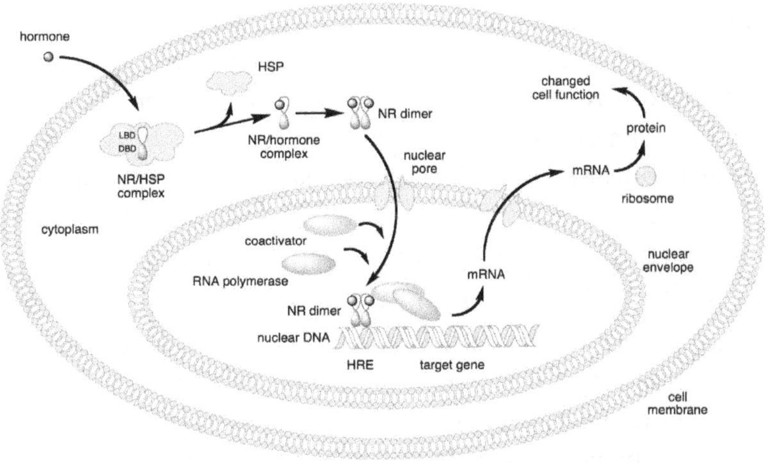

Figure 5. Activation of a cytosolic nuclear receptor (NR). Upon ligand binding (e.g. a hormone), the heat shock proteins (HSP) dissociate from the ligand-receptor complex, which forms a heterodimer before entering the nucleus. After recruiting other coactivating transcription factors, the activated dimer binds to the hormone response element (HRE). RNA polymerase binds to this complex and starts transcription of mRNA, which is excreted from the nucleus into the cytosol and transcribed in corresponding proteins. Source: https://upload.wikimedia.org/wikipedia/commons/3/3f/Nuclear_receptor_action.png

Further Reading

Boelsterli, U.A. (2009). Mechanistic Toxicology (2nd edition). Informa Healthcare, New York, London.

Denison, M.S., Nagy, S.R. (2003). Activation of the aryl hydrocarbon receptor by structurally diverse exogenous and endogenous chemicals. Annual Reviews of Pharmacology and Toxicology 43, 309–334.

Le Ferrec, E., Øvrevik J. (2018). G-protein coupled receptors (GPCR) and environmental exposure. Consequences for cell metabolism using the b-adrenoceptors as example. Current Opinion in Toxicology 8, 14-19. https://courses.washington.edu/conj/bess/gpcr/gpcr.htm

4.3. TOXICITY TESTING

Introduction

Laboratory toxicity tests may provide insight into the potential of chemicals to bioaccumulate in organisms and into their hazard, the latter usually being expressed as toxicity values derived from concentration-response relationships. Section 4.3.1 on Bioaccumulation testing describes how to perform tests to assess the bioaccumulation potential of chemicals in aquatic and terrestrial organisms, and under static and dynamic exposure conditions. Basic to toxicity testing is the establishment of a concentration-response relationship, which relates the endpoint measured in the test organisms to exposure concentrations. Section 4.3.2 on Concentration-response relationships elaborates on the calculation of the relevant toxicity parameters like the median lethal concentration (LC50) and the medium effective concentration (EC50) from such toxicity tests. It also discusses the pros and cons of different methods for analyzing data from toxicity tests.

Several issues have to be addressed when designing toxicity tests that should enable assessing the environmental or human health hazard of chemicals. This concerns among others the selection of test organisms (see section 4.3.4 on the Selection of test organisms for ecotoxicity testing), exposure media, test conditions, test duration and endpoints, but also requires clear criteria for checking the quality of toxicity tests performed (see below). Different whole organism endpoints that are commonly used in standard toxicity tests, like survival, growth, reproduction or avoidance behavior, are discussed in section 4.3.3 on Endpoints. The sections 4.3.4 to 4.3.7 are focusing on the selection and performance of tests with organisms representative of aquatic and terrestrial ecosystems. This includes microorganisms (section 4.3.6), plants (section 4.3.5), invertebrates (section 4.3.4) and vertebrate test organisms (e.g. fish: section 4.3.4 on ecotoxicity tests, and birds: section 4.3.7). Testing of vertebrates, including fish (section 4.3.4) and birds (section 4.3.7), is subject to strict regulations, aimed at reducing the use of test animals. Data on the potential hazard of chemicals to human health therefore preferably have to be obtained in other ways, like by using in vitro test methods (section 4.3.8), by using data from post-registration monitoring of exposed humans (section 4.3.9 on Human toxicity testing), or from epidemiological analysis on exposed humans (section 4.3.10).

Inclusion of Novel Endpoints in Toxicity Testing

Traditionally, toxicity tests focus on whole organism endpoints, with survival, growth and reproduction being the most measured parameters (section 4.3.3). In case of vertebrate toxicity testing, also other endpoints may be used addressing effects at the level of organs or tissues (section 4.3.9 on human toxicity testing). Behavioural (e.g. avoidance behavior) and biochemical endpoints, like enzyme activity, are also regularly included in toxicity testing with vertebrates and invertebrates (sections 4.3.3, 4.3.4, 4.3.7, 4.3.9).

With the rise of molecular biology, novel techniques have become available that may provide additional information on the effects of chemicals. Molecular tools may, for instance, be applied in molecular epidemiology (section 4.3.11) to find causal relationships between health effects and the exposure to chemicals. Toxicity testing may also use gene expression responses (transcriptomics; section 4.3.12) or changes in metabolism (metabolomics; section 4.3.13) in relation to chemical exposures to help unraveling the mechanism(s) of action of chemicals. A major challenge still is to explain whole organism effects from such molecular responses.

Standardization of Tests

The standardization of tests is organized by international bodies like the Organization for Economic Co-operation and Development (OECD), the International Standardization Organization (ISO), and ASTM International (formerly known as the American Society for Testing and Materials). Standardization aims at reducing variation in test outcomes by carefully describing the methods for culturing and handling the test organisms, the procedures for performing the test, the properties and composition of test media, the exposure conditions and the analysis of the data. Standardized test guidelines are usually based on extensive testing of a method by different laboratories in a so-called round-robin test.

Regulatory bodies generally require that toxicity tests supporting the registration of new chemicals are performed according to internationally standardized test guidelines. In Europe, for instance, all toxicity tests submitted within the framework of REACH have to be performed according to the OECD guidelines for the testing of chemicals (see section on Regulation of chemicals).

Quality Control of Toxicity Tests

Since toxicity tests are performed with living organisms, this inevitably leads to (biological) variation in outcomes. Coping with this variation requires the use of sufficient replication, careful test designs and good choice of endpoints (section 4.3.3) to enable proper estimates of relevant toxicity data.

In order to control the quality of the outcome of toxicity tests, several criteria have been developed, which mainly apply to the performance of the test organisms in the non-exposed controls. These criteria may e.g. require a minimum % survival of control organisms, a minimum growth rate or number of offspring being produced by the controls and limited variation (e.g. <30%) of the replicate control growth or reproduction data (sections 4.3.4, 4.3.5, 4.3.6, 4.3.7). When tests do not meet these criteria, the outcome is prone to doubts, as for instance a poor control survival will make it hard to draw sound conclusions on the effect of the test chemical on this endpoint. As a consequence, tests that do not meet these validity criteria may not be accepted by other scientists and by regulatory authorities.

In case the test chemical is added to the test medium using a solvent, toxicity tests should also include a solvent control, in addition to a regular non-exposed control (see section 4.3.4 on the selection of test organisms for ecotoxicity testing). In case the response in the solvent control differs significantly from that in the negative control, the solvent control will be used as the control for analyzing the effects of the test chemical. The negative control will then only be used to check if the validity criteria have been met and to monitor the condition of the test organisms. In case the responses in the negative control and the solvent control do not differ significantly, both controls can be pooled for the data analysis.

Most test guidelines also require frequent testing of a *positive control*, a chemical with known toxicity, to check if the long-term culturing of the test organisms does not lead to changes in their sensitivity.

4.3.1. Bioaccumulation Testing

Bioaccumulation is defined as the uptake of chemicals in organisms from the environment. The degree of bioaccumulation is usually indicated by the bioconcentration factor (BCF) in case the exposure is via water, or the biota-to-soil/sediment accumulation factor (BSAF) for exposure in soil or sediment (see section on Bioaccumulation).

Because of the potential risk for food-chain transfer, experimental determination of the bioaccumulation potential of chemicals is usually required in case of a high lipophilicity ($\log_{Kow} > 3$), unless the chemical has a very low persistency. For very persistent chemicals, experimental determination of bioaccumulation potential may already be triggered at $\log_{Kow} > 2$. The experimental determination of BCF and BSAF values makes use of static or dynamic exposure systems.

In *static tests*, the medium is dosed once with the test chemical, and organisms are exposed for a certain period of time after which both the organisms and the test medium are analyzed for the test chemical. The BCF or BSAF are calculated from the measured concentrations. There are a few concerns with this way of bioaccumulation testing.

First, exposure concentrations may decrease during the test, e.g. due to (bio) degradation, volatilization, sorption to the walls of the test container, or uptake of the test compound by the test organisms. As a consequence, the concentration in the test medium measured at the start of the test may not be indicative of the actual exposure during the test. To take this into account, exposure concentrations can be measured at the start and the end of the test and also at some intermediate time points. Body concentrations in the test organisms may then be related to *time-weighted-average (TWA) exposure concentrations*. Alternatively, to overcome the problem of decreasing concentrations in aquatic test systems, continuous flow systems or passive dosing techniques can be applied. Such methods, however, are not applicable to soil or sediment tests, where repeated transfer of organisms to freshly spiked medium is the only way to guarantee more or less constant exposure concentrations in case of rapidly degrading compounds. To avoid that the uptake of the test chemical in test organisms leads to decreasing exposure concentrations, the amount of biomass per volume or mass of test medium should be sufficiently low.

Second, it is uncertain whether at the end of the exposure period steady state or equilibrium is reached. If this is not the case, the resulting BSAF or BCF values may underestimate the bioaccumulation potential of the chemical. To tackle this problem, a dynamic test may be run to assess the uptake and elimination rate constants to derive a BSAF or BCF values using uptake and elimination rate constants (see below).

Such uncertainties also apply to BCF and BSAF values obtained by analyzing organisms collected from the field and comparing body concentrations with exposure levels in the environment. Using data from field-exposed organisms on one hand have large uncertainty as it remains unclear whether equilibrium was reached, on the other hand they to do reflect exposure over time under fluctuating but realistic exposure conditions.

Dynamic tests, also indicated as uptake/elimination or toxicokinetic tests, may overcome some, but not all, of the disadvantages of static tests. In dynamic tests, organisms are exposed for a certain period of time in spiked medium to assess the uptake of the chemical, after which they are transferred to clean medium for determining the elimination of the chemical. During both the uptake and the elimination phase, at different points in time, organisms are sampled and analyzed for the test chemical. The medium is also sampled frequently to check for a possible decline of the exposure concentration during the uptake phase. Also in dynamic tests, keeping exposure concentrations constant as much as possible is a major challenge, requiring frequent renewal (see above).

Toxicokinetic tests should also include controls, consisting of test organisms incubated in the clean medium and transferred to clean medium at the same time the organisms from the treated medium are transferred. Such controls may help identifying possible irregularities in the test, such as poor health of the test organisms or unexpected (cross)contamination occurring during the test.

The concentrations of the chemical measured in the test organisms are plotted against the exposure time, and a first-order one-compartment model is fitted to the data to estimate the uptake and elimination rate constants. The (dynamic) BSAF or BCF value is then determined as the ratio of the uptake and elimination rate constants (see section on Bioconcentration and kinetic models).

In a toxicokinetics test, usually replicate samples are taken at each point in time, both during the uptake and the elimination phase. The frequency of sampling may be higher at the beginning than at the end of both phases: a typical sampling scheme is shown in Figure 1. Since the analysis of toxicokinetics data using the one-compartment model is regression based, it is generally preferred to have more points in time rather than having many replicates per sampling time. From that perspective, often no more than 3-4 replicates are used per sampling time, and 5-6 sampling times for the uptake and elimination phases each.

UPTAKE PHASE

Day	0	1	2	3	4	5	6	7	8	9	10	11	12	13	14	15	16	17	18	19	20	21
Animals	3	3	3		3			3			3				3			3				3
Medium	X							X							X							X

ELIMINATION PHASE

Day	0	1	2	3	4	5	6	7	8	9	10	11	12	13	14	15	16	17	18	19	20	21
Animals	3	3	3		3			3			3				3			3				3
Medium	X																					X

Figure 1. Sampling scheme of a toxicokinetics test for assessing the uptake and elimination kinetics of chemicals in earthworms. During the 21-day uptake phase, the earthworms are individually exposed to a test chemical in soil, and at regular intervals three earthworms are sampled. After 21 days, the remaining earthworms are transferred to clean soil for the 21-day elimination period, in which again three replicate earthworms are sampled at regular points in time for measuring the body concentrations of the chemical. Also the soil is analyzed at different points in time (marked with X in the Medium row). Drawn by the author.

Preferably, replicates are independent, so destructively sampled at a specific sampling point. Especially in aquatic ecotoxicology, mass exposures are sometimes used, having all test organisms in one or few replicate test containers. In this case, at each sampling time some replicate organisms are taken from the test container(s), and at the end of the uptake phase all organisms are transferred to (a) container(s) with clean medium.

Figure 2 shows the result of a test on the uptake and elimination kinetics of molybdenum in the earthworm *Eisenia andrei*. From the ratio of the uptake rate constant (k_1) and elimination rate constant (k_2) a BSAF of approx. 1.0 could be calculated, suggesting a low bioaccumulation potential of Mo in earthworms in the soil tested.

Another way of assessing the bioaccumulation potential of chemicals in organisms includes the use of radiolabeled chemicals, which may facilitate easy detection of the test chemical. The use of radiolabeled chemicals may however, overestimate bioaccumulation potential when no distinction is made between the parent compound and potential metabolites. In case of metals, stable isotopes may also offer an opportunity to assess bioaccumulation potential. Such an approach was also applied to distinguish the role of dissolved (ionic) Zn in the bioaccumulation of Zn in earthworms from ZnO nanoparticles. Earthworms were exposed to soils spiked with mixtures of $^{64}ZnCl_2$ and ^{68}ZnO nanoparticles. The results showed that dissolution of the nanoparticles was fast and that the earthworms mainly accumulated Zn present in ionic form in the soil solution (Laycock et al., 2017).

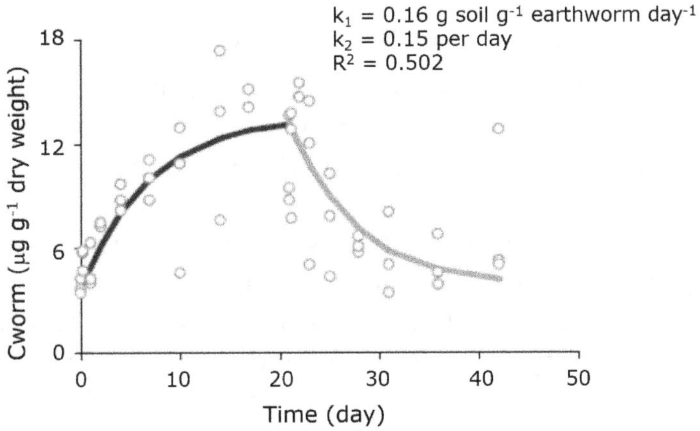

Figure 2. Uptake and elimination kinetics of molybdenum in Eisenia andrei exposed in an artificial soil spiked with a nominal Mo concentration of 10 µg g-1 dry soil. Dots represent measured internal Mo concentrations. Curves were estimated by a one-compartment model (see section on Bioconcentration and kinetic models). Parameters: k_1 = uptake rate constant [$g_{soil}/g_{worm}/d$], k_2 = elimination rate constant [d-1]. Adapted from Diez-Ortiz et al. (2010).

Standard test guidelines for assessing the bioaccumulation (kinetics) of chemicals have been published by the Organization for Economic Cooperation and Development (OECD) for sediment-dwelling oligochaetes (OECD, 2008), for earthworms/enchytraeids in soil (OECD, 2010) and for fish (OECD, 2012).

References

Diez-Ortiz, M., Giska, I., Groot, M., Borgman, E.M., Van Gestel, C.A.M. (2010). Influence of soil properties on molybdenum uptake and elimination kinetics in the earthworm *Eisenia andrei*. Chemosphere 80, 1036-1043.

Laycock, A., Romero-Freire, A., Najorka, J., Svendsen, C., Van Gestel, C.A.M., Rehkämper, M. (2017). Novel multi-isotope tracer approach to test ZnO nanoparticle and soluble Zn bioavailability in joint soil exposures. Environmental Science and Technology 51, 12756–12763.

OECD (2008). Guidelines for the testing of chemicals No. 315: Bioaccumulation in Sediment-dwelling Benthic Oligochaetes. Organization for Economic Cooperation and Development, Paris.

OECD (2010). Guidelines for the testing of chemicals No. 317: Bioaccumulation in Terrestrial Oligochaetes. Organization for Economic Cooperation and Development, Paris.

OECD (2012). Guidelines for the testing of chemicals No. 305: Bioaccumulation in Fish: Aqueous and Dietary Exposure. Organization for Economic Cooperation and Development, Paris.

4.3.2. Concentration-response Relationships

Key paradigm in human and environmental toxicology is that the dose determines the effect. This paradigm goes back to Paracelsus, stating that any chemical is toxic, but that the dose determines the severity of the effect. In practice, this paradigm is used to quantify the toxicity of chemicals. For that purpose, toxicity tests are performed in which organisms (microbes, plants, invertebrates, vertebrates) or cells are exposed to a range of concentrations of a chemical. Such tests also include incubations in non-treated control medium. The response of the test organisms is determined by monitoring selected endpoints, like survival, growth, reproduction or other parameters (see section on Endpoints). Endpoints can increase (e.g. mortality) or decrease with increasing exposure concentration (e.g. survival, reproduction, growth). The response of the endpoints is plotted against the exposure concentration, and so-called concentration-response curves (Figure 1) are fitted, from which measures of the toxicity of the chemical can be calculated.

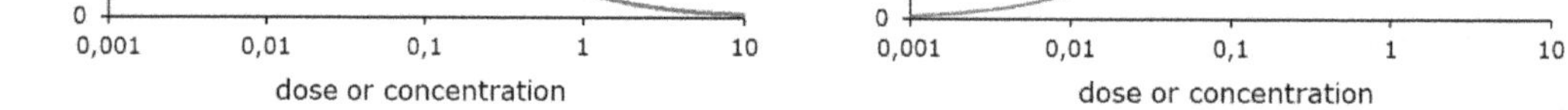

Figure 1: Concentration-response relationships. Left: response of the endpoint (e.g., survival, reproduction, growth) decreases with increasing concentration. Right: response of the endpoint (e.g., mortality, induction of enzyme activity) increases with increasing exposure concentration.

The unit of exposure, the concentration or dose, may be expressed differently depending on the exposed subject. Dose is expressed as mg/kg body weight in human toxicology and following single (oral or dermal) exposure events in mammals or birds. For other orally or dermally exposed (invertebrate) organisms, like honey bees, the dose may be expressed per animal, e.g. µg/bee. Environmental exposures generally express exposure as the concentration in mg/kg food, mg/kg soil, mg/l surface, drinking or ground water, or mg/m3 air.

Ultimately, it is the concentration (number of molecules of the chemical) at the target site that determines the effect. Consequently, expressing exposure concentrations on a molar basis (mol/L, mol/kg) is preferred, but less frequently applied.

At low concentrations or doses, the endpoint measured is not affected by exposure. At increasing concentration, the endpoint shows a concentration-related decrease or increase. From this decrease or increase, different measures of toxicity can be calculated:

- **ECx/EDx**: the "effective concentration" resp. "effective dose"; "x" denotes the percentage effect relative to an untreated control. This should always be followed by giving the selected endpoint.
- **LCx/LDx**: same, but specified for a specific endpoint: lethality.
- **EC50/ED50**: the median effect concentration or dose, with "x" set to 50%. This is the most common estimate used in environmental toxicology. This should always be followed by giving the selected endpoint.
- **LC50/LD50**: same, but specified for a specific endpoint: lethality.

The terms LCx and LDx refer to the fraction of animals responding (dying), while the ECx and EDx indicate the degree of reduction of the measured parameter. The ECx/EDx describe the overall average performance of the test organisms in terms of the parameter measured (e.g., growth, reproduction). The meaning of an LCx/LDx seems obvious: it refers to lethality of the test chemical. The use of ECx/EDx, however, always requires explicit mentioning of the endpoint it concerns.

Concentration-response models usually distinguish quantal and continuous data. Quantal data refer to constrained ("yes/no") responses and include, for instance, survival data, but may also be applicable to avoidance responses. Continuous data refer to parameters like growth, reproduction (number of juveniles or eggs produced) or biochemical and physiological measurements. A crucial difference between quantal and continuous responses is that quantal responses are population-level responses, while continuous responses can also be observed on the level of individuals. An organism cannot be half-dead, but it can certainly grow at only half the control rate.

Concentration-response models are usually sigmoidal on a log-scale and are characterized by four parameters: minimum, maximum, slope and position. The minimum response is often set to the control level or to zero. The maximum response is often set to 100%, in relation to the control or the biologically plausible maximum (e.g. 100% survival). The slope identifies the steepness of the curve, and determines the distance between the EC50 and EC10. The position parameter indicates where on the x-axis the curve is placed. The position may equal the EC50 and in that case it is named the turning point. But this in fact holds only for a small fraction of models and not for models that are not symmetrical to the EC50.

In environmental toxicology, the parameter values are usually presented with 95% confidence intervals indicating the margins of uncertainty. Statistical software packages are used to calculate these corresponding 95% confidence intervals.

Regression-based test designs require several test concentrations, and the results are dependent on the used statistical model, especially in the low-effect region. Sometimes it is simply impossible to use a regression-based design because the endpoint does not cover a sufficiently high effect range (>50% effect is typically needed for an accurate fit).

In case of quantal responses, especially survival, the slope of the concentration-response curve is an indication of the sensitivity distribution of the individuals within the population of test organisms. For a very homogenous population of laboratory test animals having the same age and body size, a steeper concentration-response curve is expected than when using field-collected animals representing a wider range of ages and body sizes (Figure 2).

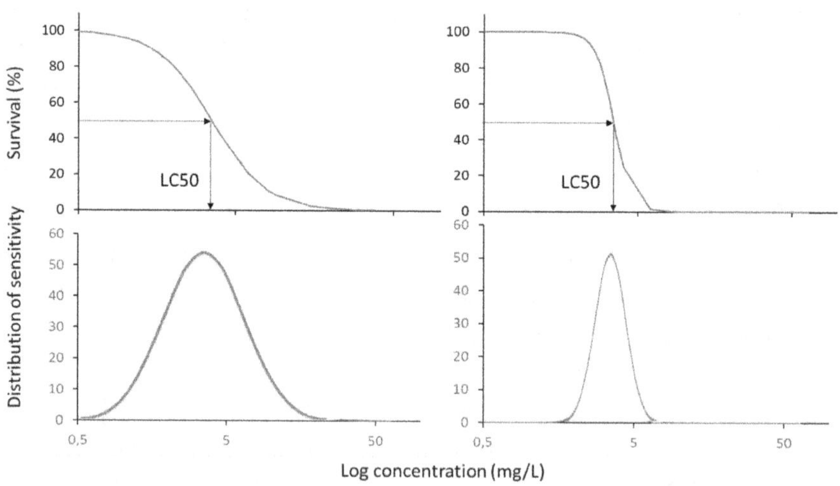

Figure 2: The steepness of the concentration-response curve for effects on survival (top) may provide insight into the sensitivity distribution of the individuals within the population of test animals (bottom). The steeper the curve the smaller the variation in sensitivity among the test organisms.

In addition to ECx values, toxicity tests may also be used to derive other measures of toxicity:

- **NOEC/NOEL:** No-Observed Effect Concentration or Effect Level
- **LOEC/LOEL:** Lowest Observed Effect Concentration or Effect Level
- **NOAEL:** No-Observed Adverse Effect Level. Same as NOEL, but focusing on effects that are negative (adverse) compared to the control.
- **LOAEL:** Lowest Observed Adverse Effect Level. Same as LOEL, but focusing on effects that are negative (adverse) compared to the control.

Where the ECx are derived by curve fitting, the NOEC and LOEC are derived by a statistical test comparing the response at each test concentration with that of the controls. The NOEC is defined as the highest test concentration where the response does not significantly differ from the control. The LOEC is the next higher concentration, so the lowest concentration tested at which the response significantly differs from the control. Figure 3 shows NOEC and LOEC values derived from a hypothetical test. Usually an Analysis of Variance (ANOVA) is used combined with a post-hoc test, e.g. Tukey, Bonferroni or Dunnett, to determine the NOEC and LOEC.

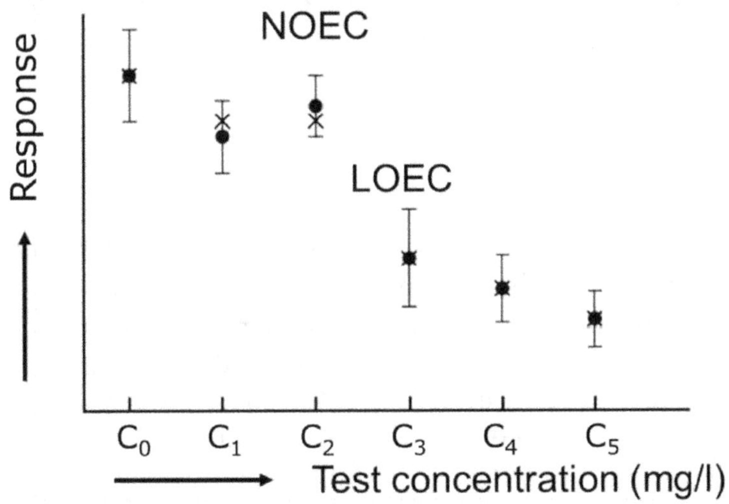

Figure 3: Derivation of NOEC and LOEC values from a toxicity test.

Most available toxicity data are NOECs, hence they are the most common values found in databases and therefore used for regulatory purposes. From a scientific point of view, however, there are quite some disadvantages related to the use of NOECs:

- Obtained by statistical test (hypothesis testing) (compared to regression analysis);

- Equal to one of the test concentrations, so not using all data from the toxicity test;
- Sensitive to the number of replicates used per exposure concentration and control;
- Sensitive to variation in response, so for differences between replicates;
- Depends on the statistical test chosen, and on the variance (σ);
- Does not have confidence intervals;
- Makes it hard to compare toxicity data between laboratories and between species.

The NOEC may, due to its sensitivity to variation and test design, sometimes be equal to or even higher than the EC50.

Because of the disadvantages of the NOEC, it is recommended to use measures of toxicity derived by fitting a concentration-response curve to the data obtained from a toxicity test. As an alternative to the NOEC, usually an EC10 or EC20 is used, which has the advantages that it is obtained using all data from the test and that it has a 95% confidence interval indicating its reliability. Having a 95% confidence interval also allows a statistical comparison of ECx values, which is not possible for NOEC values.

4.3.3. Endpoints

Introduction

Most toxicity tests performed are short-term high-dose experiments, acute tests in which mortality is often the only endpoint. Mortality, however, is a crude parameter in response to relatively high and therefore often environmentally irrelevant toxicant concentrations. At much lower and therefore environmentally more relevant toxicant concentrations, organisms may suffer from a wide variety of sublethal effects. Hence, toxicity tests gain ecological realism if sublethal endpoints are addressed in addition to mortality.

Mortality

Mortality can be determined in both acute and chronic toxicity tests. In acute tests, mortality is often the only feasible endpoint, although some acute tests take long enough to also measure sublethal endpoints, especially growth. Generally though, this is restricted to chronic toxicity tests, in which a wide variety of sublethal endpoints can be assessed in addition to mortality (Table 1).

Mortality at the end of the exposure period is assessed by simply counting the number of surviving individuals, but it can also be expressed either as percentage of the initial number of individuals or as percentage of the corresponding control.

The increasing mortality with increasing toxicant concentrations can be plotted in a dose-response relationship from which the LC50 can be derived (see section on Concentration-response relationship). If assessing mortality is non-destructive, for instance if this can be done by visual inspection, it can be scored at different time intervals during a toxicity test. Although repeated observations may take some effort, they generally do generate valuable insights in the course of the intoxication process over time.

Sublethal Endpoints in Acute Toxicity Tests

In acute toxicity tests it is difficult to assess other endpoints than mortality, since effects of toxicants on sublethal endpoints like growth and reproduction need much longer exposure times to become expressed (see section on Chronic toxicity). Incorporating sublethal endpoints in acute toxicity tests thus requires rapid responses to toxicant exposure. Photosynthesis of plants and behaviour of animals are elegant, sensitive and rapidly responding endpoints that can be incorporated into acute toxicity tests (Table 1).

Behavioural Endpoints

Behaviour is an understudied but sensitive and ecologically relevant endpoint in ecotoxicity testing, since subtle changes in animal behaviour may affect trophic interactions and ecosystem functioning. Several studies reported effects on animal behaviour at concentrations orders of magnitudes lower than lethal concentrations. Van der Geest et al. (1999) showed that changes in ventilation behaviour of fifth instar larvae of the caddisfly *Hydropsyche angustipennis* occurred at approximately 150 times lower Cu concentrations than mortality of first instar larvae. Avoidance behaviour of the amphipod *Corophium volutator* to contaminated sediments was 1,000 times more sensitive than survival (Hellou et al., 2008). Chevalier et al. (2015) tested the effect of twelve compounds covering different modes of action on the swimming behaviour of daphnids and observed that most compounds induced an early and significant swimming speed increase at concentrations near or below the 10% effective concentration (48-h EC10) of the acute immobilization test. Barata et al. (2008) reported that the short term (24 h) *D. magna* feeding inhibition assay was on average 50 times more sensitive than acute standardized tests when assessing the toxicity of a mixture of 16 chemicals in different water types combinations. These and many other examples all show that organisms may exhibit altered behaviour at relatively low and therefore often environmentally relevant toxicant concentrations.

Behavioural responses to toxicant exposure can also be very fast, allowing organisms to avoid further exposure and subsequent bioaccumulation and toxicity. A wide array of such avoidance responses have been incorporated in ecotoxicity testing (Araújo et al., 2016), including the avoidance of contaminated soil by earthworms (Eisenia fetida) (Rastetter & Gerhardt; 2018), feeding inhibition of mussels (Corbicula fluminea) (Castro et al., 2018), aversive swimming response to silver nanoparticles by the unicellular green alga Chlamydomonas reinhardtii (Mitzel et al., 2017) and by daphnids to twelve compounds covering different modes of toxic action (Chevalier et al., 2015).

Photosynthesis

Photosynthesis is a sensitive and well-studied endpoint that can be applied to identify hazardous effects of herbicides on primary producers. In bioassays with plants or algae, photosynthesis is often quantified using pulse amplitude modulation (PAM) fluorometry, a rapid measurement technique suitable for quick screening purposes. Algal photosynthesis is preferably quantified in light adapted cells as effective photosystem II (PSII) efficiency (ΦPSII) (Ralph et al., 2007; Sjollema et al., 2014). This endpoint responds most sensitively to herbicide activity, as the most commonly applied herbicides either directly or indirectly affect PSII (see section on Herbicide toxicity).

Sublethal Endpoints in Chronic Toxicity Tests

Besides mortality, growth and reproduction are the most commonly assessed endpoints in ecotoxicity tests (Table 1). Growth can be measured in two ways, as an increase in length and as an increase in weight. Often only the length or weight at the end of the exposure period is determined. This, however, includes both the growth before and during exposure. It is therefore more distinctive to measure length or weight at the beginning as well as at the end of the exposure, and then subtract the individual or average initial length or weight from the final individual length or weight. Growth during the exposure period may subsequently be expressed as percentage of the initial lengths or weight. Ideally the initial length or weight is measured from the same individuals that will be exposed. When organisms are sacrificed to measure the initial length or weight, which is especially the case for dry weight, this is not feasible. In that case a subsample from the individuals is taken apart at the beginning of the test.

Reproduction is a sensitive and ecological relevant endpoint in chronic toxicity tests. It is an integrated parameter, incorporating many different aspects of the process, that can be assessed one by one. The first reproduction parameter is the day of first reproduction. This is an ecologically very relevant parameter, as delayed reproduction obviously has strong implications for population growth. The next reproduction parameter is the amount of offspring. In this case the number of eggs, seeds, neonates or juveniles can be counted. For organisms that produce egg ropes or egg masses, both the number of egg masses as well as the number of eggs per mass can be determined. Lastly the quality of the offspring can be quantified. This can be achieved by determining their physiological status (e.g. fat content), their size, survival and finally their chance or reaching adulthood.

Table 1. Whole organism endpoints often used in toxicity tests. Quantal refers to a yes/no endpoint, while graded refers to a continuous endpoint (see section on Concentration-response relationship).

Endpoint	Acute/Chronic	Quantal/Graded
mortality	both	quantal
behaviour	acute	graded
avoidance	acute	quantal
photosynthesis	acute	graded
growth (length and weight)	mostly chronic	graded
reproduction	chronic	graded

A wide variety of other, less commonly applied sublethal whole organism endpoints can be assessed upon chronic exposure. The possibilities are endless, with some specific endpoints being designed for the effect of a single compound only, or species specific endpoints, sometimes described for only one organism. Sub-organismal endpoints are described in a separate chapter (see section on Molecular endpoints in toxicity tests).

References

Araujo, C.V.M., Moreira-Santos, M., Ribeiro, R. (2016). Active and passive spatial avoidance by aquatic organisms from environmental stressors: A complementary perspective and a critical review. Environment International 92-93, 405-415.

Barata, C., Alanon, P., Gutierrez-Alonso, S., Riva, M.C., Fernandez, C., Tarazona, J.V. (2008). A Daphnia magna feeding bioassay as a cost effective and ecological relevant sublethal toxicity test for environmental risk assessment of toxic effluents. Science of the Total Environment 405(1-3), 78-86.

Castro, B.B., Silva, C., Macario, I.P.E., Oliveira, B., Concalves, F., Pereira, J.L. (2018). Feeding inhibition in Corbicula fluminea (OF Muller, 1774) as an effect criterion to pollutant exposure: Perspectives for ecotoxicity screening and refinement of chemical control. Aquatic Toxicology 196, 25-34.

Chevalier, J., Harscoët, E., Keller, M., Pandard, P., Cachot, J., Grote, M. (2015). Exploration of Daphnia behavioral effect profiles induced by a broad range of toxicants with different modes of action. Environmental Toxicology and Chemistry 34, 1760-1769.

Hellou J., Cheeseman, K., Desnoyers, E., Johnston, D., Jouvenelle, M.L., Leonard, J., Robertson, S., Walker, P. (2008). A non-lethal chemically based approach to investigate the quality of harbor sediments. Science of the Total Environment 389, 178-187.

Mitzel, M.R., Lin, N., Whalen, J.K., Tufenkji, N. (2017). Chlamydomonas reinhardtii displays aversive swimming response to silver nanoparticles Environmental Science: Nano 4, 1328-1338.

Ralph, P.J., Smith, R.A., Macinnis-Ng, C.M.O., Seery, C.R. (2007). Use of fluorescence-based ecotoxicological bioassays in monitoring toxicants and pollution in aquatic systems: Review. Toxicological and Environmental Chemistry 89, 589–607.

Rastetter, N., Gerhardt, A. (2018). Continuous monitoring of avoidance behaviour with the earthworm Eisenia fetida. Journal of Soils and Sediments 18, 957-967.

Sjollema, S.B., Van Beusekom, S.A.M., Van der Geest, H.G., Booij, P., De Zwart, D., Vethaak, A.D., Admiraal, W. (2014). Laboratory algal bioassays using PAM fluorometry: Effects of test conditions on the determination of herbicide and field sample toxicity. Environmental Toxicology and Chemistry 33, 1017–1022.

Van der Geest, H.G., Greve, G.D., De Haas, E.M., Scheper, B.B., Kraak, M.H.S., Stuijfzand, S.C., Augustijn, C.H., Admiraal, W. (1999). Survival and behavioural responses of larvae of the caddisfly Hydropsyche angustipennis to copper and diazinon. Environmental Toxicology and Chemistry 18, 1965-1971.

4.3.4. Selection of Test Organisms - Eco Animals

Introduction

Standardized laboratory ecotoxicity tests require constant test conditions, standardized endpoints (see section on Endpoints) and good performance in control treatments. Actually, in reliable, reproducible and easy to perform toxicity tests, the test compound should be the only variable. This sets high demands on the choice of the test organisms.

For a proper risk assessment, it is crucial that test species are representative of the community or ecosystem to be protected. Criteria for selection of organisms to be used in toxicity tests have been summarized by Van Gestel et al. (1997). They include: 1. Practical arguments, including feasibility, cost-effectiveness and rapidity of the test, 2. Acceptability and standardisation of the tests, including the generation of reproducible results, and 3. Ecological significance, including sensitivity, biological validity etc. The most practical requirement is that the test organism should be easy to culture and maintain, but equally important is that the test species should be sensitive towards different stressors. These two main requirements are, however, frequently conflicting. Species that are easy to culture are often less sensitive, simply because they are mostly generalists, while sensitive species are often specialists, making it much harder to culture them.

For scientific and societal support of the choice of the test organisms, preferably they should be both ecologically and economically relevant or serve as flagship species, but again, these are opposite requirements. Economically relevant species, like crops and cattle, hardly play any role in natural ecosystems, while ecologically highly relevant species have no obvious economic value. This is reflected by the research efforts on these species, since much more is known about economically relevant species than about ecologically relevant species.

There is no species that is most sensitive to all pollutants. Which species is most sensitive depends on the mode of action and possibly also other properties of the chemical, the exposure route, its availability and the properties of the organism (e.g., presence of specific targets, physiology, etc.).

It is therefore important to always test a number of species, with different life traits, functions, and positions in the food web. According to Van Gestel et al. (1997) such a battery of test species should be:

- Representative of the ecosystem to protect, so including organisms having different life-histories, representing different functional groups, different taxonomic groups and different routes of exposure;
- Representative of responses relevant for the protection of populations and communities; and
- Uniform, so all tests in a battery should be applicable to the same test media and applying to the same test conditions, e.g. the same range of pH values.

Representation of Environmental Compartments

Each environmental compartment, water, air, soil and sediment, requires its specific set of test organisms. The most commonly applied test organisms are daphnids (*Daphnia magna*) for water, chironomids (*Chironomus riparius*) for sediments and earthworms (Eisenia fetida) for soil. For air, in the field of inhalation toxicology, humans and rodents are actually the most studied organism. In ecotoxicology, air testing is mostly restricted to plants, concerning studies on toxic gasses. Besides the most commonly applied organisms, there is a long list of other standard test organisms for which test protocols are available (Table 1; OECD site).

Table 1. Non-exhaustive list of standard ecotoxicity test species.

Environmental compartment(s)	Organism group	Test species
Water	Plant	*Myriophyllum spicatum*
Water	Plant	*Lemna*
Water	Algae	Species of choice
Water	Cyanobacteria	Species of choice
Water	Fish	*Danio rerio*
Water	Fish	*Oryzias latipes*
Water	Amphibian	*Xenopus laevis*
Water	Insect	*Chironomus riparius*
Water	Crustacean	*Daphnia magna*
Water	Snail	*Lymnaea stagnalis*
Water	Snail	*Potamopyrgus antipodarum*
Water-sediment	Plant	*Myriophyllum spicatum*
Water-sediment	Insect	*Chironomus riparius*
Water-sediment	Oligochaete worm	*Lumbriculus variegatus*
Sediment	Anaerobic bacteria	Sewage sludge
Soil	Plant	Species of choice
Soil	Oligochaete worm	*Eisenia fetida or E. andrei*
Soil	Oligochaete worm	*Enchytraeus albidus or E. crypticus*
Soil	Collembolan	*Folsomia candida or F. fimetaria*
Soil	Mite	*Hypoaspis (Geolaelaps) aculeifer*
Soil	Microorganisms	Natural microbial community
Dung	Insect	*Scathophaga stercoraria*
Dung	Insect	*Musca autumnalis*
Air-soil	Plant	Species of choice
Terrestrial	Bird	Species of choice
Terrestrial	Insect	*Apis mellifera*
Terrestrial	Insect	*Bombus terrestris/B. impatiens*
Terrestrial	Insect	*Aphidius rhopalosiphi*
Terrestrial	Mite	*Typhlodromus pyri*

Non-standard Test Organisms

The use of standard test organisms in standard ecotoxicity tests performed according to internationally accepted protocols strongly reduces the uncertainties in ecotoxicity testing. Yet, there are good reasons for deviating from these protocols. The species in Table 1 are listed according to their corresponding environmental compartment, but ignores differences between ecosystems and habitats. Soils may differ extensively in composition, depending on e.g. the sand, clay or silt content, and properties, e.g. pH and water content, each harbouring different species. Likewise, stagnant and current water have few species in common. This implies that based on ecological arguments there may be good reasons to select non-standard test organisms. Effects of compounds in streams can be better estimated with riverine insects rather than with the stagnant water inhabiting daphnids, while the compost worm Eisenia fetida is not necessarily the most appropriate species for sandy soils. The list of non-standard test organisms is of course endless, but if the methods are well documented in the open literature, there are no limitations to employ these alternative species. They do involve, however, experimental challenges, since non-standard test organisms may be hard to culture and to maintain under laboratory conditions and no protocols are available for the ecotoxicity test. Thus increasing the ecological relevance of ecotoxicity tests also increases the logistical and experimental constraints (see chapter 6 on Risk assessment).

Increasing the Number of Test Species

The vast majority of toxicity tests is performed with a single test species, resulting in large margins of uncertainty concerning the hazardousness of compounds. To reduce these uncertainties and to increase ecological relevance it is advised to incorporate more test species belonging to different trophic levels, for water e.g. algae, daphnids and fish. For deriving environmental quality standards from Species Sensitivity Distributions (see section on SSDs) toxicity data is required for minimal eight species belonging to different taxonomical groups. This obviously causes tension between the scientific requirements and the available financial resources.

References

OECD site. https://www.oecd-ilibrary.org/environment/oecd-guidelines-for-the-testing-of-chemicals-section-2-effects-on-biotic-systems_20745761.

Van Gestel, C.A.M., Léon, C.D., Van Straalen, N.M. (1997). Evaluation of soil fauna ecotoxicity tests regarding their use in risk assessment. In: Tarradellas, J., Bitton, G., Rossel, D. (Eds). Soil Ecotoxicology. CRC Press, Inc., Boca Raton: 291-317.

4.3.5. Selection of Test Organisms - Eco plants

Introduction

Photo-autotrophic primary producers use chlorophyll to convert CO_2 and H_2O into organic matter through photosynthesis under (sun)light. These primary producers are the basis of the food web and form an essential component of ecosystems. Besides serving as a food source, multicellular photo-autotrophs also form habitat for other primary producers (epiphytes) and many fauna species. Primary producers are a very diverse group, ranging from tiny unicellular pico-plankton up to gigantic trees. For standardized ecotoxicity tests, primary producers are represented by (micro)algae, aquatic macrophytes and terrestrial plants. Since herbicides are the largest group of pesticides used globally to maintain high crop production in agriculture, it is important to assess their impact on primary producers (Wang & Freemark, 1995). However, concerning testing intensity, primary producers are understudied in comparison to animals.

Standardized laboratory ecotoxicity tests with primary producers require good control over test conditions, standardized endpoints (Arts et al., 2008; see the Section on Endpoints) and growth in the controls (i.e. doubling of cell counts, length and/or biomass within the experimental period). Since the metabolism of primary producers is strongly influenced by light conditions, availability of water and inorganic carbon (CO_2 and/or HCO_3^- and CO_3^{2-}), temperature and dissolved nutrient concentrations, all these conditions should be monitored closely. The general criteria for selection of test organisms are described in the previous chapter (see the section on the Selection of ecotoxicity test organisms). For primary producers, the choice is mainly based on the available test guidelines, test species and the environmental compartment of concern.

Standardized Ecotoxicity Testing with Primary Producers

There are a number of ecotoxicity tests with a variety of primary producers standardized by different organizations including the OECD and the USEPA (Table 1). Characteristic for most primary producers is that they are growing in more than one environmental compartment (soil/sediment; water; air). As a result of this, toxicant uptake for these photo-autotrophs might be diverse, depending on the chemical and the compartment where exposure occurs (air, water, sediment/soil).

For both marine and freshwater ecosystems, standardized ecotoxicity tests are available for microalgae (unicellular micro-organisms sometimes forming larger colonies) including the prokaryotic Cyanobacteria (blue-green algae) and the eukaryotic Chlorophyta (green algae) and Bacillariophyceae (diatoms). Macrophytes (macroalgae and aquatic plants) are multicellular organisms, the latter consisting of differentiated tissues, with a number of species included in standardized ecotoxicity tests. While macroalgae grow in the water compartment only, aquatic plants are divided into groups related to their growth form (emergent; free-floating; submerged and sediment-rooted; floating and sediment-rooted) and can extend from the sediment (roots and root-stocks) through the water into the air. Both macroalgae and aquatic plants contain a wide range of taxa and are present in both marine and freshwater ecosystems.

Terrestrial higher plants are very diverse, ranging from small grasses to large trees. Plants included in standardized ecotoxicity tests consist of crop and non-crop species. An important distinction in terrestrial plants is reflected in dicots and monocots, since both groups differ in their metabolic pathways and might reflect a difference in sensitivity to contaminants.

Table 1. Open source standard guidelines for testing the effect of compounds on primary producers. All tests are performed in (micro)cosms except marked with *

Primary producer	Species	Compartment	Test number	Organisation
Microalgae & cyano-bacteria	various species	Freshwater	201	OECD 2011
	Anabaena flos-aque	Freshwater	850.4550	USEPA 2012
	Pseudokirchneriella subcapitata, Skeletonema costatum	Freshwater, Marine water	850.4500	USEPA 2012
Floating macrophytes	Lemna spp.	Freshwater	221	OECD 2006
	Lemna spp.	Freshwater	850.4400	USEPA 2012
Submerged macrophytes	Myriophyllum spicatum	Freshwater	238	OECD 2014
	Myriophyllum spicatum	Sediment (Freshwater)	239	OECD 2014
Aquatic plants*	not specified	Freshwater	850.4450	USEPA 2012
Terrestrial plants	wide variety of species	Air	227	OECD 2006
	wide variety of species	Air	850.4150	USEPA 2012
	wide variety of species (crops and non-crops)	Soil & Air	850.4230	USEPA 2012
	legumes and rhizobium symbiont	Soil & Air	850.4600	USEPA 2012
	wide variety of species (crops and non-crops)	Soil	208	OECD 2006
	wide variety of species (crops and non-crops)	Soil	850.4100	USEPA 2012
	various crop species	Soil & Air	850.4800	USEPA 2012
Terrestrial plants*	not specified	Terrestrial	850.4300	USEPA 2012

Representation of Environmental Compartments

Since primary producers can take up many compounds directly by cells and thalli (algae) or by their leaves, stems, roots and rhizomes (plants), different environmental compartments need to be included in ecotoxicity testing depending on the chemical characteristics of the contaminants. Moreover, the chemical characteristics of the compound under consideration determine if and how the compound might enter the primary producers and how it is transported through organisms.

For all aquatic primary producers, exposure through the water phase is relevant. Air exposure occurs in the emergent and floating aquatic plants, while rooting plants and algae with rhizoids might be exposed through sediment. Sediment exposure introduces additional challenges for standardized testing conditions, since changes in redox conditions and organic matter content of sediments can alter the behavior of compounds in this compartment.

All terrestrial plants are exposed through air, soil and water (soil moisture, rain, irrigation). Air exposure and water deposition (rain or spraying) directly exposes aboveground parts of terrestrial plants, while belowground plant parts and seeds are exposed through soil and soil moisture. Soil exposure introduces additional challenges for standardized testing conditions, since changes in water or sediment organic matter content of soils can alter the behavior of compounds in this compartment.

Test Endpoints

Bioaccumulation after uptake and translocation to specific cell organelles or plant tissue can result in incorporation of compounds in primary producers. This has been observed for heavy metals, pesticides and other organic chemicals. The accumulated compounds in primary producers can then enter the food chain and be transferred to higher trophic levels (see the section on Biomagnification). Although concentrations in primary producers are indicative of the presence of bioavailable compounds, these concentrations do not necessarily imply adverse effects on these organisms. Bioaccumulation measurements can therefore be best combined with one or more of the following endpoint assessments.

Photosynthesis is the most essential metabolic pathway for primary producers. The mode of action of many herbicides is therefore photosynthesis inhibition, whereby different metabolic steps can be targeted (see the section on Herbicide toxicity). This endpoint is relevant for assessing acute effects on the chlorophyll electron transport using Pulse-Amplitude-Modulation (PAM) fluorometry or as a measure of oxygen or carbon production by primary producers.

Growth represents the accumulation of biomass (microalgae) or mass (multicellular primary producers). Growth inhibition is the most important endpoint in test with primary producers since this endpoint integrates responses of a wide range of metabolic effects into a whole organism or a population response of primary producers. However, it takes longer to assess, especially for larger primary producers. Cell counts, increase in size over time for either leaves, roots, or whole organisms, and (bio)mass (fresh weight and dry weight) are the growth endpoints mostly used.

Seedling emergence reflects the germination and early development of seedlings into plants. This endpoint is especially relevant for perennial and biannual plants depending on seed dispersal and successful germination to maintain healthy populations.

Other endpoints include elongation of different plant parts (e.g. roots), necrosis of leaves, or disturbances in plant-microbial symbiont relationships.

Current Limitations and Challenges for Using Primary Producers in Ecotoxicity Tests

For terrestrial vascular plants, many crop and non-crop species can be used in standardized tests, however, for other environmental compartments (aquatic and marine) few species are available in standardized test guidelines. Also not all environmental compartments are currently covered by standardized tests for primary producers. In general, there are limited tests for aquatic sediments and there is a total lack of tests for marine sediments. Finally, not all major groups of primary producers are represented in standardized toxicity tests, for example mosses and some major groups of algae are absent.

Challenges to improve ecotoxicity tests with plants would be to include more sensitive and early response endpoints. For soil and sediment exposure of plants to contaminants, development of endpoints related to root morphology and root metabolism could provide insights into early impact of substances to exposed plant parts. Also the development of ecotoxicogenomic endpoints (e.g. metabolomics) (see the section on Metabolomics) in the field of plant toxicity tests would enable us to determine effects on a wider range of plant metabolic pathways.

References

Arts, G.H.P., Belgers, J.D.M., Hoekzema, C.H., Thissen, J.T.N.M. (2008). Sensitivity of submersed freshwater macrophytes and endpoints in laboratory toxicity tests. Environmental Pollution 153, 199-206.

Wang, W.C., Freemark, K. (1995) The use of plants for environmental monitoring and assessment. Ecotoxicology and Environmental Safety 30: 289-301.

4.3.6. Selection of Test Organisms - Microorganisms

The Importance of Microorganisms

Most organisms are microorganisms, which means they are generally too small to see with the naked eye. Nevertheless, microorganisms affect almost all aspects of our lives. Viruses are the smallest of microorganisms, the prokaryotic bacteria and archaea are bigger (in the micrometer range), and the sizes of eukaryotic microorganisms range from three to hundred micrometers. The microscopic eukaryotes have larger cells with a nucleus and come in different shapes like green algae, protists and fungi. Cyanobacteria and eukaryotic algae perform photosynthesis in the oceans, seas, brackish and freshwater ecosystems. They fix carbon dioxide into biomass and form the basis of the largest aquatic ecosystems. Bacteria and fungi degrade complex organic molecules into carbon dioxide and minerals, which are needed for plant growth.

Plants often live in symbiosis with specialized microorganisms on their roots, which facilitate their growth by enhancing uptake of water and nutrients, speeding up plant growth. Invertebrate and vertebrate animals, including humans, have bacteria and other microorganisms in their intestines to facilitate the digestion of food. Cows for example cannot digest grass without the microorganisms in their rumen. Also, termites would not be able to digest lignin, a hard to digest wood polymer, without the aid of gut fungi. Leaf cutter ants transport leaves into their nest to feed the fungi which they depend on.

Also, humans consume many foodstuffs with yeasts, fungi or bacteria for preservation of the food and a pleasant taste. Beer, wine, cheese, yogurt, sauerkraut, vinegar, bread, tempeh, sausage and may other foodstuffs need the right type of microorganisms to be palatable. Having the right type of microorganisms is also vital for human health. Human mother's milk contains oligosaccharides, which are indigestible for the newborn child. These serve as a major food source for the intestinal bacteria in the baby, which reduce the risk of dangerous infections.

This shows that the interaction between specific microorganisms and higher organisms are often highly specific. Marine viruses are very abundant and can limit algal blooms promoting a more diverse marine phytoplankton. Pathogenic viruses, bacteria, fungi and protists enhance the biodiversity of plants and animals by the following mechanism: The densest populations are more susceptible to diseases since the transmission of the disease becomes more frequent. When the most abundant species become less frequent, there is more room for the other species and biodiversity is enhanced. In agriculture, this enhanced biodiversity is unwanted since the livestock and crop are the most abundant species. That is why disease control becomes more important in high intensity livestock farming and in large monocultures of crops. Microorganisms are at the base of all ecosystems and are vital for human health and the environment.

The microbiological society has a nice video explaining why microbiology matters.

Protection Goals

The functioning of natural ecosystems on earth is threatened by many factors, such as habitat loss, habitat fragmentation, global warming, species extinction, over fertilization, acidification and pollution. Natural and man-made chemicals can exhibit toxic effects on the different organisms in natural ecosystems. Toxic chemicals released in the environment may have negative effects on biodiversity or microbial processes. In the ecosystem strongly affected by such changes, the abundance of different species could be smaller. The loss of biodiversity of the species in a specific ecosystem can be used as a measure for the degradation of the ecosystem. Humans benefit from the presence of properly functioning ecosystems. These benefits can be quantified as ecosystem services. Microbial processes contribute heavily to many ecosystem services. Groundwater for example, is often a suitable source of drinking water since microorganisms have removed pollutants and pathogens from the infiltrating water. See Section on Ecosystem services and protection goals.

Environmental Toxicity Tests

Most environmental toxicity tests are single species tests. Such tests typically determine toxicity of a chemical to a specific biological species like for example the bioluminescence by the Allivibrio fisheri bacteria in the Microtox test or the growth inhibition test on freshwater algae and cyanobacteria (see Section on Selection of test organisms – Eco plants). These tests are relatively simple using a specific toxic chemical on a specific biological species in an optimal setting. The OECD guidelines for the testing of chemicals, section 2, effects on biotic systems gives a list of standard tests. Table 1 lists different tests with microorganisms standardized by the Organization for Economic Cooperation and Development (OECD).

Table 1. Generally accepted environmental toxicity tests using microorganisms, standardized by the Organization for Economic Cooperation and Development (OECD).

OECD test No	Title	Medium	Test type
201	Freshwater algae and cyanobacteria, growth inhibition test (chapter reference)	Aquatic	Single species
209	Activated sludge, respiration inhibition test	Sediment	Process
224 (draft guideline)	Determination of the inhibition of the activity of anaerobic bacteria	Sediment	Process
217	Soil microorganisms: carbon transformation test	Soil	Process
216	Soil microorganisms: nitrogen transformation test	Soil	Process

The outcome of these tests can be summarized as EC10 values (see Section on Concentration-response relationships), which can be used in risk assessment (see Sections on Predictive risk assessment approaches and tools and on Diagnostic risk assessment approaches and tools) Basically, there are three types of tests. Single species tests, community tests and tests using microbial processes.

Single Species Tests

The ecological relevance of a single species test can be a matter of debate. In most cases it is not practical to work with ecologically relevant species since these can be hard to maintain under laboratory conditions. Each ecosystem will also have its own ecologically relevant species, which would require an extremely large battery of different test species and tests, which are difficult to perform in a reproducible way. As a solution to these problems, the test species are assumed to exhibit similar sensitivity for toxicants as the ecological relevant species. This assumption was confirmed in a number of cases. If the sensitivity distribution of a given toxicant for a number of test species would be similar to the sensitivity distribution of the relevant species in a specific ecosystem, one could use a statistic method to estimate a safe concentration for most of the species.

Toxicity tests with short incubation times are often disputed since it takes time for toxicants to accumulate in the test animals. This is not a problem in microbial toxicity tests since the small size of the test organisms allows a rapid equilibrium of the concentrations of the toxicant in the water and in the test organism. On the contrary, long incubation times under conditions that promote growth, can lead to the occurrence of resistant mutants, which will decrease the apparent sensitivity of the test organism. This selection and growth of resistant mutants cannot, however, be regarded as a positive thing since these mutants are different from the parent strain and might also have different ecological properties. In fact, the selection of antibiotic resistant microorganisms in the environment is considered to be a problem since these might transfer to pathogenic (disease promoting) microorganisms which gives problems for patients treated with antibiotics.

The OECD test no 201, which uses freshwater algae and cyanobacteria, is a well-known and sensitive single species microbial ecotoxicity test. These are explained in more detail in the Section on Selection of test organisms – Eco plants.

Community Tests

Microorganisms have a very wide range of metabolic diversity. This makes it more difficult to extrapolate from a single species test to all possible microbial species including fungi, protists, bacteria, archaea and viruses. One solution is to test a multitude of species (a whole community) exposed in a single toxicity experiment, it becomes more difficult to attribute the decline or increase of species to toxic effects. The rise and decline of species can also be caused by other factors, including species interactions. The method of Pollution-induced community tolerance is used for the detection of toxic effects on communities. Organisms survive in polluted environments only when they can tolerate toxic chemical concentrations in their habitat. During exposure to pollution the sensitive species become extinct and tolerant species take over their place and role in the ecosystem (Figure 1).

This takeover can be monitored by very simple toxicity tests using a part of the community extracted from the environment. Some tests use the incorporation of building blocks for DNA (thymidine) and protein (leucine). Other tests use different substrates for microbial growth. The observation that this part of the community becomes more tolerant as measured by these simple toxicity tests reveals that the pollutant really affects the microbial community. This is especially helpful when complex and diverse environments like biofilms, sediments and soils are studied.

Tests Using Microbial Processes

The protection of ecosystem services is fundamentally different from the protection of biodiversity. When one wants to protect biodiversity all species are equally important and are worth protecting. When one wants to protect ecosystem services only the species that perform the process have to be protected. Many contributing species can be intoxicated without having much impact on the process. An example is nitrogen transformation, which is tested by measuring the conversion of ammonium into nitrite and nitrate (see box).

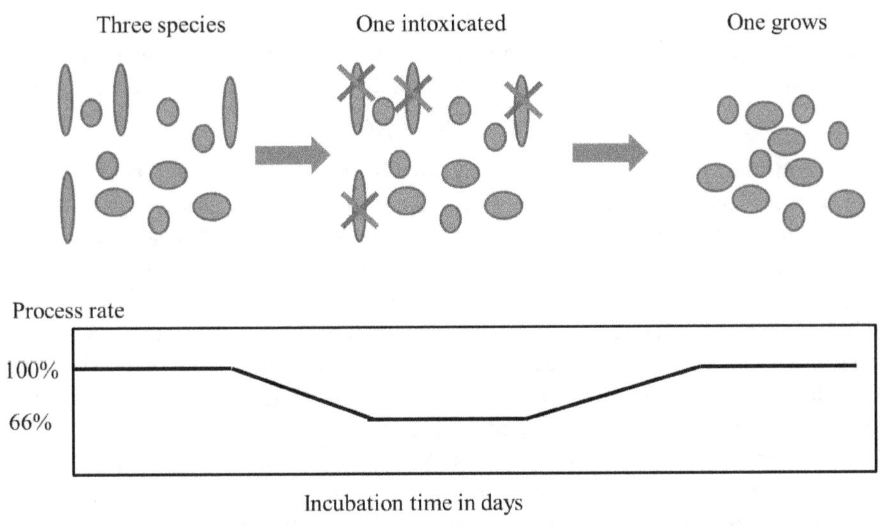

Figure 1. The effect of growth on an intoxicated process performed by different species of microorganisms. The intoxication of some species may temporarily decrease process rate, but due to growth of the tolerant species this effect soon disappears and process rate is restored. Source: Patrick van Beelen.

The inactivation of the most sensitive species can be compensated by the prolonged activity or growth of less sensitive species. The test design of microbial process tests aims to protect the process and not the contributing species. Consequently, the process tests from Table 1 seldom play a decisive role in reducing the maximum tolerable concentration of a chemical. Reason is that the single species toxicity tests generally are more sensitive since they use a specific biological species as test organism instead of a process.

Box: Nitrogen transformation test

The OECD test no. 216 Soil Microorganisms: Nitrogen Transformation Test is a very well-known toxicity test using the soil process of nitrogen transformation. The test for non-agrochemicals is designed to detect persistent adverse effects of a toxicant on the process of nitrogen transformation in soils. Powdered clover meal contains nitrogen mainly in the form of proteins which can be degraded and oxidized to produce nitrate. Soil is amended with clover meal and treated with different concentrations of a toxicant. The soil provides both the test organisms and the test medium. A sandy soil with a low organic carbon content is used to minimize sorption of the toxicant to the soil. Sorption can decrease the toxicity of a toxicant in soil. According to the guideline, the soil microorganisms should not be exposed to fertilizers, crop protection products, biological materials or accidental contaminations for at least three months before the soil is sampled. In addition, the soil microorganisms should at least form 1% of the soil organic carbon. This indicates that the microorganisms are still alive. The soil is incubated with clover meal and the toxicant under favorable growth conditions (optimal temperature, moisture) for the microorganisms. The quantities of nitrate formed are measured after 7 and 28 days of incubation. This allows for the growth of microorganisms resistant to the toxicant during the test, which can make the longer incubation time less sensitive. The nitrogen in the proteins of clover meal will be converted to ammonia by general degradation processes. The conversion of clover meal to ammonia can be performed by a multitude of species and is therefore not very sensitive to inhibition by toxic compounds.

$\text{clovermeal} \rightarrow \text{proteins} \rightarrow \text{aminoacids} \rightarrow NH^+_4 \rightarrow NO^-_2 \rightarrow NO^-_3$

The conversion of ammonia to nitrate generally is performed in two steps. First, ammonia oxidizing bacteria or archaea, oxidize ammonia into nitrite. Second, nitrite is oxidized by nitrite oxidizing bacteria into nitrate. These latter two steps are generally much slower than ammonium production, since they require specialized microorganisms. These specialized microorganisms also have a lower growth rate than the common microorganisms involved in the general degradation of proteins into amino acids. This makes the nitrogen transformation test much more sensitive compared to the carbon transformation test, which uses more common microorganisms. Under the optimal conditions in the nitrogen transformation test some minor ammonia or nitrite oxidizing species might seem unimportant since they do not contribute much to the overall process. Nevertheless these minor species can become of major importance under less optimal conditions. Under acid conditions for example, only the archaea oxidize ammonia into nitrite while the ammonia oxidizing bacteria become inhibited. The nitrogen transformation test has a minimum duration of 28 days at 20°C under optimal moisture conditions, but can be prolonged to 100 days. Shorter incubation times would make the test more sensitive.

4.3.7. Selection of Test Organisms - Birds

Introduction

Birds are seen as important models in ecotoxicology for a number of reasons:

- they are a diverse, abundant and widespread order inhabiting many human altered habitats like agriculture;

- they have physiological features that make them different from other vertebrate classes that may affect their sensitivity to chemical exposure;
- they play a specific role ecologically and fulfill essential roles in ecosystems (e.g. in seed dispersal, as biological control agents through eating insects, and removal of carcasses e.g. by vultures);
- protection goals are frequently focused on iconic species that appeal to the public.

A few specific physiological features will be discussed here. Birds are oviparous, laying eggs with hard shells. This leads to concentrated exposure (as opposed to exposure via the bloodstream as in most other vertrebrate species) to maternally transferred material, and where relevant, its metabolites. It also means that offspring receive a single supply of nutrients (and not a continuous supply through the blood stream). This makes birds sensitive to contaminants in a different way than non-oviparous vertebrates, since the embryos develop without physiological maternal interference. The bird embryo starts to regulate its own hormone homeostasis early on in its development in contrast to mammalian embryos. As a result contaminants deposited in the egg by the female bird may cause disturbance of the regulation of these embryonic processes (Murk et al., 1996). Birds have a higher body temperature (40.6 ºC) and a relatively high metabolic rate, which can impact their response to chemicals. As chicks, birds generally have a rapid growth rate, compared to many vertebrate species. Chicks of precocial (or nidifugous) species leave the nest upon hatching and, while they may follow the parents around, they are fully feathered and feed independently. They typically need a few months to grow to full size. Altricial species are naked, blind and helpless at hatch and require parental care until they fledge the nest. They often grow faster – passerines (such as swallows) can reach full size and fledge 14 days after hatching. Many bird species migrate seasonally over long distances and adaptation to this, changes their physiology and biochemical processes. Internal concentrations of organic contaminants, for example, may increase significantly due to the use of lipids stores during migration, while changes in biochemistry may increase the sensitivity of birds to the chemical.

Birds function as good biological indicators of environmental quality largely because of their position in the foodchain and habitat dependence. Protection goals are frequently focused on iconic species, for example the Atlantic puffin, the European turtle dove and the common barn owl (Birdlife International, 2018).

It was recognized early on that exposure of birds to pesticides can take place through many routes of dietary exposure. Given their association with a wide range of habitats, exposure can take place by feeding on the crop itself, on weeds, or (treated) weed seeds, on ground dwelling or foliar dwelling invertebrates, by feeding on invertebrates in the soil, such as earthworms, by drinking water from contaminated streams or by feeding on fish living in contaminated streams (Figure 1, Brooks et al., 2017). Following the introduction of persistent and highly toxic synthetic pesticides in the 1950s and prior to safety regulations, use of many synthetic organic pesticides led to wildlife losses – of birds, fish, and other wildlife (Kendall and Lacher, 1994). As a result, national and international guidelines for assessing first acute and subacute effects of pesticides on birds were developed in the 1970s. In the early 1980s tests were developed to study long-term or reproductive effects of pesticides. Current bird testing guidelines focused primarily on active ingredients used in plant protection products, veterinary medicines and biocides. In Europe the industrial chemicals regulation REACH only requires information on long-term or reproductive toxicity for substances manufactured or imported in quantities of at least 1000 tonnes per annum. These data may be needed to assess the risks of secondary poisoning by a substance that is likely to bioaccumulate and does not degrade rapidly. Secondary poisoning may occur, for example when raptors consume contaminated fish. In the United States no bird tests are required under the industrial chemicals legislation.

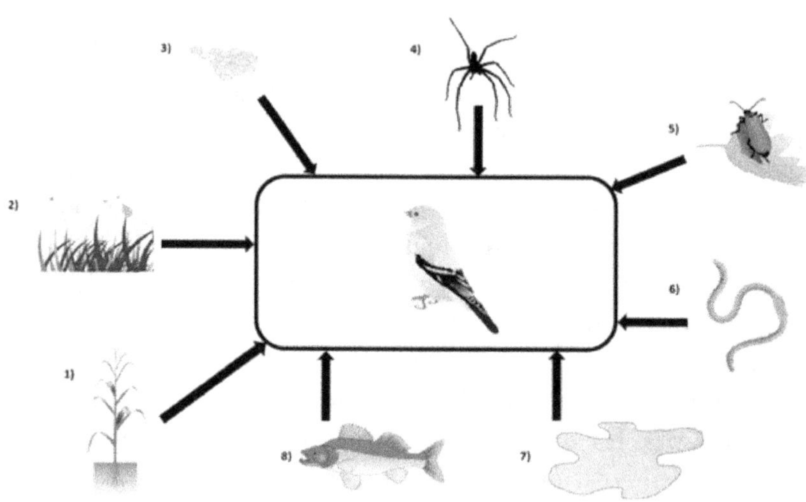

Figure 1. Potential routes of dietary exposure for birds feeding in agricultural fields sprayed with a crop protection product (pesticide). Most of the pesticide will land up in the treated crop area, but some of it may land in neighbouring surface water. Exposure to birds can therefore take place through many routes: by feeding on the crop itself (1), on weeds (2), or weed seeds (3), on ground-dwelling (4) or foliar-dwelling (5) invertebrates. Birds may also feed on earthworms living in the treated soil (6). Exposure may also occur by drinking from contaminated puddles within the treated crop area (7) or birds may feed on fish living in neighbouring contaminated surface waters (8). Based on Brooks et al. (2017).

The objective of performing avian toxicity tests is to inform an avian effects assessment (Hart et al., 2001) in order to:

- provide scientifically sound information on the type, size, frequency and pattern over time of effects expected from defined exposures of birds to chemicals.
- reduce uncertainty about potential effects of chemicals on birds.
- provide information in a form suitable for use in risk assessment.
- provide this information in a way that makes efficient use of resources and avoids unnecessary use and suffering of animals.

Bird Species Used in Toxicity Testing

Selection of bird species for toxicity testing occurs primarily on the basis of their ecological relevance, their availability and ability to adjust to laboratory conditions for breeding and testing. This means most test species have been domesticated over many years. They should have been shown to be relatively sensitive to chemicals through previous experience or published literature and ideally have available historical control data.

The bird species most commonly used in toxicity testing have all been domesticated:

- the waterfowl species mallard duck (*Anas platyrynchos)* is in the mid range of sensitivity to chemicals, an omnivorous feeder, abundant in many parts of the world, a precocial species; raised commercially and test birds show wild type plumage;
- the ground dwelling game species bobwhite quail (*Colinus virginianus)* is common in the USA and is of similar in sensitivity to mallards; feeds primarily on seeds and invertebrates; a precocial species; raised commercially and test birds show wild type plumage;
- the ground dwelling species Japanese quail (*Coturnix coturnix japonica)* occurs naturally in East Asia; feeds on plants material and terrestrial invertebrates. Domesticated to a far great extent than mallard or bobwhite quail and birds raised commercially (for eggs or for meat) are further removed genetically from the wild type. This species is unique in that the young of the year mature and breed (themselves) within 12 months;
- the passerine, altricial species zebra finch *(Taeniopygia guttata)* occurs naturally in Australia and Indonesia; they eat seeds; are kept and sold as pets; are not far removed from wild type;

- the budgerigar (*Melopsittacus undulates*) is also altricial); occurs naturally in Australia; eats seeds eating; is bred in captivity and kept and sold as pets.

Other species of birds are sometimes used for specific, often tailor-designed studies. These species include:

- the canary (*Serinus canaria domestica).*
- the rock pigeon (*Columba livia)*
- the house sparrow (*Passer domesticus),*
- red-winged blackbird (*Agelaius phoeniceus)* - US only
- the ring-necked pheasant, (*Phasianus colchicus),*
- the grey partridge (*Perdix perdix)*

Most Common Avian Toxicity Tests:

Table 1 provides an overview of all the avian toxicity tests that have been developed over the past approximately 40 years, the most commonly used guidelines, the recommended species, the endpoints recorded in each of these tests, the typical age of birds at the start of the test, the test duration and the length of exposure.

Most Common Avian Toxicity Tests:

Table 1 provides an overview of all the avian toxicity tests that have been developed over the past approximately 40 years, the most commonly used guidelines, the recommended species, the endpoints recorded in each of these tests, the typical age of birds at the start of the test, the test duration and the length of exposure.

Table 1: Most common avian toxicity tests with their recommended species and key characteristics.

Avian toxicity test	Guideline	Recommended species	Endpoints	Age at start of test	Length of study	Length of exposure
Acute oral gavage–sequential testing – average 26 birds	OECD 223	bobwhite quail, Japanese quail, zebra finch, budgerigar	mortality, clinical signs, body weight, food consumption, gross necropsy	Young birds not yet mated, at least 16 weeks old at start of test.	At least 14 days	Single oral dose at beginning of test
Acute oral gavage – 60 bird design	USEPA OCSPP 850.2100	Bobwhite quail single passerine species recommended.	See above	Young birds not yet mated, at least 16 weeks old at start of test.	14 days	Single oral dose at beginning of test
Sub-acute dietary toxicity *	OCSPP 850.2200	Bobwhite quail, mallard	See above	Mallard: 5 days old Bobwhite quail: 10-14 days old	8 days	5 days
One-generation reproduction	OECD 206 OCSPP 850.2200	Bobwhite quail, mallard, Japanese quail**	Adult body weight,food consumption, egg production, fertility, embryo survivial, hatchrate, chick survival.	Approaching first breeding season: Mallard (6 to 12 months old) Bobwhite quail: 20 -24 weeks	20 – 22 weeks	10 weeks
Avoidance testing (pen trials)	OECD Report	As closely related to species at risk as possible; eg: sparrow rock dove, pheasant, grey partridge	Food intake, mortality, Sublethal effects	Young adults if possible (depends on study design)	One to several days, depending on the study design.	One to several days, depending on the study design.

Two-generation endocrine disruptor test	OCSPP 890.2100	Japanese quail	In addition to endpoints listed for one-generation study: male sexual behaviour, biochemical, histological, and morphological endpoints	4 weeks post hatch	38 weeks	8 weeks – adult (F0) generation +14 weeks F1 generation.
Field studies to refine food residues in higher tier Bird risk assessments.	Appendix N of the EFSA Bird and Mammal Guidance.	Depends on the species at risk in the are of pesticide use.	Depends on the study design developed.	Uncontrollable in a field study	Depends on the study design developed.	Depends on the study design developed.

* This study is hardly every asked for anymore.

** Only in OECD Guideline

Acute Toxicity Testing

To assess the short-term risk to birds, acute toxicity tests must be performed for all pesticides (the active ingredient thereof) to which birds are likely to be exposed, resulting in an LD50 (mg/kg body/day) (see section on Concentration-response relationships). The acute oral toxicity test involves gavage or capsule dosing at the start of the study (Figure 2). Care must be taken when dosing birds by oral gavage. Some species can readily regurgitate leading to uncertainty in the the dose given. These include mallard duck, pigeons and some passerine species. Table 1 gives the birds species recommended in the OECD and USEPA guidelines, respectively. Gamebirds and passerines are a good combination to take account of phylogeny and a good starting point to better understand the distribution of species sensitivity.

The OECD guideline 223 uses on average 26 birds and is a sequential design (Edwards et al., 2017). Responses of birds to each stage of the test are combined to estimate and improve the estimate of the LD50 and slope. The testing can be stopped at any stage once the accuracy of the LD50 estimate meets the requirements for the risk assessment, hence using far fewer birds for compliance with the 3Rs (reduction, refinement and replacement). If toxicity is expected to be low, 5 birds are dosed at the limit dose of 2000 mg/kg (which is the highest acceptable dose to be given by oral gavage, from a humane point of view). If there is no mortality in the limit test after 14 days the study is complete and the LD50 >2000 mg/kg body weight. If there is mortality a single individual is treated at each of 4 different doses in Stage 1. With these results a working estimate of the LD50 is determined to select 10 further dose levels for a better estimate of the LD50 in Stage 2. If a slope is required a further Stage 3 is required using 10 more birds in a combination of doses selected on the basis of a provisional estimate of the slope.

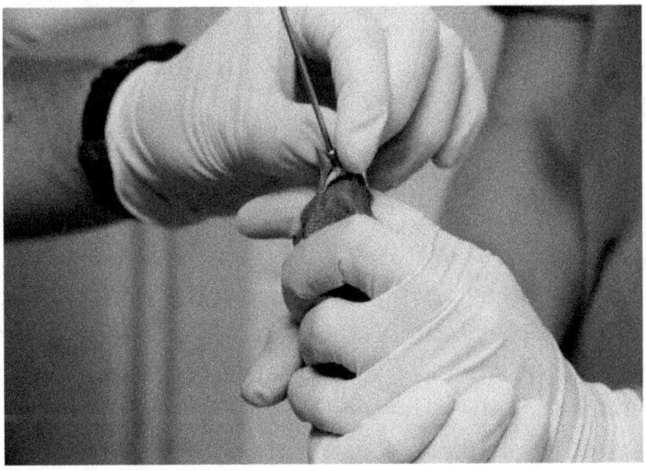

Figure 2. Gavage dosing of a zebrafinch – Eurofins Agroscience Services, Easton MD, USA.

The USEPA guideline is a single stage design preceeded by a range finding test (used only to set the concentrations for the main test). The LD50 test uses 60 birds (10 at each of five test concentrations and 10 birds in the control group). Despite the high numbers of birds used, the ability to estimate a slope is poor compared to OECD223 (the ability to calculate the LD50 is similar to the OECD 223 guideline).

Dietary Toxicity Testing

For the medium-term risk assessment an avian dietary toxicity test was regularly performed in the past exposing juvenile (chicks) of bobwhite quail, Japanese quail or mallard to a treated diet. This test determines the median lethal concentration (LC50) of a chemical in response to a 5-day dietary exposure. Given the scientific limitations and animal welfare concerns related to this test (EFSA, 2009) current European regulations recommend to only perform this test when it is expected that the LD50 value measured by the medium-term study will be lower than the acute LD50 i.e. if the chemical is cumulative in its effect.

Reproduction Testing

One-generation reproduction tests in bobwhite quail and/or mallard are requested for the registration of all pesticides to which birds are likely to be exposed during the breeding season. Table 1 presents the two standard studies: OECD Test 206 and the US EPA OCSPP 850.2100 study. The substance to be tested is mixed into the diet from the start of the test. The birds are fed ad libitum for a recommended period of 10 weeks before they begin laying eggs in response to a change in photoperiod. The egg-laying period should last at least ten weeks. Endpoints include adult body weight, food consumption, macroscopic findings at necropsy and reproductive endpoints, with the number of 14-day old surviving chicks/ducklings as an overall endpoint.

The OECD guideline states that the Japanese quail (Coturnix coturnix japonica), is also acceptable.

Avoidance (or Repellancy) Testing

Avoidance behaviour by birds in the field could be seen as reducing the risk of exposure to a pesticide and therefore could be considered in the risk assessment. However, the occurrence of avoidance in the laboratory has a confounding effect on estimates of toxicity in dietary studies (LD_{50}). Avoidance tests thus far have greatest relevance in the risk assessment of seed treatments. A number of factors need to be taken into account including the feeding rate and dietary concentration which may determine whether avoidance or mortality is the outcome. The following comprehensive OECD report provides an overview of guideline development and research activities that have taken place to date under the OECD flag. Sometimes these studies are done as semi-field (or pen) studies.

Endocrine Disruptor Testing

Endocrine-disrupting substances can be defined as materials that cause effects on reproduction through the disruption of endocrine-mediated processes. If there is reason to suspect that a substance might have an endocrine effect in birds, a two-generation avian test design aimed specifically at the evaluation of endocrine effects could be performed. This test has been developed by the USEPA (OCSPP 890.210). The test has not, however, been accepted as an OECD test to date. It uses the Japanese quail as the preferred species. The main reasons that Japanese quail were selected for this test were: 1) Japanese quail is a precocial species as mentioned earlier. This means that at hatch Japanese quail chicks are much further in their sexual differentiation and development than chicks of altricial species would be. Hormonal processes occurring in Japanese quail in these early stages of development can be disturbed by chemicals maternally deposited in the egg (Ottinger and Dean, 2011). Conversely altricial species undergo these same sexual development stages post-hatch and can be exposed to chemicals in food that might impact these same hormonal processes. 2) as mentioned above, the young of the year mature and breed (themselves) within 12 months which makes the test more efficient that if one used bobwhite quail or mallard.

It is argued among avian toxicologists, that it is necessary to develop a zebra finch endocrine assay system, alongside the Japanese quail system, as this will allow a more systematic determination of differences between responses to EDC's in altricial and precocial species, there by allowing a better evaluation and subsequent risk assessment of potential endocrine effects in birds. Differences in parental care, nesting behaviour and territoriality are examples of aspects that could be incorporated in such an approach (Jones et al., 2013).

Field Studies

Field studies can be used to test for adverse effects on a range of species simultaneously, under conditions of actual exposure in the environment (Hart et al, 2001). The numbers of sites and control fields and methods (corpse searches, censusing and radiotracking) need careful consideration for optimal use of field studies in avian toxicology. The field site will define the species studied and it is important to consider the relevance of that species in other locations. For further reading about techniques and methods to be used in avian field research Sutherland et al and Bibby et al. (2000) are recommended.

References

Bibby, C., Jones, M., Marsden, S. (2000). Expedition Field Techniques Bird Surveys. Birdlife International.

Birdlife International (2018). The Status of the World's Birds. https://www.birdlife.org/sites/default/files/attachments/BL_ReportENG_V11_spreads.pdf

Brooks, A.C., Fryer, M., Lawrence, A., Pascual, J., Sharp, R. (2017). Reflections on bird and mammal risk assessment for plant protection products in the European Union: Past, present and future. Environmental Toxicology and Chemistry 36, 565-575.

Edwards, P.J., Leopold, A., Beavers, J.B., Springer, T.A., Chapman, P., Maynard, S.K., Hubbard, P. (2017). More or less: Analysis of the performance of avian acute oral guideline OECD 223 from empirical data. Integrated Environmental Assessment and Management 13, 906-914.

Hart, A., Balluff, D., Barfknecht, R., Chapman, P.F., Hawkes, T., Joermann, G., Leopold, A., Luttik, R. (Eds.) (2001). Avian Effects Assessment: A Framework for Contaminants Studies. A report of a SETAC workshop on 'Harmonised Approaches to Avian Effects Assessment', held with the support of the OECD, in Woudschoten, The Netherlands, September 1999. A SETAC Book.

Jones, P.D., Hecker, M., Wiseman, S., Giesy, J.P. (2013). Birds. Chapter 10 In: Matthiessen, P. (Ed.) Endocrine Disrupters - Hazard Testing and Assessment Methods. Wiley & Sons.

Kendall, R.J., Lacher Jr, T.E. (Eds.) (1994). Wildlife Toxicology and Population Modelling – Integrated Studies of Agrochecosystems. Special Publication of SETAC.

Murk, A.J., Boudewijn, T.J., Meininger, P.L., Bosveld, A.T.C., Rossaert, G., Ysebaert, T., Meire, P., Dirksen, S. (1996). Effects of polyhalogenated aromatic hydrocarbons and related contaminants on common tern reproduction: Integration of biological, biochemical, and chemical data. Archives of Environmental Contamination and Toxicology 31, 128–140.

Ottinger, M.A., Dean, K. (2011). Neuroendocrine Impacts of Endocrine-Disrupting Chemicals in Birds: Life Stage and Species Sensitivities. Journal of Toxicology and Environmental Health, Part B: Critical Reviews. 26 July 2011.

Sutherland, W.J., Newton, I., Green, R.E. (Eds.) (2004). Biological Ecology and Conservation. A Handbook of Techniques. Oxford University Press

4.3.8. In Vitro Toxicity Testing

Introduction

In vitro bioassays refer to testing methods making use of tissues, cells, or proteins. The term "in vitro" (meaning "in glass") refers to the test tubes or petri dishes made from glass that were traditionally used to perform these types of toxicity tests. Nowadays, in vitro bioassays are more often performed in plastic microtiter wells-plates containing multiple (6, 12, 24, 48, 96, 384, or 1536) test containers (called "wells") per plate (Figure 1). In vitro bioassays are usually performed to screen individual substances or samples for specific bioactive properties. As such, in vitro toxicology refers to the science of testing substances or samples for specific toxic properties using tissues, cells, or proteins.

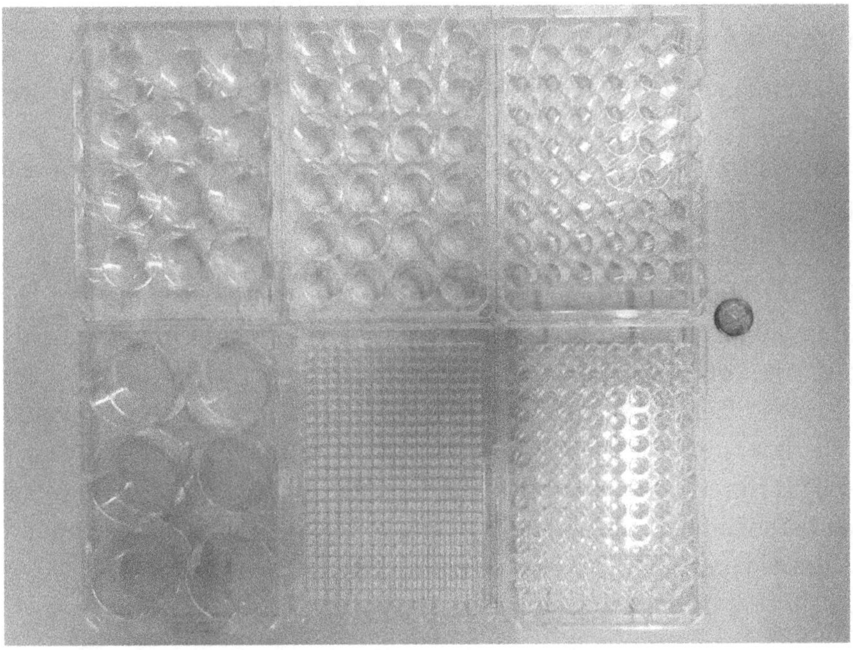

Figure 1. Six different microtiter well-plates, consisting of multiple small-volume test containers. In clockwise direction starting from the LOWER left: 6-wells plate, 12-wells plate, 24-wells plate, 48-wells plate, 96 wells plate, 384 wells plate.

Most in vitro bioassays show a mechanism-specific response, which is for instance indicative of the inhibition of a specific enzyme or the activation of a specific molecular receptor. Moreover, in vitro bioassays are usually performed in small test volumes and have short test durations (usually incubation periods range from 15 minutes to 48 hours). As a consequence, multiple samples can be tested simultaneously in a single experiment and multiple experiments can be performed in a relatively short test period. This "medium-throughput" characteristic of in vitro bioassays can even be increased to high-throughput" if the time-limiting steps in the test procedure (e.g. sample preparation, cell culturing, pipetting, read-out) are further automated.

Toxicity tests making use of bacteria are also often performed in small volumes, allowing short test-durations and high-throughput. Still, such tests make use of intact organisms and should therefore strictly be considered as in vivo bioassays. This holds especially true if bacteria are used to study endpoints like survival or population growth. However, bacteria test systems studying specific toxic mechanisms, such as the Ames test used to screen substances for mutagenic properties (see section on Carcinogenicity and Genotoxicity), are often considered as in vitro bioassays, because of the similarity in test characteristics when compared to in vitro toxicity tests with cells derived from higher organisms.

Protein-based Assays

The simplest form of an in vitro binding assay consists of a purified protein that is incubated with a potential toxic substance or sample. Purified proteins are usually obtained by isolation from an intact organism or from cultures of recombinant bacteria, which are genetically modified to express the protein of interest.

Ligand binding assays are used to determine if the test substance is capable of binding to the protein, thereby inhibiting the binding capacity of the natural (endogenous) ligand to that protein (see section on Protein Inactivation). Proteins of interest are for instance receptor proteins or transporter proteins. Ligand binding assays often make use of a natural ligand that has been labelled with a radioactive isotope. The protein is incubated with the labelled ligand in the presence of different concentrations of the test substance. If protein-binding by the test substance prevents ligand binding to the protein, the free ligand shows a concentration-dependent increase in radioactivity (See Figure 2).

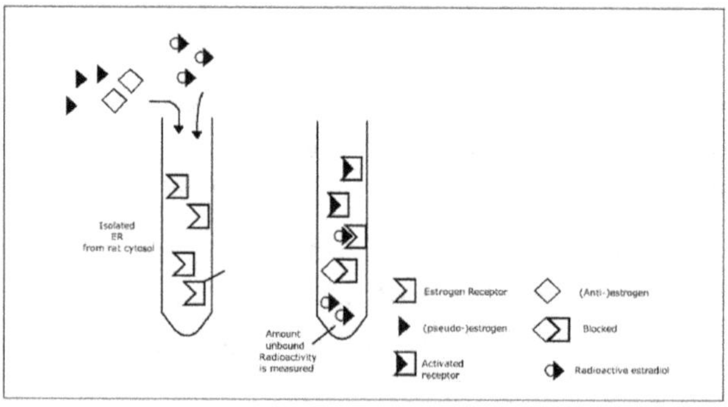

Figure 2. Principle of a radioactive ligand binding assay to determine binding of (anti-)estrogenic compounds to the estrogen receptor (ER). The ER is incubated with radiolabeled estradiol in combination with different concentrations of the test compound. If the compound is capable of binding to the ER, it will displace estradiol from the receptor. After separation of the free and bound estradiol, the amount of unbound radioactivity is measured. Increasing test concentrations of (anti-)estrogenic ER-binders will cause an increase in unbound radioactivity (and consequently a decrease in bound radioactivity). Redrawn from Murk et al. (2002) by Wilma Ijzerman.

Consequently, the ligand-protein complex shows a concentration-dependent decrease in radioactivity. Alternatively, the natural ligand may be labelled with a fluorescent group. Binding of such a labelled ligand to the protein often causes an increase in fluorescence. Consequently, a decrease in fluorescence is observed if a test substance prevents ligand binding to the protein.

Enzyme inhibition assays are used to determine if a test substance is capable to inhibit the enzymatic activity of a protein. Enzymatic activity is usually determined as the conversion rate of a substrate into a product. Enzyme inhibition is determined as a decrease in conversion rate, corresponding to lower concentrations of product and higher concentrations of substrate after different periods of incubation. Quantitative measures of substrate disappearance or product formation can be done by chemical analysis of the substrate or the product. Preferably, however, the reaction rate is measured by spectrophotometry or by fluorescence. This is achieved by performing the reaction with a substrate that has a specific colour or fluorescence by itself or that yields a product with a specific colour or fluorescence, in some cases after reaction with an additional indicator compound. A well-known example of an enzyme inhibition assay is the acetylcholinesterase inhibition assay (see section on Diagnosis - In vitro bioassays).

Cell Cultures

Cell-based bioassays make use of cell cultures that are maintained in the laboratory. Cell culturing starts with mechanical or enzymatic isolation of single cells from a tissue (obtained from an animal or a plant). Subsequently, the cells are grown in cell culture medium, i.e. a liquid that contains all essential nutrients required for optimal cell growth (e.g. growth factors, vitamins, amino acids) and regulates the physicochemical environment of the cells (e.g. pH buffer, salinity). Typically, several types of cell cultures can be distinguished (Figure 3).

Primary cell cultures consist of cells that are directly isolated from a donor organism and are maintained in vitro. Typically, such cell cultures consist of either a cell suspension of non-adherent cells or a monolayer of adherent cells attached to a substrate (i.e. often the bottom of the culture vessel). The cells may undergo several cell divisions until the cell suspension becomes too dense or the adherent cells grow on top of each other. The cells can then be further subcultured by transferring part of the cells from the primary culture to a new culture vessel containing fresh medium. This progeny of the primary cell culture is called a cell line, whereas the event of subculturing is called a passage. Typically, cell lines derived from primary cells undergo senescence and stop proliferating after a limited number (20-60) of cell divisions. Consequently, such a finite cell line can undergo only a limited number of passages. Primary cell cultures and their subsequent finite cell lines have the advantage that they closely resemble the physiology of the cells in vivo. The disadvantage of such cell cultures for toxicity testing is that they divide relatively slowly, require specific cell culturing conditions, and are finite. New cultures can only be obtained from new donor organisms, which is time-consuming, expensive, and may introduce genetic variation.

Alternatively, continuous cell lines have been established, which have an indefinite life span because the cells are immortal. Due to genetic mutations cells from a continuous cell line can undergo an indefinite number of cell divisions and behave like cancer cells. The immortalizing mutations may have been present in the original primary cell culture, if these cells were isolated from a malign cancer tumour tissue. Alternatively, the original finite cell line may have been transformed into a continuous cell line by introducing a viral or chemical induced mutation. The advantage of continuous cell lines is that the cells proliferate quickly and are easy to culture and to manipulate (e.g. by genetic modification). The disadvantage is that continuous cell lines have a different genotype and phenotype than the original healthy cells in vivo (e.g. have lost enzymatic capacity) and behave like cancer cells (e.g. have lost their differentiating capacities and ability to form tight junctions).

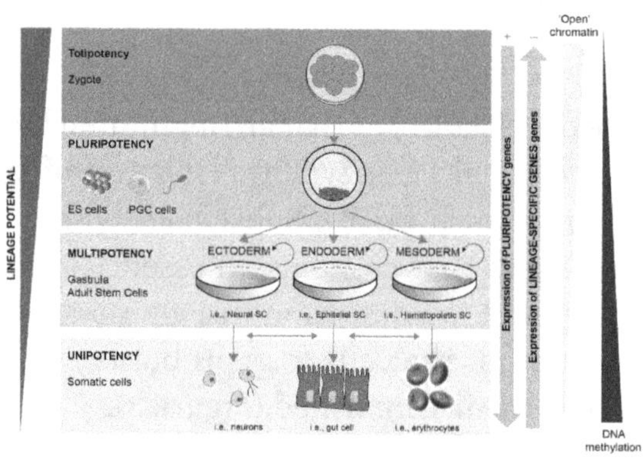

Figure 3. Different types of cell culturing, showing the establishment of a primary cell culture, a finite cell line, and a continuous cell line. See text for further explanation.

Differentiation Models

To study the toxic effects of compounds in vitro, toxicologists prefer to use cell cultures that resemble differentiated, healthy cells rather than undifferentiated cancer cells. Therefore, differentiation models have gained increasing attention in in vitro toxicology in recent years. Such differentiation models are based on stem cells, which are cells that possess the potency to differentiate into somatic cells. Stem cells can be obtained from embryonic tissues at different stages of normal development, each with their own potency to differentiate into somatic cells (Figure 4). In the very early embryonic stage, cells from the "morula stage" (i.e. after a few cell divisions of the zygote) are totipotent, meaning that they can differentiate in all cell types of an organism. Later in development, cells from the inner cell mass of the trophoblast are pluripotent, meaning that they can differentiate in all cell types, except for extra-embryonic cells. During gastrulation, cells from the different germ layers (i.e. ectoderm, mesoderm, and endoderm) are multipotent, meaning that they can differentiate into a restricted number of cell types. Further differentiation results in precursor cells that are unipotent, meaning that they are committed to differentiate into a single ultimate differentiated cell type.

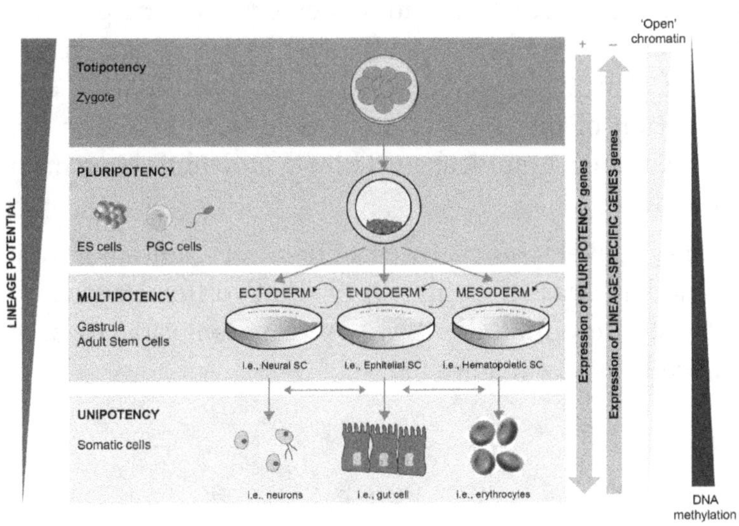

Figure 4. Lineage restriction of human developmental potency. Totipotent cells at the morula stage have the ability to self-renew and differentiate into all of the cell types of an organism, including extraembryonic tissues. Pluripotent cells – for example, in vitro embryonic stem (ES) cells established at the blastocyst stage and primordial germ cells (PGCs) from the embryo – lose the capacity to form extraembryonic tissues like placenta. Restriction of differentiation is imposed during normal development, going from multipotent stem cells (SCs), which can give rise to cells from multiple but not all lineages, to the well-defined characteristics of a somatic differentiated cell (unipotent). Specific chromatin patterns and epigenetic marks can be observed during human development since they are responsible for controlling transcriptional activation and repression of tissue-specific and pluripotency-related genes, respectively. Global increases of heterochromatin marks and DNA methylation occur during differentiation. Redrawn from Berdasco and Esteller (2011) by Evelin Karsten-Meessen.

While remaining undifferentiated, in vitro embryonic stem cell (ESC) cultures can divide indefinitely, because they do not suffer from senescence. However, an ESC cell line cannot be considered as a continuous (or immortalized) cell line, because the cells contain no genetic mutations. ESCs can be differentiated into the cell type of interest by manipulating the cell culture conditions in such a way that specific signalling pathways are stimulated or inhibited in the same sequence as happens during in vivo cell type differentiation. Manipulation may consist of addition of growth factors, transcription factors, cytokines, hormones, stress factors, etc. This approach requires good understanding of which factors affect decision steps in the cell lineage of the cell type of interest.

Differentiation of ESCs into differentiated cells is not only applicable in in vitro toxicity testing, but also in drug discovery, regenerative medicine, and disease modelling. Still, the destruction of a human embryo for the purpose of isolation of – mainly pluripotent – human ESCs (hESCs) raises ethical issues. Therefore, alternative sources of hESCs have been explored. The isolation and subsequent in vitro differentiation of multipotent stem cells from amniotic fluid (collected during caesarean sections), umbilical cord blood, and adult bone marrow is a very topical field of research.

A revolutionary development in the field of non-embryonic stem cell differentiation models was the discovery that differentiated cells can be reprogrammed to undifferentiated cells with pluripotent capacities, called induced pluripotent stem cells (iPSCs) (Figure 5).

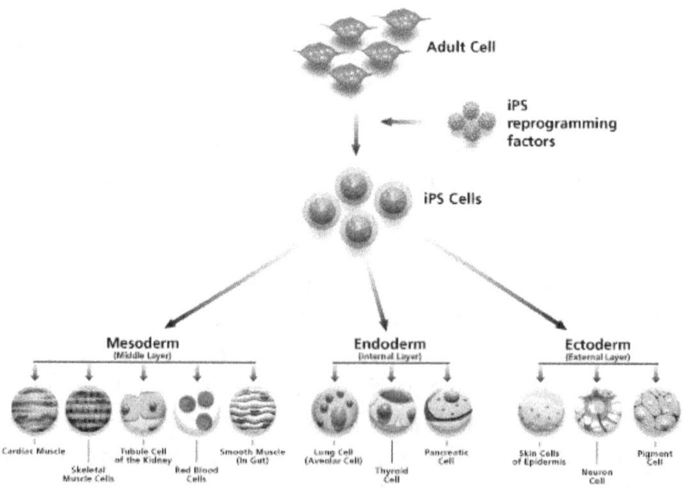

Figure 5. Principle of generating induced pluripotent stem cells (iPSCs) that can differentiate into any type of somatic cell. Source: https://beyondthedish.wordpress.com/2015/08/08/new-york-stem-cell-foundation-invents-robotic-platform-for-making-induced-pluripotent-stem-cells//

In 2012, the Nobel Prize in Physiology or Medicine was awarded to John B. Gurdon and Shinya Yamanaka for this ground-breaking discovery. Reprogramming of differentiated cells isolated from an adult donor is obtained by exposing the cells to a mixture of reprogramming factors, consisting of transcription factors typical for pluripotent stem cells. The obtained iPSCs can be differentiated again (similar as ESCs) into any type of differentiated cells, for which the required conditions for cell lineage are known and can be simulated in vitro.

Whereas iPSC based differentiation models require a complete reprogramming of a differentiated somatic cell back to the stem cell level, transdifferentiation (or lineage reprogramming) is an alternative technique by which differentiated somatic cells can be transformed into another type of differentiated somatic cells, without undergoing an intermediate pluripotent stage. Especially fibroblast cell lines are known for their capacity to be transdifferentiated into different cell types, like neurons or adipocytes (Figure 6).

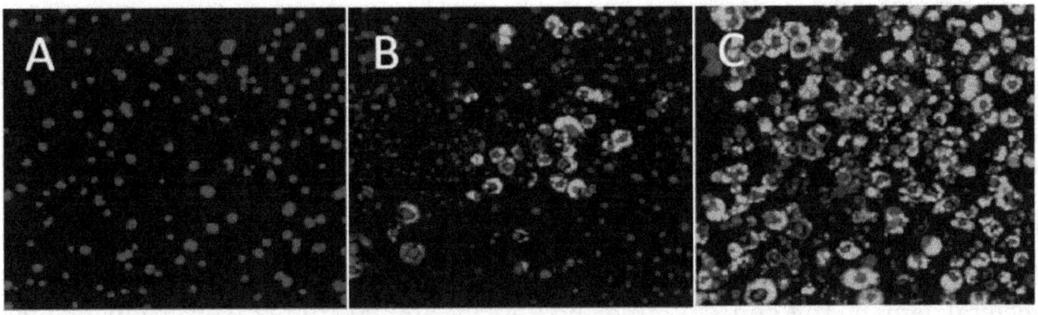

Figure 6. In vitro trans-differentiation of fibroblast cells from the 3T3-L1 cell line into mature adipocytes containing lipid vesicles (green). Each individual cell is visualized by nuclear staining (blue). A: undifferentiated control cells, B:cells exposed to an adipogenic cocktail consisting 3-isobutyl-1-methylxanthine, dexamethasone and insulin (MDI), C: cells exposed to MDI in combination with the PPAR gamma agonist troglitazone, an antidiabetic drug. Source: Vrije Universiteit Amsterdam-Dept. Environment & Health.

Cell-based Bioassays

In cell-based in vitro bioassays, the cell cultures are exposed to test compounds or samples and their response is measured. In principle, all types of cell culture models discussed above can be used for in vitro toxicity testing. For reasons of time, money, and comfort, continuous cell lines are commonly used, but more and more often primary cell lines and iPSC-derived cell lines are used, for reasons of higher biological relevance. Endpoints that are measured in in vitro cell cultures exposed to toxic compounds typically range from effects on cell viability (measured as decreased mitochondrial functioning, increased membrane damage, or changes in cell metabolism; see section on Cytotoxicity) and cell growth to effects on cell kinetics (absorption, elimination and biotransformation of cell substrates), changes in the cell transcriptome, proteome or metabolome, or effects on cell-type dependent functioning. In addition, cell differentiation models can be used not only to study effects of compounds on differentiated cells, but also to study the effects on the process of cell differentiation per se by exposing the cells during differentiation.

A specific type of cell-based bioassays are the reporter gene bioassays, which are often used to screen individual compounds or complex mixtures extracted from environmental samples for their potency to activate or inactivate receptors that play a role in the expression of genes that play an important role in a specific path. Reporter gene bioassays make use of genetically modified cell lines or bacteria that contain an incorporated gene construct encoding for an easily measurable protein (i.e. the reporter protein). This gene construct is developed in such a way that its expression is triggered by a specific interaction between the toxic compound and a cellular receptor. If the receptor is activated by the toxic compound, transcription and translation of the reporter protein takes place, which can be easily measured as a change in colour, fluorescence, or luminescence (see section on Diagnosis – In vitro bioassays).

Future Developments

Although there is a societal need for a non-toxic environment, there is also a societal demand to Reduce, Refine and Replace animal studies (three R principles). Replacement of animal studies by in vitro tests requires that the obtained in vitro results are indicative and predictive for what happens in the in vivo situation. It is obvious that a cell culture consisting of a single cell type is not comparable to a complex organism. For instance, toxicokinetic aspects are hardly taken into account in cell-based bioassays. Although some cells might have metabolic capacities, processes like adsorption, distribution, and elimination are not represented as exposure is usually directly on the cells. Moreover, cell cultures often lack repair mechanisms, feedback loops, and any other interaction with other cell types/tissues/organs as found in intact organisms. To expand the scope of in vitro – in vivo extrapolation (IVIVE), more complex in vitro models are developed nowadays that have a closer resemblance to the in vivo situation. For instance, whereas cell culturing was traditionally done in 2D monolayers (i.e. in layers of 1 cell thickness), 3D cell culturing is gaining ground. The advantage of 3D culturing is that it represents a more realistic type of cell growth, including cell-cell interactions, polarization, differentiation, extracellular matrix, diffusion gradients, etc. For epithelial cells (e.g. lung cells), such 3D cultures can even be grown at the air-liquid interphase reflecting the in vivo situation. Another development is cell co-culturing where different cell types are cultured together in a cell culture. For instance, two cell types that interact in an organ can be co-cultured. Alternatively, a differentiated cell type that has poor metabolic capacity can be co-cultured with a liver cell in order to take possible detoxification or bioactivation after biotransformation into account. The latest development in increasing complexity in in vitro test systems are so-called organ-on-a-chip devices, in which different cell types are co-cultured in miniaturized small channels. The cells can be exposed to different flows representing for instance the blood stream, which may contain toxic compounds (see for instance video clips at https://wyss.harvard.edu/technology/human-organs-on-chips/). Based on similar techniques, even human body-on-a-chip devices can be constructed. Such chips contain different miniaturized compartments containing cell co-cultures representing different organs, which are all interconnected by different channels representing a microfluid circulatory system (Figure 7). Although such devices are in their infancies and regularly run into impracticalities, it is to be expected that these innovative developments will play their part in the near future of toxicity testing.

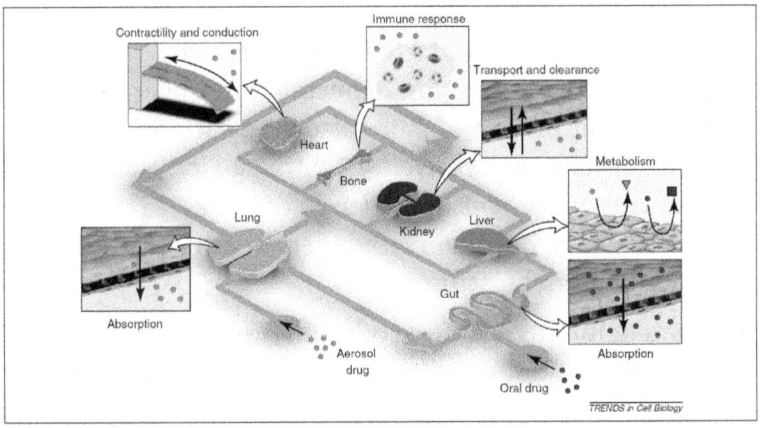

Figure 7. The human-on-a-chip device, showing miniaturized compartments (or biomimetic microsystems) containing (co-)cultures representing different organs, interconnected by a microfluidic circulatory system. Compartments are connected in a physiologically relevant manner to reflect complex, dynamic ADME processes and to allow toxicity evaluation. In this example, an integrated system of microengineered organ mimics (lung, heart, gut, liver, kidney and bone) is used to study the absorption of inhaled aerosol substances (red) from the lung to microcirculation, in relation to their cardiotoxicity (e.g. changes in heart contractility or conduction), transport and clearance in the kidney, metabolism in the liver, and immune-cell contributions to these responses. To investigate the effects of oral administration, substances can also be introduced into the gut compartment (blue).

4.3.9. Human Toxicity Testing - I. General Aspects

Introduction

Toxicity is the capacity of a chemical to cause injury to a living organism. Small doses of a chemical can in theory be tolerated due to the presence of systems for physiological homeostasis (i.e., the ability to maintain physiological stability) or compensation (i.e., physiological adaptation). Above a given chemical-specific threshold, however, the ability of organisms to compensate for toxic stress becomes saturated, leading to loss of homeostasis and adverse effects, which may be reversible or irreversible, and ultimately fatal.

Toxicity testing serves two main aims, i.e. to identify the potential adverse effects of a chemical on humans (i.e., hazard identification), and to establish the relationship between the dose or concentration and the incidence and severity of an effect. The data from toxicity testing thus needs to be suitable for classification and labelling and should allow toxicologists to determine safe levels of human exposure (section 6.3.3), to predict and evaluate the risks of these chemicals to humans and to prioritize chemicals for further risk assessment (section 6.1) and risk management (section 6.6).

Toxicologists gather toxicity data from the scientific literature and selected databases or produce these data in experimental settings, mostly involving experimental animals, but more and more also alternative test systems with cells/ cell lines, tissues or organs (see section 4.3.9.II). Toxicity data are also obtained from real-life exposures of humans in epidemiological research (section 4.3.10.I) or in experiments with human volunteers under strict ethical rules. This chapter will focus on experimental animal testing.

The scope of toxicity testing depends on the anticipated use, with route, duration and frequency of administration as representative as possible for human exposure to the chemical during normal use. The oral, dermal or inhalation routes are the routes of preference, and the time scale can vary from single exposures up to repeated or continuous exposure over parts or the whole of the lifetime of the experimental organism. In toxicity testing, specific toxicity endpoints such as irritation, sensitization, carcinogenicity , mutagenicity, reproductive toxicity, immunotoxicity and neurotoxicity need to be addressed (see respective subchapters in section 4.2, and section 4.3.9.III). These toxicity endpoints can be investigated at different time scales, ranging from acute exposure (e.g., single dose oral testing) up to chronic exposure (e.g., lifelong testing for carcinogenicity) (see also under 'test duration' below).

Other useful tests are tests designed to investigate the mechanisms of action at the tissue, cellular, subcellular and receptor levels (section 4.2), and toxicokinetic studies, investigating the uptake, distribution, metabolism and excretion of the chemical. Such data helps in the design of the testing strategy (which tests, which route of exposure, the order of the tests, the dose levels) and the interpretation of the results.

International Cooperation and Harmonization

The regulation of chemicals is more and more an international affair, not in the least to facilitate trade, transport and use of chemicals at a global scale. This requires strong international cooperation and harmonization. For instance, guidelines for protocol testing and assessment of chemicals have been developed by the World Health Organization (WHO) and the Organisation for Economic Co-operation and Development (OECD). These WHO and OECD guidelines often are the basis for regulatory requirements at regional (e.g., EU) and national scales (e.g., USA, Japan).

Of prime Importance for harmonization is the OECD Mutual Acceptance of Data (MAD) system. This system is built on two instruments for ensuring harmonized data generation and data quality: the OECD Guidelines for the Testing of Chemicals and the OECD Principles of Good Laboratory Practice (GLP). Under MAD, laboratory test results related to the safety of chemicals that are generated in an OECD member country in accordance with these instruments are to be accepted in all OECD member countries and a range of other countries adhering to MAD.

The OECD test guidelines are accepted internationally as standard methods for safety testing by industries, academia, governments and independent laboratories. They cover tests for physical-chemical properties, effects on biotic systems (ecotoxicity), environmental fate (degradation and accumulation) and health effects (toxicity). These guidelines are regularly updated, and new test guidelines are developed and added, based on specific regulatory needs. This happens in cooperation with experts from regulatory agencies, academia, industry, environmental and animal welfare organizations.

The OECD GLP principles provide quality assurance concepts concerning the organization of test laboratories and the conditions under which laboratory studies are planned, performed, monitored, and reported.

Alternative Testing

The use of animal testing for risk assessment has been a matter of debate for a long time, first of all for ethical reasons, but also because of the costs of animal testing and the difficulties in translating the results of animal tests to the human situation. Therefore, there is political and societal pressure to develop and implement alternative methods to replace, reduce and refine animal testing. In some legal frameworks such as the EU cosmetics regulation, the use of experimental animals is already banned. Under the EU chemicals legislation REACH, experimental animal testing is a last resort option. In 2017, the number of animals used for the first time for research and testing in the EU was just below 10 million. Twenty-three percent of these animals were for all regulatory uses, of which approximately one-third was for toxicity, pharmacology and other safety testing (850,000 animals) for industrial chemicals, food and feed chemicals, plant protection products, biocides, medicinal products and medical devices (European Commission, 2020).

Alternative methods include the use of (quantitative) structure-activity relationships ((Q)SARs; i.e., theoretical models to predict the physicochemical and biological (e.g. toxicological) properties of molecules from the knowledge of chemical structure), in vitro tests (section 4.3.9.II; preferably with cells/cell lines, organs or tissues of human origin) and read-across methods (using toxicity data on structurally related chemicals to predict the toxicity of the chemical under investigation). In Europe, the European Union Reference Laboratory for alternatives to animal testing (EURL_ECVAM) has an important role in the development, validation and uptake of alternative methods. It is an important contributor to the OECD Test Guideline Programme; a number of OECD test guidelines are now based on non-animal tests.

Since alternative methods do not always fit easily in current regulatory risk assessment and standard setting approaches, there is also a huge effort to develop testing strategies in which the results of alternative tests are combined with toxicokinetic information and information on the mechanism of action, adverse outcome pathways (AOPs), genetic information (OMICS), read-across and in vitro in vivo extrapolation (IVIVE). Such methods are also called: Integrated Approaches to Testing and Assessment (IATA) or intelligent testing strategies (ITS). These will help in making alternative methods more acceptable for regulatory purposes.

Core Elements of Toxicity Testing

Currently, there are around 80 OECD Test guidelines for human health effects, including both in vivo and in vitro tests. The in vivo tests relate to acute (single exposures) and repeated dose toxicity (28 days, 90 days, lifetime) for all routes of exposure (oral, dermal, inhalation), reproductive toxicity (two generations, (extended) one generation, developmental (neuro)toxicity), genotoxicity, skin and eye irritation, skin sensitization, carcinogenicity, neurotoxicity, endocrine disruption, skin absorption and toxicokinetics. The in vitro tests concern skin absorption, skin and eye irritation and corrosion, phototoxicity, skin sensitization, genotoxicity and endocrine disruption.

Important elements of these test guidelines include the identity, purity and chemical properties of the test substance, route of administration, dose selection,

selection and care of animals, test duration, environmental variables such as caging, diet, temperature and humidity, parameters studied, presentation and interpretation of results. Other important issues are: good laboratory practice (GLP), personnel requirements and animal welfare.

Test Substance

The test substance should be accurately characterized. Important elements here are: chemical structure(s), composition, purity, nature and quantity of impurities, stability, and physicochemical properties such as lipophilicity, density, vapor pressure.

Route of Administration

The three main routes of administration used in experimental animal testing are oral, dermal and inhalation. The choice of the route of administration depends on the physical and chemical characteristics of the test substance and the predominant route of exposure of humans.

Dose and Dose Selection

The selection of the dose level depends on the type of study. In general, studies require careful selection and spacing of the dose levels in order to obtain the maximum amount of information possible. The dose selection should also consider and ensure that the data generated is adequate to fulfill the regulatory requirements across OECD countries as appropriate (e.g., hazard and risk assessment, classification and labelling, endocrine disruption assessment, etc.).

To allow for the determination of a dose-response relationship, the number of dose levels is usually at least three (low, mid, high) in addition to concurrent control group(s). Increments between doses generally vary between factors of 2 and 10. The high dose level should produce sufficient evidence of toxicity, however without severe suffering of the animals and without excess mortality (above 10%) or morbidity. The mid dose should produce slight toxicity and the low dose no toxicity. Toxicokinetic data and tests already performed, such as range-finding studies and other toxicity studies, can help in dose selection. Measurement of dose levels and concentrations in media (air, drinking water, feed) is often recommended, in order to know the exact exposure and to detect mistakes in the dosing.

Animal Species

Interspecies and intraspecies variation is a fact of life even when exposure route and pattern are the same. Knowledge of and experience with the laboratory animal to be used is of prime importance. It provides the investigator with the inherent strengths and weaknesses of the animal model, for instance, how much the model resembles humans. Although the guiding principle in the choice of species is that it should resemble humans as closely as possible in terms of absorption, distribution, metabolic pattern, excretion and effect(s) at the site, small laboratory rodents (mostly rats) of both sexes are usually used for economic and logistic reasons. They additionally provide the possibility of obtaining data on a sufficient number of animals for valid statistical analysis. For specialized toxicity testing guinea pigs, rabbits, dogs and non-human primates may be used as well. Most test guidelines specify the minimum number of animals to be tested.

Test Duration

The response of an organism to exposure to a potentially toxic substance will depend on the magnitude and duration of exposure. Acute or single-dose toxicity refers to the adverse effects occurring within a short time (usually within 14 days) after the administration of a single dose (or exposure to a given concentration) of a test substance, or multiple doses given within 24 hours. In contrast, repeated dose toxicity comprises the adverse effects observed following exposure to a substance for a smaller or bigger part of the expected lifespan of the experimental animal. For example, standard tests with rats are the 28-day subacute test, the 90-day semi-chronic (sub-chronic) test and the 2-year lifetime/chronic test.

Diet

Composition of the diet or the nature of a vehicle in which the substance is administered influences physiology and as a consequence, the response to a chemical substance. The test substance may also change the palatability of the diet or drinking water, which may affect the observations, too.

Other Environmental Variables

Housing conditions, such as caging, grouping and bedding, temperature, humidity, circadian-rhythm, lighting and noise, may all influence animal response to toxic substances. OECD and WHO have made valid suggestions in the relevant guidelines for maintaining good standards of housing and care. The variables referred to should be kept constant and controlled.

Parameters Studied

Methods of investigation have changed dramatically in the past few decades. A better understanding of physiology, biochemistry and pathology has led to more and more parameters being studied in order to obtain information about functional and morphological states. In general, more parameters are studied in the more expensive *in vivo* tests for longer durations such as reproductive toxicity tests, chronic toxicity tests and carcinogenicity tests. Nowadays, important parameters to be assessed in routine toxicity testing are biochemical organ function, physiological measurements, metabolic and haematological information and extensive general and histopathological examination. Some other important parameters that lately gained more interest, such as endocrine parameters or atherogenic indicators, are not or not sufficiently incorporated in routine testing.

Presentation and Evaluation of Results

Toxicity studies must be reported in great detail in order to comply with GLP regulations and to enable in-depth evaluation by regulating agencies. Electronic data processing systems have become indispensable in toxicity testing and provide the best way of achieving the accuracy required by the internationally accepted GLP regulations. A clear and objective interpretation of the results of toxicity studies is important: this requires a clear definition of the experimental objectives, the design and proper conduct of the study and a careful and detailed presentation of the results. As there are many sources of uncertainty in the toxicity testing of substances, these should also be carefully considered.

Toxicity studies aim to derive insight into adverse effects and possible target organs, to establish dose-response relationships and no observed adverse effect levels (NOAELs) or other intended outcomes such as benchmark doses (BMDs). Statistics are an important tool in this evaluation. However, statistical significance and toxicological/biological significance should always be evaluated separately.

Good Laboratory Practice

Non-clinical toxicological or safety assessment studies that are to be part of a safety submission for the marketing of regulated products, are required to be carried out according to the principles of GLP, including both quality control (minimizing mistakes or errors and maximizing the accuracy and validity of the collected data) and quality assurance (assuring that procedures and quality control were carried out according to the regulations).

Personnel Requirements and Animal Welfare

GLP regulations require the use of qualified personnel at every level. Teaching on the subject of toxicity has improved tremendously over the last two decades and accreditation procedures have been implemented in many industrialized countries. This is also important because every toxicologist should feel the responsibility to reduce the number of animals used in toxicity testing, to reduce stress, pain and discomfort as much as possible, and to seek for alternatives, and this requires proper qualifications and experience.

Relevant Sources and Recommendations for Further Reading:

European Commission (2020). 2019 Report on the statistics on the use of animals for scientific purposes in the Member States of the European Union in 2015-2017, Brussels, Belgium, COM(2020) 16 final

OECD Test guidelines for chemicals. https://www.oecd.org/env/ehs/testing/oecdguidelinesforthetestingofchemicals.htm

OECD Integrated Approaches to Testing and Assessment (IATA). http://www.oecd.org/chemicalsafety/risk-assessment/iata-integrated-approaches-to-testing-and-assessment.htm

Van Leeuwen, C.J., Vermeire, T.G. (eds) (2007). Risk assessment of chemicals: an introduction, Second edition. Springer Dordrecht, The Netherlands. ISBN 978-1-4020-6101-1 (handbook), ISBN 978-1-4020-6102-8 (e-book), Chapters 6 (Toxicity testing for human health risk assessment), 11 (Intelligent Testing Strategies) and 16 (The OECD Chemicals Programme). DOI 10.1007/978-1-4020-6102-8

WHO, International Programme on Chemical Safety (1978). Principles and methods for evaluating the toxicity of chemicals. Part I. World Health Organization, Environmental Health Criteria 6. IPCS, Geneva, Switzerland. https://apps.who.int/EHC_6

World Health Organization & Food and Agriculture Organization of the United Nations (2009). Principles and methods for the risk assessment of chemicals in food. World Health Organization, Environmental Health Criteria 240, Chapter 4. IPCS, Geneva, Switzerland. https://apps.who.int/EHC_240_4

4.3.9. Human Toxicity Testing - II. In Vitro Tests

Introduction

Toxicity tests are required to assess potential hazards of new compounds to humans. These tests reveal species-, organ- and dose- specific toxic effects of the compound under investigation. Toxicity can be observed by either in vitro studies using cells/cell lines (see section on in vitro bioassay) or by in vivo exposure on laboratory animals; and involves different durations of exposure (acute, subchronic, and chronic). In line with Directive 2010/63/EC on the protection of animals used for scientific purposes, it is encouraged to use alternatives to animal testing (OECD: alternative methods for toxicity testing). The first step towards replacing animals is to use in vitro methods which can be used to predict acute toxicity. In this chapter, we present acute in vitro cytotoxicity tests (= quality of being toxic to cells) and skin corrosive, irritant, phototoxic, and sensitivity tests as skin is the largest organ of the body.

Cytotoxicity Tests

The cytotoxicity test is one of the biological evaluation and screening tests *in vitro* to observe cell viability. Viability levels of cells are good indicators of cell health. Conventionally used tests for cytotoxicity include dye exclusion or uptake assays such as Trypan Blue Exclusion (TBE) and Neutral Red Uptake (NRU).

The **TBE test** is used to determine the number of viable cells present in a cell suspension. Live cells possess intact cell membranes that exclude certain dyes, such as trypan blue, whereas dead cells do not. In this assay, a cell suspension incubated with serial dilutions of a test compound under study is mixed with the dye and then visually examined. A viable cell will have a clear cytoplasm whereas a nonviable cell will have a blue cytoplasm. The number of viable and/or dead cells per unit volume is determined by light microscopy using a hemacytometer counting chamber (Figure 1). This method is simple, inexpensive and a good indicator of membrane integrity but counting errors (~10%) can occur due to poor dispersion of cells, or improper filling of counting chamber.

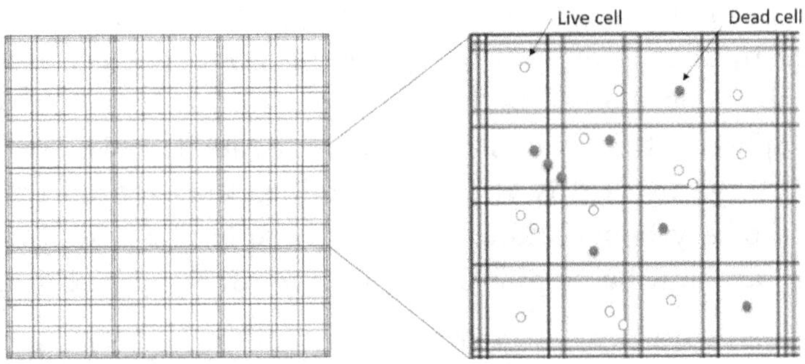

Figure 1. Graphical view hemacytometer counting chamber and illustration of viable and non-viable cells.

The NRU assay is an approach that assesses the cellular uptake of a dye (Neutral Red) in the presence of a particular substance under study (see e.g.Figure 1 in Repetto et al., 2008). This test is based on the ability of viable cells to incorporate and bind neutral red in the lysosomes, a process based on universal structures and functions of cells (e.g. cell membrane integrity, energy production and metabolism, transportation of molecules, secretion of molecules). Viable cells can take up neutral red via active transport and incorporate the dye into their lysosomes while non-viable cells cannot. After washing, viable cells can release the incorporated dye under acidified-extracted conditions. The amount of released dye can be measured by the use of spectrophotometry.

Nowadays, colorimetric assays to assess cell viability have become popular. For example, the MTT assay (3-(4,5-dimethylthiazol-2-yl)-2,5-diphenyltetrazolium bromide) tests cell viability by assessing activity of mitochondrial enzymes. NAD(P)H-dependent oxidoreductase enzymes, which under defined conditions reflect the number of viable cells, are capable of reducing the yellow formazan salt dye into an insoluble purple crystalline product. After solubilizing the end product using dimethyl sulfoxide (DMSO), the product can be quantitatively measured by light absorbance at a specific wavelength. This method is easy to use, safe and highly reproducible. One disadvantage is that MTT formazan is insoluble so DMSO is required to solubilize the crystals.

Skin Corrosion and Irritation

Skin corrosion refers to the production of irreversible damage to the skin; namely, visible necrosis (= localized death of living cells, see section on cell death) through the epidermis and into the dermis occurring after exposure to a substance or mixture. Skin irritation is a less severe effect in which a local inflammatory reaction is observed onto the skin after exposure to a substance or mixture. Examples of these substances are detergents and alkalis which commonly affect hands.

The identification and classification of irritant substances has conventionally been achieved by means of skin or eye observation *in vivo*. Traditional animal testing used rabbits because of its thin skin. In the Draize test, for example, the test substance is applied to the eye or shaved skin of a rabbit, and covered for 24h. After 24 and 72h, the eye or skin is visually examined and graded subjectively based on appearance of erythema and edema. As these *in vivo* tests have been heavily criticized, they are now being phased out in favor of *in vitro* alternatives.

The Skin Corrosion Test (SCT) and Skin Irritation Test (SIT) are in vitro assays that can be used to identify whether a chemical has the potential to corrode or irritate skin. The method uses a three-dimensional (3D) human skin model (Episkin model) which comprises of a main basal, supra basal, spinous and granular layers and a functional stratum corneum (the outer barrier layer of skin). It involves topical application of a test substance and subsequent assessment of cell viability (MTT assay). Test compounds considered corrosive or irritant are identified by their ability to decrease cell viability below the defined threshold level (lethal dose by which 50% of cells are still viable - LD50).

Skin Phototoxicity

Phototoxicity (photoirritation) is defined as a toxic response that is elicited after the initial exposure of skin to certain chemicals and subsequent exposure to light (e.g. chemicals that absorb visible or ultraviolet (UV) light energy that induces toxic molecular changes).

The 3T3 NRU PT assay is based on an immortalised mouse fibroblast cell line called Balb/c 3T3. It compares cytotoxicity of a chemical in the presence or absence of a non-cytotoxic dose of simulated solar light. The test expresses the concentration-dependent reduction of the uptake of the vital dye neutral red when measured 24 hours after treatment with the chemical and light irradiation. The exposure to irradiation may alter cell surface and thus may result in decreased uptake and binding of neutral red. These differences can be measured with a spectrophotometer.

Skin Sensitisation

Skin sensitisation is the regulatory endpoint aiming at the identification of chemicals able to elicit an allergic response in susceptible individuals. In the past, skin sensitisation has been detected by means of guinea pigs (e.g. guinea pig maximisation test and the Buehler occluded patch tests) or murine (e.g. murine local lymph node assay). The latter is based upon quantification of T-cell proliferation in the draining lymph nodes behind the ear (auricular) of mice after repeated topical application of the test compound.

The key biological events (Figure 2) underpinning the skin sensitisation process are well established and include:

- haptenation, the covalent binding of the chemical compounds (haptens) to skin proteins (key event 1);
- signaling, the release of pro-inflammatory cytokines and the induction of cyto-protective pathways in keratinocytes (key event 2);
- the maturation, and mobilisation of dendritic cells, immuno-competent cells in the skin (key event 3);
- migration of dendritic cells, movement of dendritic cells bearing hapten-protein comples from skin to draining local lymph node;
- the antigen presentation to naïve T-cells and proliferation (clonal expansion) of hapten-peptide specific T-cells (key event 4)

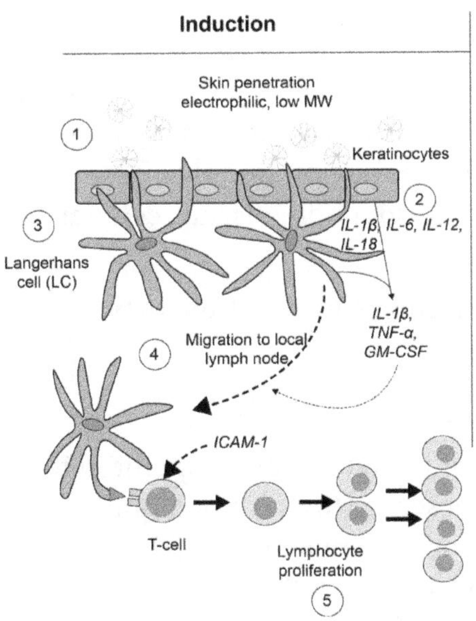

Figure 2. Key biological events in skin sensitisation. Figure adapted from D. Sailstad by Evelin Karsten-Meessen.

Today a number of non-animal methods addressing each a specific key mechanism of the induction phase of skin sensitisation can be employed. These include the Direct Peptide Reactivity Assay (DPRA), ARE-Nrf2 Luciferase Test Method: KeratinoSens, Human Cell Line Activation Test (h-CLAT), U937 cell line activation test (U-SENS), and Interleukin-8 Reporter Gene assay (IL-8 Luc assay). Detailed information of these methods can be found on OECD site: skin sensitization.

References

EUR-lex site.https://eur-lex.europa.eu/legal-content/EN/ALL/?uri=CELEX:32010L0063.

OECD site. Alternative methods for toxicity testing. https://ec.europa.eu/jrc/en/eurl/ecvam/alternative-methods-toxicity-testing.

Episkin model. http://www.episkin.com/Episkin

OECD site. Skin sensitization. https://ec.europa.eu/jrc/en/eurl/ecvam/alternative-methods-toxicity-testing/validated-test-methods/skin-sensitisation.

Repetto, G., Del Peso, A., Zurita, J.L. (2008). Neutral red uptake assay for the estimation of cell viability/cytotoxicity.Nature protocols 3(7), 1125.

4.3.9. Human Toxicity Testing - III. Carcinogenicity Assays

Introduction

The term "carcinogenicity" refers to the property of a substance to induce or increase the incidence of cancer after inhalation, ingestion, injection or dermal application.

Traditionally, carcinogens have been classified according to their mode of action (MoA). Compounds directly interacting with DNA, resulting in DNA-damage or chromosomal aberrations are classified as genotoxic (GTX) carcinogens. Non-genotoxic (NGTX) compounds do not directly affect DNA and are believed to affect gene expression, signal transduction, disrupt cellular structures and/or alter cell cycle regulation.

The difference in mechanism of action between GTX and NGTX require a different approach in many cases.

Genotoxic Carcinogens

Genotoxicity itself is considered to be an endpoint in its own right. The occurrence of DNA-damage can be observed/determined quit easily by a variety of methods based on both bacterial and mammalian cells. Often a tiered testing method is used to evaluate both heritable germ cell line damage and carcinogenicity.

Currently eight in vitro assays have been granted OECD guidelines, four of which are commonly used.

- The Ames test

The gold standard for genotoxicity testing is the Ames test, a test that has been developed in the early seventies. The test evaluates the potential of a chemical to induce mutations (base pair substitutions, frame shift induction, oxidative stress, etc.) in *Salmonella typhimurium*. During the safety assessment process, it is the first test performed unless deemed unsuitable for specific reasons (e.g. during testing for antibacterial substances). With a sensitivity of 70-90% it is a relatively good predictor of genotoxicity.

The principle of the test is fairly simple. A bacterial strain, with a genetic defect, is placed on minimal medium containing the chemical in question. If mutations are induced, the genetic defect in some cells will be restored, thus rendering the cells able to synthesize the deficient amino acid caused by the defect in the original cells.

- Escherichia coli reverse mutation assay

The Ames assay is basically a bacterial reverse mutation assay. In this case different strains of *E. coli*, which are deficient in both DNA-repair and an amino acid, are used to identify genotoxic chemicals. Often a combination of different bacterial strains is used to increase the sensitivity as much as possible.

- *In vitro* mammalian chromosome aberration assay

Chromosomal mutations can occur in both somatic cells and in germ cells, leading to neoplasia or birth and developmental abnormalities respectively. There are two types of chromosomal mutations:

- **Structural changes:** stable aberrations such as translocations and inversions, and unstable aberrations such as gaps and breaks.
- **Numerical changes:** aneuploidy (loss or gain of chromosomes) and polyploidy (multiples of the diploid chromosome complement).

To perform the assay, mammalian cells are exposed *in vitro* to the potential carcinogen and then harvested. Through microscopy the frequency of aberrations is determined. The chromosome aberration can be and is performed with both rodent and human cells which is an interesting feat regarding the translational power of the assay.

- *In vitro* mammalian cell gene mutation test

This mammalian genotoxicity assay utilizes the HPRT gene, a X-chromosome located reporter gene. The test relies on the fact that cells with an intact HPRT gene are susceptible to the toxic effects of 6-thioguanine, while mutants are resistant to this purine analogue. Wild-type cells are sensitive to the cytostatic effect of the compound while mutants will be able to proliferate in the presence of 6-thioguanine.

- Micronucleus test

Next to the four mentioned assays, there is a more recent developed test that already has proven to be a valuable resource for genotoxicity testing. The test provides an alternative to the chromosome aberration assay but can be evaluated faster. The test allows for automated measurement as the analysis of the damage proves to be less subjective. Micronuclei are "secondary" nuclei formed as a result of aneugenic or clastogenic damage.

It is important to note that these assays are all described from an *in vivo* perspective. However, an *in vivo* approach can always be used, see two-year rodent assay. In that case, live animals are exposed to the compound after which specific cells are harvested. The advantage of this approach is the presence of the natural niche in which susceptible cells normally grow, resulting in a more relevant range of effects. The downside of *in vivo* assays is the current ethical pressure on these kind of methods. Several instances actively promote the development and usage of *in vitro* or simply non-animal alternative methods.

- Two-year rodent carcinogenicity assay

For over 50 years, 2-year rodent carcinogenicity assay has been the golden standard for carcinogenicity testing. The assay relies on the exposure to a compound during a major part of an organism's lifespan. During the further development of the assay, a 2-species/2-gender setup became the preferred method, as some compounds showed different results in e.g. rats and mice and even between male and female individuals.

For this approach model organisms are exposed for two years to a compound. Depending on the possible mode of exposure (i.e. inhalation, ingestion, skin/eye/... contact) when the compounds enters the relevant industry, a specific mode of exposure towards the model is chosen. During this time period the health of the model organism is documented through different parameters. Based on this a conclusion regarding the compound is drawn.

Non-genotoxic Carcinogens

Carcinogens not causing direct DNA-damage are classified as NGTX compounds. Due to the fact that there are a large number of potential malign pathways or effects that could be induced, the identification of NGTX carcinogens is significantly more difficult compared to GTX compounds.

The two-year rodent carcinogenicity assay is one of the assays capable of accurately identify NGTX compounds. The use of transgenic models has greatly increased the sensitivity and specificity of this assay towards both groups of carcinogens while also improving the refinement of the assay by shortening the required time to formulate a conclusion regarding the compound.

An *in vitro* method to identify NGTX compounds is rare. Not many alternative assays are able to cope with the vast variety of possible effects caused by the compounds resulting in many false negatives. However, cell morphology based methods such as the cell transformation assay, can be a good start in developing methods for this type of carcinogens.

References

Stanley, L. (2014). Molecular and Cellular Toxicology. p. 434.

4.3.10. Environmental epidemiology - I Basic Principles and study designs

Definitions of epidemiology

Epidemiology (originating from Ancient Greek: Epi -upon, demos - people, logos - the study of) is the study of the distribution and determinants of health-related states or events in specified populations, and the application of this study to the prevention and control of health problems (Last, 2001). Epidemiologists study human populations with measurements at one or more points in time. When a group of people is followed over time, we call this a cohort (originating from Latin: cohors (Latin), a group of Roman soldiers). In epidemiology, the relationship between a determinant or risk factor and health outcome - variable is investigated. The outcome variable mostly concerns morbidity: a disease, e.g. lung cancer, or a health parameter, e.g. blood pressure, or mortality: death. The determinant is defined as a collective or individual risk factor (or set of factors) that is (causally) related to a health condition, outcome, or other defined characteristic. In human health – and, specifically, in diseases of complex etiology – sets of determinants often act jointly in relatively complex and long-term processes (International Epidemiological Association, 2014).

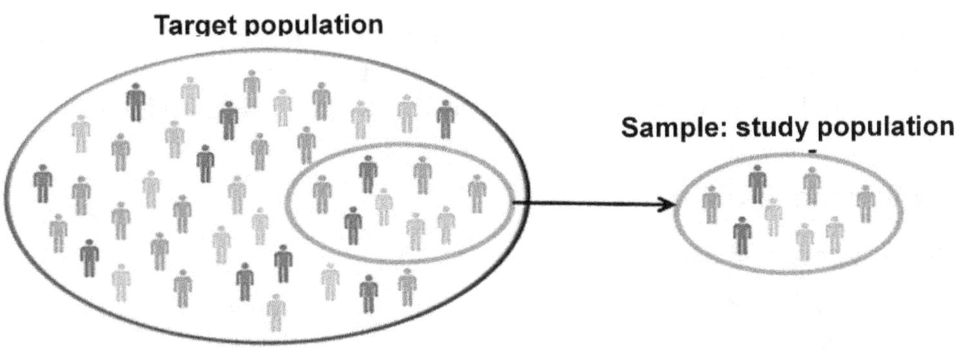

Figure 1: On the left, the target population is presented and from this population, a representative sample is drawn, including all types of individuals from the target population.

The people that are subject of interest are the target population. In most cases, it is impossible and unnecessary to include all people from the target population and therefore, a sample will be taken from the target population, which is called the study population. The sample is ideally representative of the target population (Figure 1). To get a representative sample, it is possible to recruit subjects at random.

Study Designs

Epidemiologic research can either be observational or experimental (Figure 2). Observational studies do not include interference (e.g. allocation of subjects into exposed / non-exposed groups), while experimental studies do. With regard to observational studies analytical and descriptive studies can be distinguished. Descriptive studies describe the determinant(s) and outcome without making comparisons, while analytical studies compare certain groups and derive inferences.

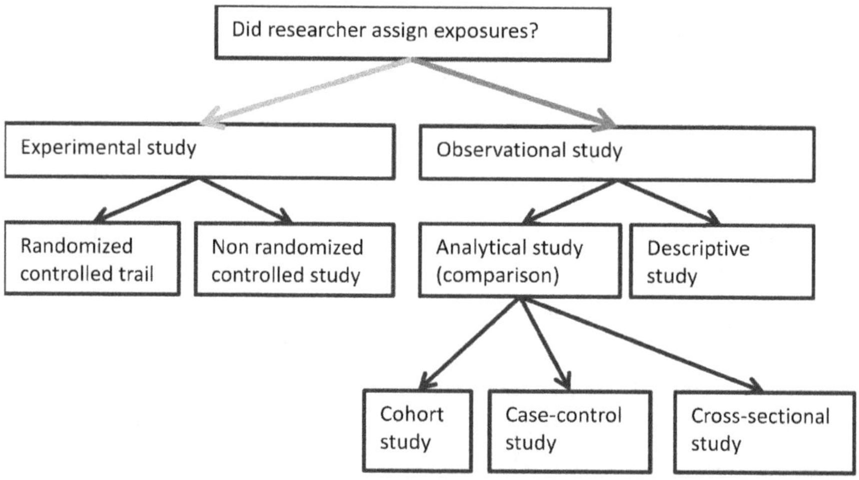

Figure 2: The types of study designs, the branch on the left includes an exposure/ intervention assigned by the researcher, while the branch on the right is observational, and does not include an exposure/intervention assigned by the researcher.

2.1 Observational studies

2.1.1. Cross-sectional Study

In a cross-sectional study, determinant and outcome are measured at the same time. For example, pesticide levels in urine (determinant) and hormone levels in serum (outcome) are collected at one point in time. The design is quick and cheap because all measurements take place at the same time. The drawback is that the design does not allow to conclude about causality, that is whether the determinant precedes the outcome, it might be the other way around or caused by another factor (lacking Hill's criterion for causality temporality, Box 1). This study design is therefore mostly hypothesis generating.

2.1.2 Case-control Study

In a case-control study, the sample is selected based on the outcome, while the determinant is measured in the past. In contrast to a cross-sectional study, this design can include measurements at several time points, hence it is a longitudinal study. First, people with the disease (cases) are recruited and then matched controls (people not affected by the disease), comparable with regard to e.g. age, gender and geographical region, are involved into the study. Important is that controls have the same risk to develop the disease as the cases. The determinant is collected retrospectively, meaning that participants are asked about exposure in the past.

The retrospective character of the design poses a risk for recall bias when people are asked about events that happened in the past, they might not remember them correctly. Recall bias is a form of information bias, when a measurement error results in misclassification. Bias is defined as a systematic deviation of results or inferences from the truth (International Epidemiological Association, 2014). One should be cautious to draw conclusions about causality with the case-control study design. According to Hill's criterion temporality (see Box 1), the exposure precedes the outcome, but because the exposure is collected retrospectively, the evidence may be too weak to draw conclusions about a causal relationship. The benefits are that the design is suitable for research on diseases with a low incidence (in a prospective cohort study it would result in a low number of cases), and for research on diseases with a long latency period, that is the time that exposure to the determinant can result in the disease (in a prospective cohort study, it would take many years to follow-up participants until the disease develops).

An example of a case-control study in environmental epidemiology offman et al. (2017) investigated papillary thyroid cancer (PTC) and exposure to flame retardant chemicals (FRs) in the indoor environment. FRs are chemicals which are added to household products in order to limit the spread of fire, but can leach to house dust where residents can be exposed to the contaminated house dust. FRs are associated with thyroid disease and thyroid cancer. In this case-control study, PTC cases and matched cases were recruited (outcome), and FR exposure (determinant) was assessed by measuring FRs in the house dust of the participants. The study showed that participants with higher exposure to FRs (bromodiphenyl ether-209 concentrations above the median level) had 2.3 more odds (see section Quantifying disease and associations) on having PTC, compared to participants with lower exposure to FRs (bromodiphenyl ether-209 concentrations below the median level).

2.1.3 Cohort Study

A cohort study, another type of a longitudinal study, includes a group of individuals that are followed over time in the future (prospective) or that will be asked about the past (retrospective). In a prospective cohort study, the determinant is measured at the start of the study and the incidence of the disease is calculated after a certain time period, the follow-up. The study design needs to start with people who are at risk for the disease, but not yet affected by the disease.

Therefore, the prospective study design allows to conclude that there may be a causal relationship, since the health outcome follows the determinant in time (Hill's criterion temporality). However, interference of other factors is still possible, see paragraph 3 about confounding and effect modification.

It is possible to look at more than 1 health outcome, but the design is less suitable for diseases with a low incidence or with a long latency period, because then you either need a large study population to have enough cases, or need to follow the participants for a long time to measure cases. A major issue with this study design is attrition (loss to follow-up), it means to what extent do participants drop out during the study course. Selection bias can occur when a certain type of participants drops out more often, and the research is conducted with a selection of the target population. Selection bias can also occur at the start of a study, when some members of the target population are less likely to be included in the study population in comparison to other members and the sample therefore is not representative of the target population.

An example of a prospective cohort study

De Cock et al. (2016) present a prospective cohort study investigating early life exposure to chemicals and health effects in later life, the LInking EDCs in maternal Nutrition to Child health (LINC study). For this, over 300 pregnant women were recruited during pregnancy. Prenatal exposure to chemicals was measured in, amongst others, cord blood and breast milk and the children were followed over time, measuring, amongst others, height and weight status.

For example, prenatal exposure to dichlorodiphenyl-dichloroethylene (DDE), a metabolite of the pesticide dichlorodiphenyl-trichloroethane (DDT), was assessed by measuring DDE in umbilical cord blood, collected at delivery. During the first year, the body mass index (BMI), based on weight and height, was monitored. DDE levels in umbilical cord blood were divided into 4 equal groups, called quartiles.

Boys with the lowest DDE concentrations (the first quartile) had a higher BMI growth curve in the first year, compared to boys with the highest concentrations DDE (the fourth quartile) (De Cock et al., 2016).

2.1.4 Nested Case-control study

When a case-control study is carried out within a cohort study, it is called a nested case-control study. Cases in a cohort study are selected, and matching non-cases are selected as controls. This type of study design is useful in case of a low amount of cases in a prospective cohort study.

An example of a nested case-control study

Engel et al. (2018) investigated attention-deficit hyperactivity disorder (ADHD) in children in relation to prenatal phthalate exposure. Phthalates are added to various consumer products to soften plastics. Exposure occurs during ingestion, inhalation or dermal absorption and sources are for example plastic packaging of food, volatile household products and personal care products (Benjamin et al., 2017). Engel et al. (2018) carried out a nested case-control study within the Norwegian Mother and Child Cohort (MoBa).

The cohort included 112,762 mother-child pairs of which only a small amount of cases with a clinical ADHD diagnosis. A total of 297 cases were randomly sampled from registrations of clinically ADHD diagnoses. In addition, 553 controls without ADHD were randomly sampled from the cohort. Phthalate metabolites were measured in maternal urine collected at midpregnancy and concentrations were divided into 5 equal groups, called quintiles. Children of mothers in the highest quintile of the sum of metabolites of the phthalate bis(2-ethylhexyl) phthalate (DEHP) had 2.99 (95%CI: 1.47-5.49) more odds (see chapter Quantifying disease and associations) of an ADHD diagnosis in comparison to the lowest quintile.

2.1.5 Ecological Study Design

All previously discussed study designs deal with data from individual participants. In the ecological study design data at aggregated level is used. This study design is applied when individual data is not available or when large-scale comparisons are being made, such as geographical comparisons of the prevalence of disease and exposure. Published statistics are suitable to use which makes the design relatively cheap and fast. Within environmental epidemiology, ecological study designs are frequently used in air pollution research.

For example, time trends of pollution can be detected using aggregated data over several time points and can be related to the incidence of health outcomes. Caution is necessary with interpreting the results: groups that are being compared might be different in other ways that are not measured. Moreover, you do not know whether, within the groups you are comparing, the people with the outcome you are interested in are also the people who have the exposure. This study design is, therefore, hypothesis-generating.

2.2 Experimental studies

A randomized controlled trial (RCT) is an experimental study in which participants are randomly assigned to an intervention group or a control group. The intervention group receives an intervention or treatment, the control group receives nothing, usual care or a placebo. Clinical trials that test the effectiveness of medication are an example of an RCT. If the assignment of participants to groups is not randomized, the design is called a non-randomized controlled trial. The latter design provides less strength of evidence.

When groups of people instead of individuals, are randomized, the study design is called a cluster-randomized controlled trial. This is, for example, the case when classrooms with children at school are randomly assigned to the intervention- and control group. Variations are used to switch groups between the intervention and control group. For example, a crossover design makes it possible that people are both intervention group and control group in different phases of the study. In order to not restrain the benefits of the intervention to the control group, a waiting list design makes the intervention available to the control group after the research period.

An example of an experimental study

An example of an experimental study design within environmental research is the study of Bae and Hong (2015). In a randomized crossover trial, participants had to drink beverages either from a BPA containing can, or a BPA-free glass bottle. Besides BPA levels in urine, blood pressure was measured after exposure. The crossover design included 3 periods, with either drinking only canned beverages, both canned and glass-bottled beverages or only glass-bottled beverages. BPA concentration was increased with 1600% after drinking canned beverages in comparison to drinking from glass bottles.

3. Confounding and effect modification

Confounding occurs when a third factor influences both the outcome and the determinant (see Figure 3). For example, the number of cigarettes smoked is positively associated with the prevalence of esophageal cancer. However, the number of cigarettes smoked is also positively associated with the amount of standard glasses alcohol consumption. Besides, alcohol consumption is a risk factor for esophageal cancer. Alcohol consumption is therefore a confounder in the relationship smoking and esophageal cancer. One can correct for confounders in the statistical analysis, e.g. using stratification (results are presented for the different groups separately).

Effect modification occurs when the association between exposure/determinant and outcome is different for certain groups (Figure 3). For example, the risk of lung cancer due to asbestos exposure is about ten times higher for smokers than for non-smokers. A solution to deal with effect modification is stratification as well.

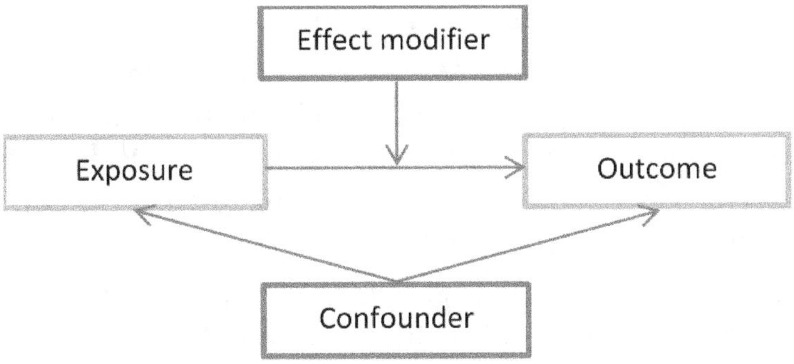

Figure 3: Confounding and effect modification in an association between exposure and outcome. A confounder has associations with both the exposure/determinant and the outcome. An effect modifier alters the association between the exposure/determinant and the outcome.

Box 1: Hill's Criteria for Causation

With epidemiological studies it is often not possible to determine a causal relationship. That is why epidemiological studies often employ a set of criteria, the Hill's criteria of causation, according to Sir Austin Bradford Hill, that need to be considered before conclusions about causality are justified (Hill, 1965).

- **Strength:** stronger associations are more reason for causation.
- **Consistency:** causation is likely when observations from different persons, in different populations and circumstances are consistent.
- **Specificity:** specificity of the association is reason for causation.
- **Temporality:** for causation the determinant must precede the disease.
- **Biological gradient:** is there biological gradient between the determinant and the disease, for example, a dose-response curve?
- **Plausibility:** is it biological plausible that the determinant causes the disease?
- **Coherence:** coherence between findings from laboratory analysis and epidemiology.
- **Experiment:** certain changes in the determinant, as if it was an experimental intervention, might provide evidence for causal relationships.
- **Analogy:** consider previous results from similar associations.

References

Bae, S., Hong, Y.C. (2015). Exposure to bisphenol a from drinking canned beverages increases blood pressure: Randomized crossover trial. Hypertension 65, 313-319. https://doi.org/10.1161/HYPERTENSIONAHA.114.04261

Benjamin, S., Masai, E., Kamimura, N., Takahashi, K., Anderson, R.C., Faisal, P.A. (2017). Phthalates impact human health: Epidemiological evidences and plausible mechanism of action. Journal of Hazardous Materials 340, 360-383. https://doi.org/10.1016/j.jhazmat.2017.06.036

De Cock, M., De Boer, M.R., Lamoree, M., Legler, J., Van De Bor, M. (2016). Prenatal exposure to endocrine disrupting chemicals and birth weight-A prospective cohort study. Journal of Environmental Science and Health - Part A Toxic/ Hazardous Substances and Environmental Engineering 51, 178-185. https:// doi.org/10.1080/10934529.2015.1087753

De Cock, M., Quaak, I., Sugeng, E.J., Legler, J., Van De Bor, M. (2016). Linking EDCs in maternal Nutrition to Child health (LINC study) - Protocol for prospective cohort to study early life exposure to environmental chemicals and child health. BMC Public Health 16: 147. https://doi.org/10.1186/s12889-016-2820-8

Engel, S.M., Villanger, G.D., Nethery, R.C., Thomsen, C., Sakhi, A.K., Drover, S.S.M., ... Aase, H. (2018). Prenatal phthalates, maternal thyroid function, and risk of attention-deficit hyperactivity disorder in the Norwegian mother and child cohort. Environmental Health Perspectives. https://doi.org/10.1289/ EHP2358

Hill, A.B. (1965). The Environment and Disease: Association or Causation? Journal of the Royal Society of Medicine 58, 295–300. https://doi. org/10.1177/003591576505800503

Hoffman, K., Lorenzo, A., Butt, C.M., Hammel, S.C., Henderson, B.B., Roman, S.A., ... Sosa, J.A. (2017). Exposure to flame retardant chemicals and occurrence and severity of papillary thyroid cancer: A case-control study. Environment International 107, 235-242. https://doi.org/10.1016/j.envint.2017.06.021

International Epidemiological Association. (2014). Dictionary of epidemiology. Oxford University Press. https://doi.org/10.1093/ije/15.2.277

Last, J.M. (2001). A Dictionary of Epidemiology. 4th edition, Oxford, Oxford University Press.

4.3.10. Environmental epidemiology - II

Quantifying Disease and Associations

1. Measures of Disease

Prevalence is the proportion of a population with an outcome at a certain time point (e.g. currently, 40% of the population is affected by disease Y) and can be calculated in cross-sectional studies.

Incidence concerns only new cases, and the cumulative incidence is the proportion of new cases in the population over a certain time span (e.g. 60% new cases of influenza per year). The (cumulative) incidence can only be calculated in prospective study designs, because the population needs to be at risk to develop the disease and therefore participants should not be affected by the disease at the start of the study.

Population Attributable Risk (PAR) is a measure to express the increase in disease in a population due to the exposure. It is calculated with this formula:

$$PAR = \frac{(incidence\,in\,the\,total\,population - incidence\,in\,the\,unexposed\,group)}{(incidence\,in\,the\,total\,population)}$$

2. Effect Sizes

2.1 In case of dichotomous outcomes (disease, yes versus no)

Risk ratio or relative risk (RR) is the ratio of the incidence in the exposed group to the incidence in the unexposed group (Table 1):

$$RR = \frac{\frac{A}{A+b}}{\frac{C}{C+D}}$$

The RR can only be used in prospective designs, because it consists of probabilities of an outcome in a population at risk. The RR is 1 if there is no risk, <1 if there is a decreased risk, and >1 if there is an increased risk. For example, researchers find an RR of 0.8 in a hypothetical prospective cohort study on the region children live in (rural vs. urban) and the development of asthma (outcome). This means that children living in rural areas have a 0.8 lower risk to develop asthma, compared to children living in the urban areas.

Risk difference (RD) is the difference between the risks in two groups (Table 1):

$$RV = \frac{A}{A+B} - \frac{C}{C+D}$$

Odds ratio (OR) is the ratio of odds on the outcome in the exposed group to the odds of the outcome in the unexposed group (Table 1).

$$OR = \frac{\frac{A}{B}}{\frac{C}{D}}$$

The OR can be used in any study design, but is most frequently used in case-control studies. (Table 1) The OR is 1 if there is no difference in odds, >1 if there is a higher odds, and <1 if there is a lower odds. For example, researchers find an OR of 2.5 in a hypothetical case-control study on mesothelioma cancer and occupational exposure to asbestos in the past. Patients with mesothelioma cancer had 2.5 higher odds on being occupational in the past exposed to asbestos compared to the healthy controls.

The OR can also be used in terms of odds on the disease instead of the exposure, the formulae is then (Table 1):

$$OR = \frac{\frac{A}{C}}{\frac{B}{D}}$$

For example, researchers find an odds ratio of 0.9 in a cross-sectional study investigating mesothelioma cancer in builders working with asbestos comparing the use of protective clothing and masks. The builders who used protective clothing and masks had 0.9 odds on having mesothelioma cancer in comparisons to builders who did not use protective clothing and masks.

Table 1: Concept table to use for calculation of the RR, RD, and OR

	Disease/outcome +	Disease/outcome -
Exposure/determinant +	A	B
Exposure/determinant -	C	D

2.2 In case of continuous outcomes (when there is a scale on which a disease can be measured, e.g. blood pressure)

Mean difference is the difference between the mean in the exposed group versus the unexposed group. This is also applicable to experimental designs with a follow-up to assess increase or decrease of the outcome after an intervention: the mean at the baseline versus the mean after the intervention. This can be standardized using the following formulae:

$$Standardized\,mean\,difference = \frac{mean\,intervention\,group - mean\,control\,group}{standard\,deviation}$$

The standard deviation (SD) is a measure of the spread of a set of values. In practice, the SD must be estimated either from the SD of the control group, or from an 'overall' value from both groups. The best-known index for effect size is Cohens 'd'. The standardized mean difference can have both a negative and a positive value (between -2.0 and +2.0). With a positive value, the beneficial effect of the intervention is shown, with a negative value, the effect is counterproductive. In general, an effect size of for example 0.8 means a large effect.

3. Statistical Significance and Confidence Interval

Effect measurements such as the relative risk, the odds ratio and the mean difference are reported together with statistical significance and/or a confidence interval. Statistical significance is used to retain or reject null hypothesis. The study starts with a null hypothesis assumption, we assume that there is no difference between variables or groups, e.g. RR=1 or the difference in means is 0. Then the statistical test gives us the probability of getting the outcome observed (e.g. OR=2.3, or mean difference=1.5) when in fact the null hypothesis is true. If the probability is smaller than 5%, we conclude that the observation is true and we may reject the null hypothesis. The 5% probability corresponds to a p-value of 0.05. A p-value cut-off of $p<0.05$ is generally used, which means that p-values smaller than 0.05 are considered statistically significant.

The 95% confidence interval. A 95% confidence interval is a range of values within which you can be 95% certain that the true mean of the population or measure of association lies. For example, in the hypothetical cross-sectional study on smoking (yes or no) and lung cancer, an OR of 2.5 was found, with an 95% CI of 1.1 to 3.5. That means, we can say with 95% certainty that the true OR lies between 1.1 and 3.5. This is regarded statistically significant, since the 1, which means no difference in odds, does not lie within the 95% CI. If researchers also studied oesophagus cancer in relation to smoking and found an OR of 1.9 with 95% CI of 0.6-2.6, this is not regarded statistically significant, since 95% CI includes 1.

4. Stratification

When two populations investigated have a different distribution of, for example, age and gender, it is often hard to compare disease frequencies among them. One way to deal with that is to analyse associations between exposure and outcome within strata (groups). This is called stratification. Example: a hypothetical study to investigate differences in health (outcome, measured with number of symptoms, such as shortness of breath while walking) between two groups of elderly, urban elderly (n=682) and rural elderly (n=143) (determinant). No difference between urban and rural elderly was found, however there was a difference in the number of women and men in both groups. The results for symptoms for urban and rural elderly are therefore stratified by gender (Table 2). Then, it appeared that male urban elderly have more symptoms than male rural elderly (p=0.01). The difference is not significant for women (p=0.07). The differences in health of elderly living an urban region are different for men and women, hence gender is an effect modifier of our association of interest.

Table 2. Number of symptoms (expressed as a percentage) for urban and rural elderly stratified by gender. Significant differences in bold.

Number of symptoms	Women		Men	
	Urban	Rural	Urban	Rural
None	16.0	30.4	16.2	43.5
One	26.4	30.4	45.2	47.8
Two or more	57.6	39.1	37.8	8.7
N	125	23	74	23
p-value	0.07		**0.01**	

4.3.11. Molecular Epidemiology - I. Human Biomonitoring

Human Biomonitoring

Human biomonitoring (HBM) involves the assessment of human exposure to natural and synthetic chemicals by the quantitative analysis of these compounds, their metabolites or reaction products in samples from human origin. Samples used in HBM can include blood, urine, faces, saliva, breast milk and sweat or other tissues, such as hair, nails and teeth.

The concentrations determined in human samples are a reflection of the exposure of an individual to the compounds analysed, also referred to as the internal dose. HBM data are collected to obtain insight into the population's exposure to chemicals, often with the objective to integrate them with health data for health impact assessment in epidemiological studies. Often, specific age groups are addressed, such as neonates, toddlers, children, adolescents, adults and elderly. Human biomonitoring is an established method in occupational and environmental exposure assessment.

In several countries, HBM studies have been conducted for decades already, such as the German Environment Survey (GerES) and the National Health and Nutrition Examination Survey (NHANES) program in the United States. HBM programs may sometimes be conducted under the umbrella of the World Health Organization (WHO).

Other examples are the Canadian Health Measures Survey, the Flemish Environment and Health Study and the Japan Environment and Children's Study, the latter specifically focuses on young children. Children are considered to be more at risk for the adverse health effects of early exposure to chemical pollutants, because of their rapid growth and development and their limited metabolic capacity to detoxify harmful chemicals.

Table 1. Information sources for Human Biomonitoring (HBM) programmes

Programme	Internet link
German Environment Survey (GerES)	www.umweltbundesamt.de/en/topics/health/assessing-environmentally-related-health-risks/german-environmental-survey-geres
National Health and Nutrition Examination Survey (NHANES)	https://www.cdc.gov/nchs/nhanes/index.htm
WHO	www.euro.who.int/en/data-and-evidence/environment-and-health-information-system-enhis/activities/human-biomonitoring-survey
Canadian Health Measures Survey (CHMS)	http://www23.statcan.gc.ca/imdb/p2SV.pl?Function=getSurvey&Id=148760
Japan Environment and Children's Study (JECS)	https://www.env.go.jp/en/chemi/hs/jecs/

Studies focusing on the impact of exposure to chemicals on health are conducted with the use of cohorts: groups of people that are enrolled in a certain study and volunteer to take part in the research program. Usually, apart from donating e.g. blood or urine samples, health measures, such as blood pressure, body weight, hormone levels, etc. but also data on diet, education, social background, economic status and lifestyle are collected, the latter through the use of questionnaires.

A cross-sectional study aims at the acquisition of exposure and health data of the whole (volunteer) group at a defined moment, whereas in a longitudinal study follow-up studies are conducted with a certain frequency (i.e. every few years) in order to follow and evaluate the changes in exposure, describe time trends as well as study health and lifestyle on the longer term (see section on Environmental Epidemiology). To obtain sufficient statistical power to derive meaningful relationships between exposure and eventual (health) effects, the number of participants in HBM studies is often very large, i.e. ranging to 100,000 participants.

Because a lot of (sometimes sensitive) data is gathered from many individuals, ethics is an important aspect of any HBM study. Before starting a certain study involving HBM, a Medical Ethical Approval Committee needs to approve it. Applications to obtain approval require comprehensive documentation of i) the study protocol (what is exactly being investigated), ii) a statement regarding the safeguarding of the privacy and collected data of the individuals, the access of researchers to the data and the safe storage of all information and iii) an information letter for the volunteers explaining the aim of and procedures used in the study and their rights (e.g. to withdraw), so that they can give consent to be included in the study.

Chemical Absorption, Distribution, Metabolism and Excretion

Because chemicals often undergo metabolic transformation (see section on Xenobiotic metabolism and defence) after entering the body via ingestion, dermal absorption and inhalation, it is important to not only focus on the parent compound (= the compound to which the individual was exposed), but also include metabolites. Diet, socio-economic status, occupation, lifestyle and the environment all contribute to the exposure of humans, while age, gender, health status and weight of an individual define the effect of the exposure. HBM data provide an aggregation of all the different routes through which the individual was exposed. For an in-depth investigation of exposure sources, however, chemical analysis of e.g. diet (including drinking water), the indoor and outdoor environment are still necessary. Another important source of chemicals to which people are exposed in their day to day life are consumer products, such as electronics, furniture, textiles, etc., that may contain flame retardants, stain repellents, colorants and dyes, preservatives, among others.

The distribution of a chemical in the body is highly dependent on its physico-chemical properties, such as lipophilicity/hydrophilicity and persistence, while also phase I and Phase II transformation (see section on Xenobiotic metabolism and defence) play a determining role, see Figure. 1. For lipophilic compounds (e.g. section on POPs) storage occurs in fat tissue, while moderately lipophilic to hydrophilic compounds are excreted after metabolic transformation, or in unchanged form. Based on these considerations, a proper choice for sampling of the appropriate matrix can be made, i.e. some chemicals are best measured in urine, while for others blood may be better suitable.

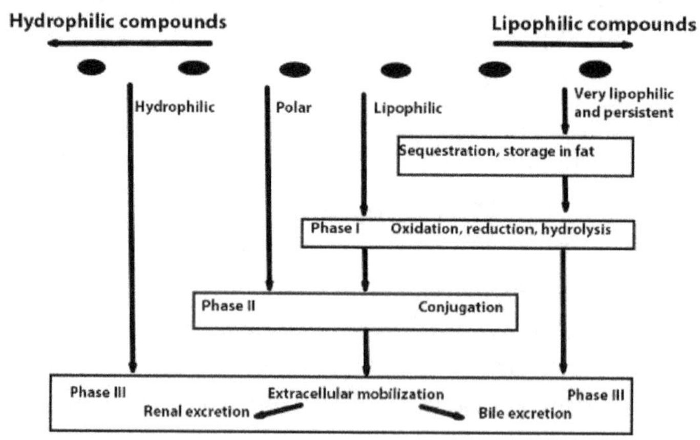

Figure 1. Distribution biotransformation of a compound (xenobiotic) in the body leading to storage or excretion.

For the design of the sampling campaign, the properties of the compounds to be analyzed should be taken into account. In case of volatility, airtight sampling containers should be used, while for light-sensitive compounds amber coloured glassware is the optimal choice.

Ideally, after collection, the samples need to be stored under the correct conditions as quickly as possible, in order the avoid degradation caused by thermal instability or by biodegradation caused by remaining enzyme activity in the sample (e.g. in blood or breast milk samples). Labeling and storage of the large quantities of samples generally included in HBM studies are important parts of the sampling campaign (see for video: https://www.youtube.com/watch?v=FQjKKvAhhjM).

Chemical Analysis of Human Samples for Exposure Assessment

Typically, for the determination of the concentrations of compounds to which people are exposed and the corresponding metabolites formed in the human body, analytical techniques such as liquid and gas chromatography (LC and GC, respectively) coupled to mass spectrometry (MS) are applied. Chromatography is used for the separation of the compounds, while MS is used to detect the compounds. Prior to the analysis using LC- or GC-MS, the sample is pretreated (e.g. particles are removed) and extracted, i.e. the compounds to be analysed are concentrated in a small volume while sample matrix constituents that may interfere with the analysis (e.g. lipids, proteins) are removed, resulting in an extract that is ready to be injected onto chromatographic system.

In Figure 2 a schematic representation is given of all steps in the analytical procedure.

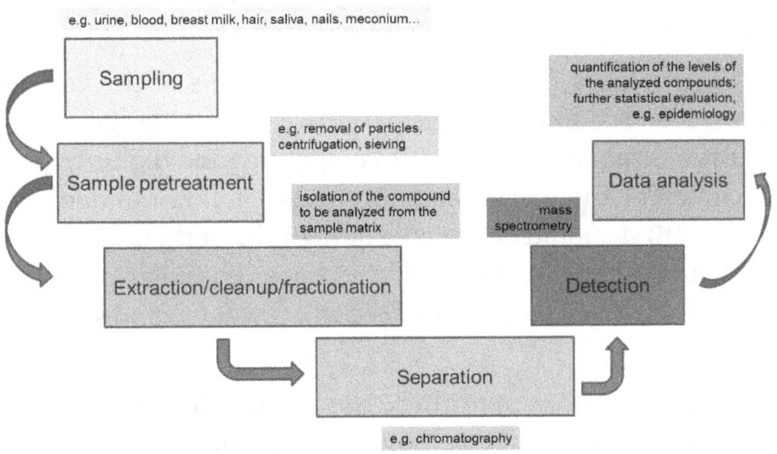

Figure 2. Schematic representation of the analytical procedure typically used for the quantitative determination of chemicals and their metabolites in human samples.

The analytical methods to quantify concentrations of chemicals in order to assess human exposure need to be of a high quality due to the specific nature of HBM studies. The compounds to be analysed are usually present in very low concentrations (i.e. in the order of pg/L for cord blood), and the sample volumes are small. For some matrices, the small sample volumes is dictated by the fact that sample availability is not unlimited, e.g. for blood. Another factor that governs the limited sample volume available is the costs that are related to the requirement of dedicated, long term storage space at conditions of -20 ^{0}C or even -80 ^{0}C to ensure sample integrity and stability.

The compounds on which HBM studies often focus are those to which we are exposed in daily life. This implies that the analytical procedure should be able to deal with contamination of the sample with the compounds to be analysed, due to the presence of the compounds in our surroundings. Higher background contamination leads to a decreased capacity to detect low concentrations, thus negatively impacting the quality of the studies.

Examples of compounds that have been monitored frequently in human urine are phthalates, such as diethyl hexyl phthalate or shortly DEHP. DEHP is a chemical used in many consumer products and therefore contamination of the samples with DEHP from the surroundings severely influences the analytical measurements. One way around this is to focus on the metabolites of DEHP after Phase I or II metabolism: this guarantees that the chemical has passed the human body and has undergone a metabolic transformation, and its detection is not due to contamination from the background, which results in a more reliable exposure metric. When the analytical method is designed for the quantitative analysis of metabolites, an enzymatic step for the deconjugation of the Phase II metabolites should be included (see section on Xenobiotic metabolism and defence).

Because the generated data, i.e. the concentrations of the compounds in the human samples, are used to determine parameters like average/median exposure levels, the detection frequency of specific compounds and highest/lowest exposure levels, the accuracy of the measurements should be high. In addition, analytical methods used for HBM should be capable of high throughput, i.e. the time needed per analysis should be low, because of the large numbers of samples that are typically analysed, in the order of a hundred to a few thousand samples, depending on the study.

Summarizing, HBM data support the assessment of temporal trends and spatial patterns in human exposure, sheds light on subpopulations that are at risk and provides insight into the effectiveness of measures to reduce or even prevent adverse health effects due to chemical exposure.

Background Information

HBM4EU project info: www.hbm4eu.eu; video: https://www.youtube.com/watch?v=DmC1v6EAeAM&feature=youtu.be.

4.3.11. Molecular Epidemiology - II. The Exposome and Internal Molecular Markers

Exposome

The exposome idea was described by Christopher Wild in 2005 as a measure of all human life-long exposures, including the process of how these exposures relate to health. An important aim of the exposome is to explain how non-genetic exposures contribute to the onset or development of important chronic disease. This concept represents the totality of exposures from three broad domains, i.e. internal, specific external and general external (Figure 1) (Wild, 2012). The internal exposome includes processes such as metabolism, endogenous circulating hormones, body morphology, physical activity, gut microbiota, inflammation, and aging. The specific external exposures include diverse agents, for example, radiation, infections, chemical contaminants and pollutants, diet, lifestyle factors (e.g., tobacco, alcohol) and medical interventions. The wider social, economic and psychological influences on the individual make up the general external exposome, including the following factors but not limited to social capital, education, financial status, psychological stress, urban-rural environment, climate, etc1.

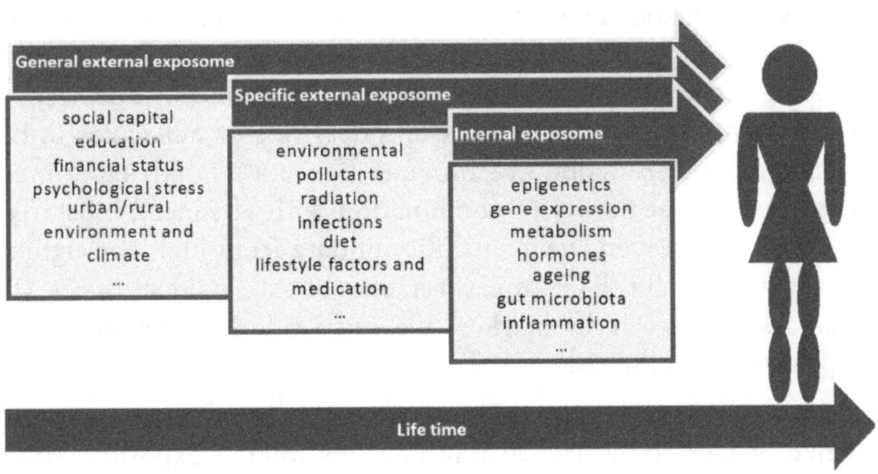

Figure 1. The exposome consists of 3 domains: the general external, the specific internal and the internal exposome.

The exposome is a theoretical concept with overlap between the three domains, however, this description serves to illustrate the full width of the exposome. The exposome model is characterized by the application of a wide range of tools in rapidly developing fields. Novel advances in monitoring exposure via wearables, modelling, internal biological measurements are recently developed and implemented to actually estimate lifelong exposures2-4. As these approaches generate extensive amounts of data, statistical and data science frameworks are warranted to analyze the exposome. Besides several bio-statistical advances combining multiple levels of exposures, biological responses and layers of personal characteristics, machine learning algorithms are developed to fully exploit collected data5,6.

The exposome concept clearly illustrates the complexity of the environment humans are exposed to nowadays, and how this can impact human health. There is a need for internal biomarkers of exposure (see section on Human biomonitoring) as well as biomarkers of effect, to disentangle the complex interplay between several exposures occurring potentially simultaneously and at different concentrations throughout life. Advances in biomedical sciences and molecular biology thereby collecting holistic information of epigenetics, transcriptome (see section on Gene expression), metabolome (see section on Metabolomics), etc. are at the forefront to identify biomarkers of exposure as well as of effect.

Internal Molecular Markers of the Exposome

Meet in the middle model

To determine the health effect of environmental exposure, markers that can detect early changes before disease arises are essential and can be implemented in preventative medicine. These types of markers can be seen as intermediate biomarkers of effect, and their discovery relies on large-scale studies at different levels of biology (transcriptomics, genomics, metabolomics). The term "omics" refers to the quantitative measurement of global sets of molecules in biological samples using high throughput techniques (i.e. automated experiments that enable large scale repetition)7, in combination with advanced biostatistics and bioinformatics tools8. Given the availability of data from high-throughput omics platforms, together with reliable measurements of external exposures, the use of omics enhances the search for markers playing a role in the biological pathway linking exposure to disease risk.

The meet-in-the-middle (MITM) concept was suggested as a way to address the challenge of identifying causal relationships linking exposures and disease outcome (Figure 2). The first step of this approach consists in the investigation of the association between exposure and biomarkers of exposure. The next step consists in the study of the relationship between (biomarkers of) exposure and intermediate omics biomarkers of early effects; and third, the relation between the disease outcome and intermediate omics biomarkers is assessed. The MITM stipulates that the causal nature of an association is reinforced if it is found in all three steps.

Molecular markers that indicate susceptibility to certain environmental exposures are starting to become uncovered and can aid in targeted prevention strategies. Therefore, this approach is heavily dependent on new developments in molecular epidemiology, in which molecular biology is merged into epidemiological studies. Below, the different levels of molecular biology currently studied to identify markers of exposure and effect are discussed in detail.

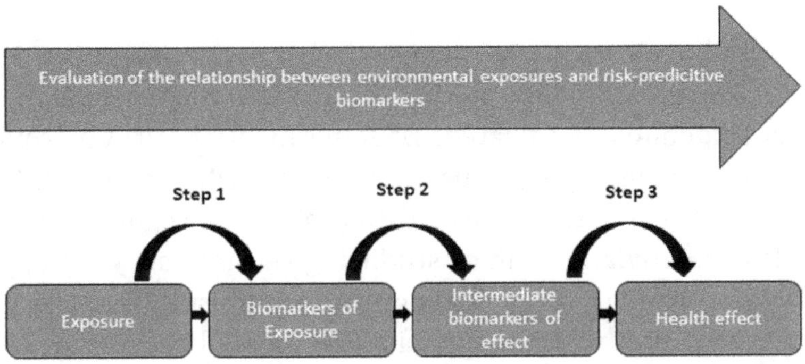

Figure 2. The meet in the middle approach. Biological samples are examined to identify molecules that represent intermediate markers of early effect. These can then be used to link exposure measures or markers with disease endpoints. Figure adapted from Vineis & Perera (2007).

Levels

Intermediate biomarkers can be identified as measurable indicators of certain biological states at different levels of the cellular machinery, and vary in their response time, duration, site and mechanism of action. Different molecular markers might be preferred depending on the exposure(s) under study.

Gene Expression

Changes at the mRNA level can be studied following a candidate approach in which mRNAs with a biological role suspected to be involved in the molecular response to a certain type of exposure (e.g. inflammatory mRNAs in the case of exposure to tobacco smoke) are selected a priori and measured using quantitative PCR technology or alternatively at the level of the whole genome by means of microarray analyses or Next Generation Sequencing technology. Changes at the transcriptome level, are studied by analysing the totality of RNA molecules present in a cell type or sample.

Both types of studies have proven their utility in molecular epidemiology. About a decade ago the first study was published reporting on candidate gene expression profiles that were associated with exposure to diverse carcinogens. Around the same time, the first studies on transcriptomics were published, including transcriptomic profiles for a dioxin-exposed population , in association with diesel-exhaust exposure, and comparing smokers versus non-smokers both in blood as well as airway epithelium cells. More recently, attention has been focused on prenatal exposures in association with transcriptomic signatures, as this fits within the scope of the exposome concept. As such, transcriptomic profiles have been described in association with exposure to maternal smoking assessed in placental tissue, as well as particulate matter exposure in cord blood samples.

Epigenetics

Epigenetics related to all heritable changes in that do not affect the DNA sequence itself directly. The most widely studied epigenetic mechanism in the field of environmental epidemiology to date is DNA methylation. DNA methylation refers to the process in which methyl-groups are added to a DNA sequence. As such, these methylation changes can alter the expression of a DNA segment without altering its sequence. DNA methylation can be studied by a candidate gene approach using a digestion-based design or, more commonly used, a bisulfite conversion followed by pyrosequencing, methylation-specific PCR or a bead array. The bisulfite treatment of DNA mediates the deamination of cytosine into uracil, and these converted residues will be read as thymine, as determined by PCR-amplification and sequencing. However, 5 mC residues are resistant to this conversion and will remain read as cytosine (Figure 3).

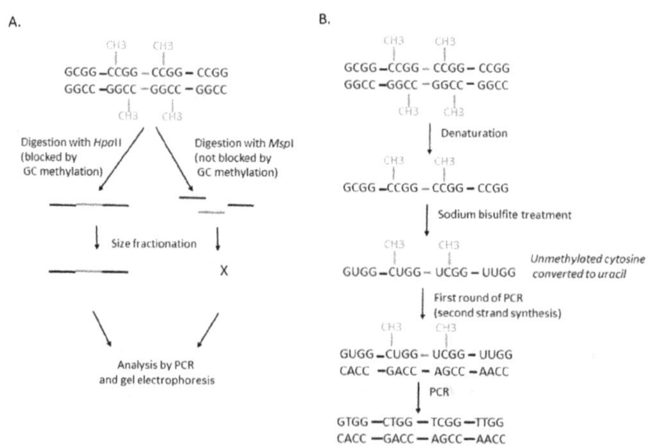

Figure 3: A. Restriction-digest based design A methylated (CH_3) region of genomic DNA is digested either with two restriction enzymes, one which is blocked by GC methylation (HpaII) and one which is not(MspI). Smaller fragments are discarded (X), enriching for methylated DNA in the HpaII treated sample. B. Bisulfite-conversion of DNA. DNA is denatured and then treated with sodium bisulfite to convert unmethylated cytosine to uracil, which is converted to thymine by PCR. An important point is that following bisulfite conversion, the DNA strands are no longer complementary, and primers are designed to assay the methylation status of a specific strand.

If an untargeted approach is desirable, several strategies can be followed to obtain whole-genome methylation data, including sequencing. Epigenotyping technologies such as the human methylation BeadChips 18 generate a methylation-state-specific 'pseudo-SNP' through bisulfite conversion; therefore, translating differences in the DNA methylation patterns into sequence differences that can be analyzed using quantitative genotyping methods.

An interesting characteristic of DNA methylation is that it can have transgenerational effects (i.e. effects that act across multiple generations). This was first shown in a study on a population that was prenatally exposed to famine during the Dutch Hunger Winter in 1944–1945. These individuals had less DNA methylation of the imprinted gene coding for insulin-like growth factor 2 (IGF2) measured 6 decades later compared with their unexposed, same-sex siblings. The association was specific for peri-conceptional exposure (i.e. exposure during the period from before conception to early pregnancy), reinforcing that very early mammalian development is a crucial period for establishing and maintaining epigenetic marks.

Post-translational modifications (i.e. referring to the biochemical modification of proteins following protein biosynthesis) recently gained more attention as they are known to be induced by oxidative stress (see sections on Oxidative stress) and specific inflammatory mediators . Besides their function in the structure of chromatin in eukaryotic cells, histones have been shown to have toxic and pro-inflammatory activities when they are released into the extracellular space . Much attention has gone to the associations between metal exposures and histone modifications, although recently a first human study on the association between particulate matter exposure and histone H_3 modification was published.

Expression of microRNAs (miRNAs are small noncoding RNAs of ~22nt in length which are involved in the regulation of gene expression at the posttranscriptional level by degrading their target mRNAs and/or inhibiting their translation) {Ambros et al,2004}{Ambros, 2004 #324}{Ambros, 2004 #324} has also been shown to serve as a valuable marker of exposure, both candidate and untargeted approaches have resulted in the identification of miRNA expression patterns that are associated with exposure to smoking , particulate matter , and chemicals such as polychlorinated biphenyls (PCBs) .

Metabolomics

Metabolomics have been proposed as a valuable approach to address the challenges of the exposome. Metabolomics, the study of metabolism at the whole-body level, involves assessment of the entire repertoire of small molecule metabolic products present in a biological sample. Unlike genes, transcripts and proteins, metabolites are not encoded in the genome. They are also chemically diverse, consisting of carbohydrates, amino acids, lipids, nucleotides and more. Humans are expected to contain a few thousand metabolites, including those they make themselves as well as nutrients and pollutants from their environment and substances produced by microbes in the gut. The study of metabolomics increases knowledge on the interactions between gene and protein expression, and the environment29. Metabolomics can be a biomarker of effect of environmental exposure as it allows for the full characterization of biochemical changes that occur during xenobiotic metabolism (see Section on Xenobiotic metabolism and defence). Recent technological developments have allowed downscaling the sample volume necessary for the analysis of the full metabolome, allowing for the assessment of system-wide metabolic changes that occur as a result of an exposure or in conjunction with a health outcome 30. As for all discussed biomarkers, both targeted metabolomics, in which specific metabolites are measured in order to characterize a pathway of interest, as well as untargeted metabolomic approaches are available. Among "omics" methodologies, metabolomics interrogates the levels of a relatively lower number of features as there are about 2900 known human metabolites versus ~30,000 genes. Therefore it has strong statistical power compared to transcriptome-wide and genome-wide studies 31. Metabolomics is, therefore, a potentially sensitive method for identifying biochemical effects of external stressors. Even though the developing field of "environmental metabolomics" seeks to employ metabolomic methodologies to characterize the effects of environmental exposures on organism function and health, the relationship between most of the chemicals and their effects on the human metabolome have not yet been studied.

Challenges

Limitations of molecular epidemiological studies include the difficulty to obtain samples to study, the need for large study populations to identify significant relations between exposure and the biomarker, the need for complex statistical methods to analyse the data. To circumvent the issue of sample collection, much effort has been focused on eliminating the need for blood or serum samples by utilizing saliva samples, buccal cells or nail clippings to read out molecular markers. Although these samples can be easily collected in a non-invasive manner, care must be taken to prove that these samples indeed accurately reflect the body's response to exposure rather than a local effect. For DNA methylation, it has been shown this is heavily dependent on the locus under study. For certain CpG sites the correlation in methylation levels is much higher than for other sites 32. For those sites that do not correlate well across tissues, it has furthermore been demonstrated that DNA methylation levels can differ in their associations with clinical outcomes 33, so care must be taken in epidemiological study design to overcome these issues.

4.3.12. Gene Expression

Synopsis

Low-dose exposure to toxicants induces biochemical changes in an organism, which aim to maintain homoeostasis of the internal environment and to prevent damage. One aspect of these changes is a high abundance of transcripts of biotransformation enzymes, oxidative stress defence enzymes, heat shock proteins and many proteins related to the cellular stress response. Such defence mechanisms are often highly inducible, that is, their activity is greatly upregulated in response to a toxicant. It is also known that most of the stress responses are specific to the type of toxicant. This principle may be reversed: if an upregulated stress response is observed, this implies that the organism is exposed to a certain stress factor; the nature of the stress factor may even be derived from the transcription profile. For this reason, microarrays, RNA sequencing or other techniques of transcriptome analysis, have been applied in a large variety of contexts, both in laboratory experiments and in field surveys. These studies suggest that transcriptomics scores high on (in decreasing order) (1) rapidity, (2) specificity, and (3) sensitivity. While the promises of genomics applications in environmental toxicology are high, most of the applications are in mode-of-action studies rather than in risk assessment.

Introduction

No organism is defenceless against environmental toxicants. Even at exposures below phenotypically visible no-effect levels a host of physiological and biochemical defence mechanisms are already active and contribute to the organism's homeostasis. These regulatory mechanisms often involve upregulation of defence mechanisms such as oxidative stress defence, biotransformation (xenobiotic metabolism), heat shock responses, induction of metal-binding proteins, hypoxia response, repair of DNA damage, etc. At the same time downregulation is observed for energy metabolism and functions related to growth and reproduction. In addition to these targeted regulatory mechanisms targeting, there are usually a lot of secondary effects and dysfunctional changes arising from damage. A comprehensive overview of all these adjustments can be obtained from analysis of the transcriptome.

In this module we will review the various approaches adopted in "omics", with an emphasis on transcriptomics. "Omics" is a container term comprising five different activities. Table 1 provides a list of these approaches and their possible contribution to environmental toxicology. Genomics and transcriptomics deal with DNA and mRNA sequencing, proteomics relies on mass spectrometry while metabolomics involves a variety of separation and detection techniques, depending on the class of compounds analysed. The various approaches gain strength when applied jointly. For example proteomics analysis is much more insightful if it can be linked to an annotated genome sequence and metabolism studies can profit greatly from transcription profiles that include the enzymes responsible for metabolic reactions. Systems biology aims to integrate the different approaches using mathematical models. However, it is fair to say that the correlation between responses at the different levels is often rather poor. Upregulation of a transcript does not always imply more protein, more protein can be generated without transcriptional upregulation and the concentration of a metabolite is not always correlated with upregulation of the enzymes supposed to produce it. In this module we will focus on transcriptomics only. Metabolomics is dealt with in a separate section.

Table 1. Overview of the various "omics" approaches

Term	Description	Relevance for environmental toxicology
Genomics	Genome sequencing and assembly, comparison of genomes, phylogenetics, evolutionary analysis	Explanation of species and lineage differences in susceptibility from the structure of targets and metabolic potential, relationship between toxicology, evolution and ecology
Transcriptomics	Genome-wide transcriptome (mRNA) analysis, gene expression profiling	Target and metabolism expression indicating activity, analysis of modes of action, diagnosis of substance-specific effects, early warning instrument for risk assessment
Proteomics	Analysis of the protein complement of the cell or tissue	Systemic metabolism and detoxification, diagnosis of physiological status, long-term or permanent effects
Metabolomics	Analysis of all metabolites from a certain class, pathway analysis	Functional read-out of the physiological state of a cell or tissue
Systems biology	Integration of the various "omics" approaches, network analysis, modelling	Understanding of coherent responses, extrapolation to whole-body phenotypic responses

Transcriptomics Analysis

The aim of transcriptomics in environmental toxicology is to gain a complete overview of all changes in mRNA abundance in a cell or tissue as a function of exposure to environmental chemicals. This is usually done in the following sequence of steps:

- Exposure of organisms to an environmental toxicant, including a range of concentrations, time-points, etc., depending on the objectives of the experiment.

- Isolation of total RNA from individuals or a sample of pooled individuals. The number of biological replicates is determined at this stage, by the number of independent RNA isolations, not by technical replication further on in the procedure.
- Reverse transcription. mRNAs are transcribed to cDNA using the enzyme reverse transcriptase that initiates at the polyA tail of mRNAs. Because ribosomal RNA lacks a poly(A)tail they are (in principle) not transcribed to cDNA. This is followed by size selection and sometimes labelling of cDNAs with barcodes to facilitate sequencing.
- Sequencing of the cDNA pool and transcriptome assembly. The assembly preferably makes use of a reference genome for the species. If no reference genome is available, the transcriptome is assembled *de novo*, which requires a greater sequencing depth and usually ends in many incomplete transcripts. A variety of corrections are applied to equalize effects of total RNA yield, library size, sequencing depth, gene length, etc.
- Gene expression analysis and estimation of fold regulation. This is done, in principle, by counting the normalized number of transcripts per gene for every gene in the genome, for each of the different conditions to which the organism was exposed. The response per gene is expressed as fold regulation, by expressing the transcripts relative to a standard or control condition. Tests are conducted to separate significant changes from noise.
- Annotation and assessment of pathways and functions as influenced by exposure. An integrative picture is developed, taking all evidence together, of the functional changes in the organism.

In the recent past, step 4 was done by microarray hybridization rather than by direct sequencing. In this technique two pools of cDNA (e.g. a control and a treatment) are hybridized to a large number of probes fixed onto a small glass plate. The probes are designed to represent the complete gene complement of the organism. Positive hybridization signals are taken as evidence for upregulated gene expression. Microarray hybridization arose in the years 1995-2005 but has now been largely overtaken by ultrafast and high-throughput next generation sequencing methods, however, due to cost-efficiency, relative simplicity of bioinformatics analysis, and standardization of the assessed genes it is still often used.

We illustrate the principles of transcriptomics analysis and the kind of data analysis that follows it, by an example from the work by Bundy et al. (2008). These authors exposed earthworms (Lumbricus rubellus) to soils experimentally amended with copper, quite a toxic element for earthworms. The copper-induced transcriptome was surveyed using a custom-made microarray and metabolic profiles were established using NMR (nuclear magnetic resonance) spectroscopy. From the 8,209 probes on the microarray, 329 showed a significant alteration of expression under the influence of copper. The data were plotted in a "heat map" diagram (Figures 1A and 1B), providing a quick overview of upregulated and downregulated genes. The expression profiles were also analysed in reduced dimensionality using principal component analysis (PCA).

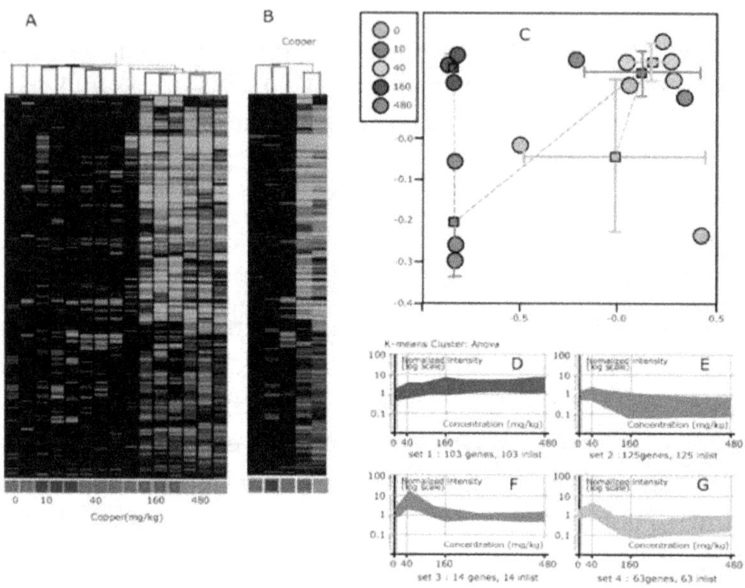

Figure 1. Example of a transcriptomics analysis aiming to understand copper toxicity to earthworms. A "heat map" of individual replicates (four in each of five copper treatments). Expression is indicated for each of the 329 differentially expressed genes (arranged from top to bottom) in red (downregulated) or green (upregulated). A cluster analysis showing the similarities is indicated above the profiles. B. The same data, but with the four replicates per copper treatment joined. The data show that at 40 mg/kg of copper in soil some of the earthworm's genes are starting to be downregulated, while at 160 mg/kg and 480 mg/kg significant upregulation and downregulation is occurring. C Principal Component Analysis of the changes in expression profile. The multivariate expression profile is reduced to two dimensions and the position of each replicate is indicated by a single point in the biplot; the confidence interval over four replicates of each copper treatment is indicated by horizontal and vertical bars. The profiles of the different copper treatments (joined by a dashed line) differ significantly from each other. D, E, F, and G. Classification of the 329 genes in four groups according to their responses to copper (plotted on the horizontal axis). Redrawn from Bundy et al. (2008) by Wilma Ijzerman.

This showed that the profiles varied considerably with treatment. Especially the highest and the penultimate highest exposures generated a profile very different from the control (see Figure 1C). The genes could be allocated to four clusters, (1) genes upregulated by copper over all exposures (Figure 1D), (2) genes downregulated by copper (see Figure 1E), (3) genes upregulated by low exposures but unaffected at higher exposures (see Figure 1F), and (4) genes upregulated by low exposure but downregulated by higher concentrations (see Figure 1G). Analysis of gene identity combined with metabolite analysis suggested that the changes were due to an effect of copper on mitochondrial respiration, reducing the amount of energy generated by oxidative phosphorylation. This mechanism was underlying the reduction of body-growth observed on the phenotypic level.

Omics in Risk Assessment

How could omics-technology, especially transcriptomics, contribute to risk assessment of chemicals? Three possible advantages have been put forward:

- Gene expression analysis is *rapid*. Gene regulation takes place on a time-scale of hours and results can be obtained within a few days. This compares very favourably with traditional toxicity testing (*Daphnia*, 48 hours, *Folsomia*, 28 days).
- Gene expression is *specific*. Because a transcription profile involves hundreds to thousands of endpoints (genes), the information content is potentially very large. By comparing a new profile generated by an unknown compound, to a trained data set, the compound can usually be identified quite precisely.
- Gene expression is *sensitive*. Because gene regulation is among the very first biochemical responses in an organism, it is expected to respond to lower dosages, at which whole-body parameters such as survival, growth and reproduction are not yet responding.

Among these advantages, the second one (specificity) has shown to be the most consistent and possibly brings the largest advantage. This can be illustrated by a study by Dom et al. (2012) in which gene expression profiles were generated for Daphnia magna exposed to different alcohols and chlorinated anilines (Figure 2).

The profiles of replicates exposed to the same compound were always clustered together, except in one case (ethanol), showing that gene expression is quite specific to the compound. It is possible to reverse this argument: from the gene expression profile the compound causing it can be deduced. In addition, the example cited showed that the first separation in the cluster analysis was between exposures that did and did not affect energy reserves and growth. So the gene expression profiles are not only indicative of the compound, but also of the type of effects expected.

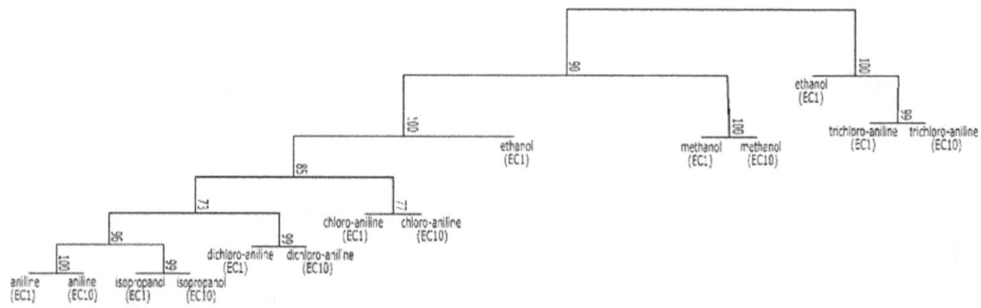

Figure 2. Clustered gene expression profiles of Daphnia magna exposed to seven different compounds. Replicates exposed to the same compound are clustered together, except for ethanol. The first split separates exposures that at the EC10 level (reproduction) did not show any effects on growth and energy reserves (right) and exposures that caused significant such effects (left). Reproduced from Dom et al. (2012) by Wilma IJzerman.

The claim of rapidity also proved true, however, the advantage of rapidity is not always borne out. It may be an issue when quick decisions are crucial (evaluating a truck loaded with suspect contaminated soil, deciding whether to discharge a certain waste stream into a lake yes or no), but for regular risk assessment procedures it proved to be less of an advantage than sometimes expected. Finally, greater sensitivity of gene expression, in the sense of lower no-observed effect concentrations than classical endpoints is a potential advantage, but proves to be less spectacular in practice. However, there are clear examples in which exposures below phenotypic effect levels were shown to induce gene expression responses, indicating that the organism was able to compensate any negative effects by adjusting its biochemistry.

Another strategy regarding the use of gene expression in risk assessment is not to focus on genome-wide transcriptomes but on selected biomarker genes. In this strategy, gene expressions are aimed for that show (1) consistent dose-dependency, (2) responses over a wide range of contaminants, and (3) correlations with biological damage. For example, De Boer *et al.* (2015) analysed a composite data set including experiments with six heavy metals, six chlorinated anilines, tetrachlorobenzene, phenanthrene, diclofenac and isothiocyanate, all previously used in standardized experiments with the soil-living collembolan, *Folsomia candida*. Across all treatments a selection of 61 genes was made, that were responsive in all cases and fulfilled the three criteria listed above. Some of these marker genes showed a very good and reproducible dose-related response to soil contamination. Two biomarkers are shown in Figure 3. This experiment, designed to diagnose a field soil with complex unknown contamination, clearly demonstrated the presence of *Cyp*-inducing organic toxicants.

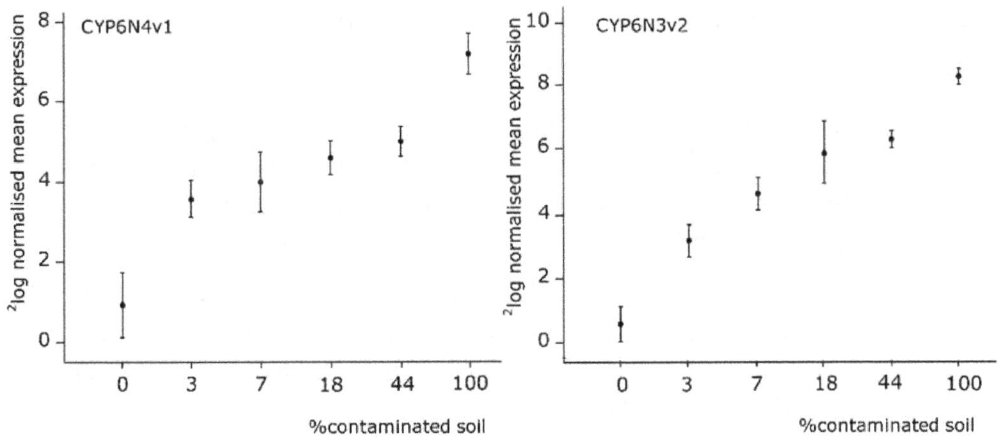

Figure 3. Showing gene expression, relative to control expression, for two selected biomarker genes (encoding cytochroms P450 phase I biotransformation enzymes) in the genome of the soil-living collembolan Folsomia candida, in response to the concentration of contaminated field soil spiked-in into a clean soil at different rates. Reproduced from Roelofs et al. (2012) by Wilma IJzerman.

Of course there are also disadvantages associated with transcriptomics in environmental toxicology, for example:

- Gene expression analysis requires a knowledge-intensive infrastructure, including a high level of expertise for some of the bioinformatics analyses. Also, adequate molecular laboratory facilities are needed; some techniques are quite expensive.
- Gene expression analysis is most fruitful when species are used that are backed up by adequate genomic resources, especially a well annotated genome assembly, although this is becoming less of a problem with increasing availability of genomic resources.
- The relationship between gene expression and ecologically relevant variables such as growth and reproduction of the animal is not always clear.

Conclusions

Gene expression analysis has come to occupy a designated niche in environmental toxicology since about 2005. It is a field highly driven by technology, and shows continuous change over the last years. It may significantly contribute to risk assessment in the context of mode of action studies and as a source of designated biomarker techniques. Finally, transcriptomics data are very suitable to feed into information regarding key events, important biochemical alterations that are causally linked up to the level of the phenotype to form an adverse outcome pathway. We refer to the section on Adverse outcome pathways for further reading.

References

Bundy,J.G.,Sidhu,J.K.,Rana,F.,Spurgeon,D.J.,Svendsen,C.,Wren,J.F.,Stürzenbaum, S.R., Morgan, A.J., Kille, P. (2008). "Systems toxicology" approach identifies coordinated metabolic responses to copper in a terrestrial non-model invertebrate, the earthworm Lumbricus rubellus. BMC Biology 6, 25.

De Boer, T.E., Janssens, T.K.S., Legler, J., Van Straalen, N.M., Roelofs, D. (2015). Combined transcriptomics analysis for classification of adverse effects as a potential end point in effect based screening. Environmental Science and Technology 49, 14274-14281.

Dom, N., Vergauwen, L., Vandenbrouck, T., Jansen, M., Blust, R., Knapen, D. (2012). Physiological and molecular effect assessment versus physico-chemistry based mode of action schemes: Daphnia magna exposed to narcotics and polar narcotics. Environmental Science and Technology 46, 10-18.

Gibson, G., Muse, S.V. (2002). A Primer of Genome Science. Sinauer Associates Inc., Sunderland.

Gibson, G. (2008). The environmental contribution to gene expression profiles. Nature Reviews Genetics 9, 575-581.

Roelofs, D., De Boer, M., Agamennone, V., Bouchier, P., Legler, J., Van Straalen, N. (2012). Functional environmental genomics of a municipal landfill soil. Frontiers in Genetics 3, 85.

Van Straalen, N.M., Feder, M.E. (2012). Ecological and evolutionary functional genomics - how can it contribute to the risk assessment of chemicals? Environmental Science & Technology 46, 3-9.

Van Straalen, N.M., Roelofs, D. (2008). Genomics technology for assessing soil pollution. Journal of Biology 7, 19.

4.3.13. Metabolomics

Introduction

Metabolomics is the systematic study of small organic molecules (<1000 Da) that are intermediates and products formed in cells and biofluids due to metabolic processes. A great variety of small molecules result from the interaction between genes, proteins and metabolites. The primary types of small organic molecules studied are endogenous metabolites (i.e., those that occur naturally in the cell) such as sugars, amino acids, neurotransmitters, hormones, vitamins, and fatty acids. The total number of endogenous metabolites in an organism is still under study but is estimated to be in the thousands. However, this number varies considerably between species and cell types. For instance, brain cells contain relative high levels of neurotransmitters and lipids, nevertheless the levels between different types of brain tissues can vary largely. Metabolites are working in a network, e.g. citric acid cycle, by the conversion of molecules by enzymes. The turnover time of the metabolites is regulated by the enzymes present and the amount of metabolites present.

The field of metabolomics is relatively new compared to genomics, with the first draft of the human metabolome available in 2007. However, the field has grown rapidly since that time due to its recognized ability to reflect molecular changes most closely associated with an organism's phenotype. Indeed, in comparison to other 'omics approaches (e.g., transcriptomics), metabolites are the downstream results from the action of genes and proteins and, as such, provide a direct link with the phenotype (Figure 1). The metabolic status of an organism is directly related to its function (e.g. energetic, oxidative, endocrine, and reproductive status) and phenotype, and is, therefore, uniquely suitable to relate chemical stress to the health status of organisms. Moreover, transcriptomics and proteomics, the identification of metabolites does not require the existence of gene sequences, making it particularly useful for the those species which lack a sequenced genome.

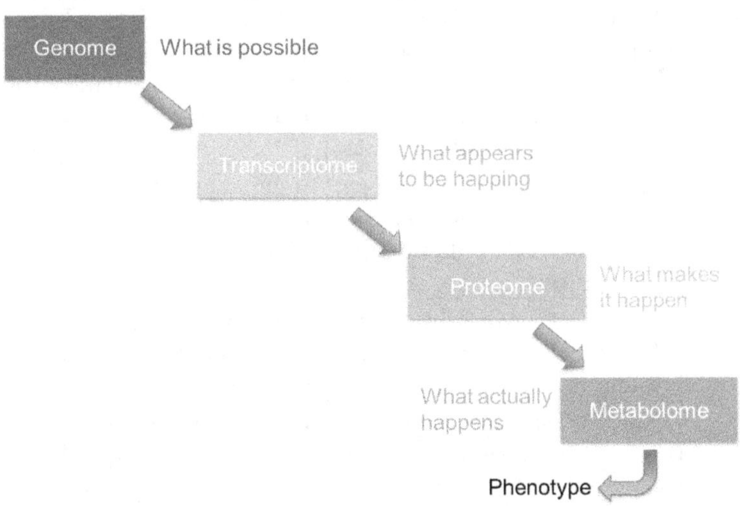

Figure 1: Cascade of different omics fields.

Definitions

The complete set of small molecules in a biological system (e.g. cells, body fluids, tissues, organism) is called the metabolome (Table 1). The term metabolomics was introduced by Oliver et al. (1998) who described it "as the complete set of metabolites/low molecular weight compounds which is context dependent, varying according to the physiology, development or pathological state of the cell, tissue, organ or organism". This quote highlights the observation that the levels of metabolites can vary due to internal as well as external factors, including stress resulting from exposure to environmental contaminants. This has resulted in the emergence and growth of the field of environmental metabolomics which is based on the application of metabolomics, to biological systems that are exposed to environmental contaminants and other relevant stressors (e.g., temperature). In addition to endogenous metabolites, some metabolomic studies also measure changes in the biotransformation of environmental contaminants, food additives, or drugs in cells, the collection of which has been termed the xenometabolome.

Table 1: Definitions of metabolomics.

Term	Definition	Relevance for environmental toxicology
Metabolomics	Analysis of small organic molecules (<1000 Da) in biological systems (e.g. cell, tissue, organism)	Functional read-out of the physiological state of a cell or tissue and directly related to the phenotype
Metabolome	Measurement of the complete set of small molecules in a biological system	Discovery of affected metabolic pathways due to contaminant exposure
Environmental metabolomics	Metabolomics analysis in biological systems that are exposed to environmental stress, such as the exposure to environmental contaminants	Metabolomics focused on environmental contaminant exposure to study for instance the mechanism of toxicity or to find a biomarker of exposure or effect
Xenometabolome	Metabolites formed from the biotransformation of environmental contaminants, food additives, or drugs	Understanding the metabolism of the target contaminant
Targeted metabolomics	Analysis of a pre-selected set of metabolites in a biological system	Focus on the effects of environmental contaminants on specific metabolic pathways
Untargeted metabolomics	Analysis of all detectable (i.e., not preselected) of metabolites in a biological system	Discovery-based analysis of the metabolic pathways affected by environmental contaminant exposure

Environmental Metabolomics Analysis

The development and successful application of metabolomics relies heavily on i) currently available analytical techniques that measure metabolites in cells, tissues, and organisms, ii) the identification of the chemical structures of the metabolites, and iii) characterisation of the metabolic variability within cells, tissues, and organisms.

The aim of metabolomics analysis in environmental toxicology can be:

- to focus on changes in the abundances of specific metabolites in a biological system

After environmental contaminant exposure: targeted metabolomics

- to provide a "complete" overview of changes in abundances of all detectable metabolites in a biological system after environmental contaminant exposure: untargeted metabolomics

In targeted metabolomics a limited number of pre-selected metabolites (typically 1-100) are quantitatively analysed (e.g. nmol dopamine/g tissue). For example, metabolites in the neurotransmitter biosynthetic pathway could be targeted to assess exposures to pesticides. Targeting specific metabolites in this way typically allows for their detection at low concentrations with high accuracy. Conversely, in untargeted metabolomics the aim is to detect as many metabolites as possible, regardless of their identities so as to assess as much of the metabolome as possible. The largest challenge for untargeted metabolomics is the identification (annotation) of the chemical structures of the detected metabolites. There is currently no single analytical method able to detect all metabolites in a sample, and therefore a combination of different analytical techniques are used to detect the metabolome. Different techniques are required due to the wide range of physical-chemical properties of the metabolites. The variety of chemical structures of metabolites are shown in Figure 2. Metabolites can be grouped in classes such as fatty acids (the classes are given in brackets in Figure 2), and within a class different metabolites can be found.

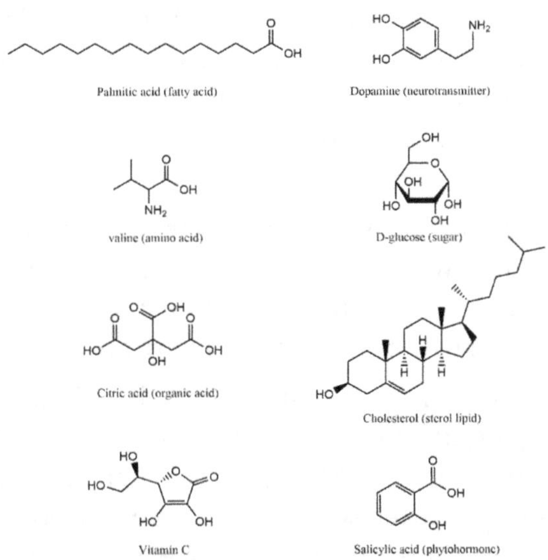

Figure 2: Examples of the chemical structures of several commonly detected metabolites. Metabolite classes are indicated in brackets. Drawn by Steven Droge.

A general workflow of environmental metabolomics analysis uses the following steps:

- of the organism or cells to an environmental contaminant. An unexposed control group must also be included. The exposures often include the use of various concentrations, time-points, etc., depending on the objectives of the study.
- Sample collection of the relevant biological material (e.g. cell, tissue, organism). It is important that the collection be done as quickly as possible so as to quench any further metabolism. Typically ice cold solvents are used.
- of the metabolites from the cell, tissue or organisms by a two-step extraction using a combination of polar (e.g. water/methanol) and apolar (e.g. chloroform) extraction solvents.
- of the polar and apolar fractions using liquid chromatography (LC) or gas chromatography (GC) combined with mass spectrometry (MS), or by nuclear magnetic resonance (NMR) spectroscopy. The analytical tool(s) used will depend on the metabolites under consideration and whether a targeted or untargeted approach is required.
- Metabolite detection (targeted or untargeted analysis).
    - Targeted metabolomics - a specific set of pre-selected metabolites are detectedand their concentrations are determined using authentic standards.
    - Untargeted metabolomics - a list of all detectable metabolites measured by MS or NMR response, and their intensities is collected. Various techniques are then used to determine the identities of those metabolites that change due to the exposure (see step 7 below).
- Statistical analysis using univariate and multivariate statistics to calculate the difference between the exposure and the control groups. The fold change (fold increase or decrease of the metabolite levels) between an exposure and control group are determined.
- For untargeted metabolomics only the chemical structure of the statistically significant metabolites are identified. The identification of the chemical structure of the metabolite can be based on the molecular weight, isotope patterns, elemental compositions, mass spectrometry fragmentation patterns, etc. Mass spectrometry libraries are used for the identification to match the above parameters in the samples with the data in libraries.

- **Data processing:** Identification of the metabolic pathways influenced by the chemical exposure. Integrative picture of the molecular and functional level of an organism due to chemical exposure. Understand the relationship between the chemical exposure, molecular pathway changes and the observed toxicity. Or to identify potential biomarkers of exposure or effect.

Box: Analytical tools for metabolomics analysis

The most frequently used analytical tools for measuring metabolites are mass spectrometry (MS) and nuclear magnetic resonance (NMR) spectroscopy. MS is an analytical tool that generates ions of molecules and then measures their mass-to-charge ratios. This information can be used to generate a "molecular finger print" for each molecule, and based on this finger print metabolites can be identified. Chromatography is typically used to separate the different metabolites of a mixture found in a sample before it enters the mass spectrometer. Two main chromatography techniques are used in metabolomics: liquid chromatography and gas chromatography. Due to its high sensitivity, MS is able to measure a large number of different metabolites simultaneously. Moreover, when coupled with a separation method such as chromatography, MS can detect and identify thousands of metabolites.

Mass spectrometry is much more sensitive than NMR, and it can detect a large range of different types of metabolites with different physical-chemical properties. NMR is less sensitive and can therefore detect a lower number of metabolites (typically 50-200). The advantages of NMR are the minimum amount of sample handling, the reproducibility of the measurements (due to high precision), and it is easier to quantify the levels of metabolites. In addition, NMR is a non-destructive technique such that a sample can often be used for further analyses after the data has been acquired.

Application of Environmental Metabolomics

Metabolomics has been widely used in drug discovery and medical sciences. More recently, metabolomics is being incorporated into environmental studies, an emerging field of research called environmental metabolomics. Environmental metabolomics is used mainly in five application domains (Table 2). Arguably the most commonly used application is for studying the mechanism of toxicity/mode of action (MoA) of contaminants. However, many studies have identified select metabolites that show promise for use as biomarkers of exposure or effect. As a result of its strength in identifying response fingerprints, metabolomics is also finding use in the regulatory toxicology field particularly for read-across studies. This application is particularly useful for rapidly screening contaminants for toxicity. Metabolomics can be also be used in dose-response studies (benchmark dosing) to derive a point of departure (POD). This is especially interesting in regulatory chemical risk assessment.

Currently, the field of systems toxicology is explored by combining data from different omics field (e.g. transcriptomics, proteomics, metabolomics) to improve our understanding of the relationship between the different omics, chemical exposure, and toxicity, and to better understand the mechanism of toxicity/ MoA.

Table 2: Application areas of metabolomics in environmental toxicology.

Application area	Description
Mechanism of toxicity/ Mode of action (MoA)	Using metabolomics to understand at the molecular level the pathways that are affected by exposure to environmental contaminants. Discovery of the mode of action of chemicals. In an adverse outcome pathway (AOP)Discovery metabolomics is used to identify the key events (KE), by linking chemical exposure at the molecular level to functional endpoints (e.g. reproduction, behaviour).
Biomarker discovery	Identification of metabolites that can be used as convenient (i.e., easy and inexpensive to measure) indicators of exposure or effect.
Read-across	In regulatory toxicology, metabolomics is used in read-across studies to provide information on the similarity of the responses between chemicals. This approach is useful to identify more environmentally toxic chemicals.
Point of departure	Metabolomics can be used in dose-response studies (benchmark dosing) to derive a point of departure (POD). This is especially interesting in regulatory chemical risk assessment. This application is currently not used yet.
Systems toxicology	Combination approach of different omics (e.g. transcriptomics, proteomics, metabolomics) to improve our understanding of the relationship between the different omics and chemical exposure, and to better understand the mechanism of toxicity/ MoA.

As an illustration of the mechanism of toxicity/mode of action application, Bundy et al. (2008) used NMR-based metabolomics to study earthworms (*Lumbricus rubellus*) exposed to various concentrations of copper in soil (0, 10, 40, 160, 480 mg copper / kg soil). They performed both transcriptomic and metabolomics studies. Both polar (sugars, amino acid, etc.) and apolar (lipids) metabolites were analysed, and fold changes relative to the control group were determined. For example, differences in the fold changes of lipid metabolites (e.g. fatty acids, triacylglycerol) as a function of copper concentration are shown as a "heatmap" in Figure 3A. Clearly the highest dose group (480 mg/kg) has a very different lipid metabolite pattern than the other groups. The polar metabolite data was analysed using principal component analysis (PCA) , a multivariate statistical tool that reduces the number of dimensions of the data. The PCA score plot shown in Figure 3B reveals that the largest differences in metabolite profiles exist between: the control and low dose (10 mg Cu/kg) groups, the 40 mg Cu/kg and 160 mg Cu/kg groups, and the highest dose (480 mg Cu/kg) group (Figure 3B). These separations indicate that the metabolite patterns in these groups were different as a result of the different copper exposures. Some of the metabolites were up- and some were down-regulated due to the copper exposure (two examples given in Figures 3c and 3D). The metabolite data were also combined with gene expression data in a systems toxicology application. This combined analysis showed that the copper exposures led to disruption of energy metabolism, particularly with regard to effects on the mitochondria and oxidative phosphorylation. Bundy et al. associated this effect on energy metabolism with a reduced growth rate of the earthworms. This study effectively showed that metabolomics can be used to understand the metabolite pathways that are affected by copper exposure and are closely linked to phenotypic changes (i.e., reduced growth rate). The transcriptome data collected simultaneously were in good accordance with the metabolome patterns, supporting Bundy et al.'s hypothesis that simultaneous measurement of the transcriptomes and metabolome can be used to validate the findings of both approaches, and in turn the value of "systems toxicology".

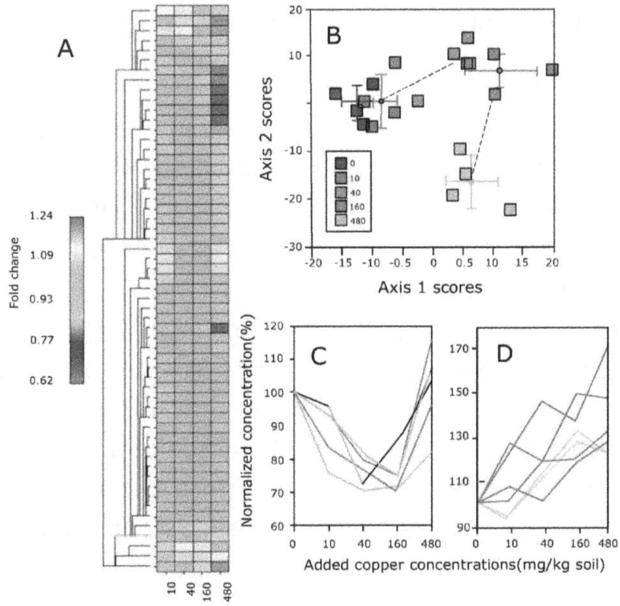

Figure 3: Example of metabolite analysis with NMR to understand the mechanism of toxicity of copper to earthworms (Bundy et al., 2008). A: Heatmap, showing the fold changes of lipid metabolites at different exposure concentrations of copper (10, 40, 160, 480 mg/kg copper in soil) and controls. B: Principal component analysis (PCA) of the polar metabolite patterns of the exposure groups. The higher dose group (480 mg/kg soil) is separate from the medium dose (40 mg Cu/kg and 160 mg Cu/kg) groups and the control and the lowest dose groups (0 mg Cu/kg and 10 mg Cu/kg soil) indicating that the metabolite patterns in these groups are different and are affected by the copper exposure. C: Down and upregulation of lipophilic amino acids (blue: aliphatics, red: aromatics). D: Upregulation of cell-membrane-related metabolites (black = betaine, glycine, HEFS, phosphoethanolamine, red: myo-inositol, scyllo-inositol). Redrawn from Bundy et al. (2008) by Wilma IJzerman.

Challenges in Metabolomics

Several challenges currently exist in the field of metabolomics. From a biological perspective, metabolism is a dynamic process and therefore very time-sensitive. Taking samples at different time-points during development of an organism, or throughout a chemical exposure can result in quite different metabolite patterns. Sample handling and storage can also be challenging as some metabolites are very unstable during sample collection and sample treatment. From an analytical perspective, metabolites possess a wide range of physico-chemical properties and occur in highly varying concentrations such that capturing the widest portion of the metabolome requires analysis with more than one analytical technique. However, the largest challenge is arguably the identification of the chemical structure of unknown metabolites. Even with state-of-the-art analytical techniques only a fraction of the unknown metabolites can be confidently identified.

Conclusions

Metabolomics is a relative new field in toxicology, but is rapidly increasing our understanding of the biochemical pathways affected by exposure to environmental contaminants, and in turn their mechanisms of action. Linking the changed molecular pathways due to the contaminant exposure to phenotypical changes of the organisms is an area of great interest.

Continual advances in state-of-the-art analytical tools for metabolite detection and identification will continue to this trend and expand the utility of environmental metabolomics for prioritizing contaminants. However, a number of challenges remain for the widespread use of metabolomics in regulatory toxicology. Fortunately, recent growth in international interest to address these challenges is underway, and is making great strides in a variety of applications.

References

Bundy,J.G., Sidhu,J.K., Rana,F., Spurgeon,D.J., Svendsen,C., Wren,J.F., Sturzenbaum, S.R., Morgan, A.J., Kille, P. (2008). 'Systems toxicology' approach identifies coordinated metabolic responses to copper in a terrestrial non-model invertebrate, the earthworm Lumbricus rubellus. BMC Biology, 6(25), 1-21.

Bundy, J.G., Matthew P. Davey, M.P., Viant, M.R. (2009). Environmental metabolomics: a critical review and future perspectives. Metabolomics, 5, 3–21.

Johnson, C.H., Ivanisevic, J., Siuzdak, G. (2016). Metabolomics: beyond biomarkers and towards mechanisms. Nature Reviews, Molecular and Cellular Biology 17, 451-459.

4.4. INCREASING ECOLOGICAL REALISM IN TOXICITY TESTING

If you want to re-use this section, for e.g. in your electronic learning environment, feel free to copy this url: https://maken.wikiwijs.nl/162480/4_4__Increasing_ecological_realism_in_toxicity_testing

Introduction

The vast majority of single-species toxicity tests reported in the literature concerns acute or short-term exposures to individual chemicals, in which mortality is often the only endpoint. This is in sharp contrast with the actual situation at contaminated sites, where organisms may be exposed to relatively low levels of mixtures of contaminants under suboptimal conditions for their entire life span. Hence there is an urgent need to increase ecological realism in single-species toxicity tests by addressing sublethal endpoints, mixture toxicity, multistress effects, chronic toxicity and multigeneration effects.

Increasing Ecological Realism in Single-species Toxicity Tests

Mortality is a crude parameter representing the response of organisms to relatively high and therefore often environmentally irrelevant toxicant concentrations. At much lower and environmentally more relevant toxicant concentrations, organisms may suffer from a wide variety of sublethal effects. Hence, the first step to gain ecological realism in single-species toxicity tests is to address sublethal endpoints instead of, or in addition to mortality (Figure 1). Yet, given the short exposure time in acute toxicity tests it is difficult to assess other endpoints than mortality. Photosynthesis of plants and behaviour of animals are elegant, sensitive and rapidly responding endpoints that can be incorporated into short-term toxicity tests to enhance their ecological realism (see section on Endpoints).

Since organisms are often exposed to relatively low levels of contaminants for their entire life span, the next step to increase ecological realism in single-species toxicity tests is to increase exposure time by performing chronic experiments (Figure 1) (see section on Chronic toxicity). Moreover, in chronic toxicity tests a wide variety of sublethal endpoints can be assessed in addition to mortality, the most common ones being growth and reproduction (see to section on Endpoints). Given the relatively short duration of the life cycle of many invertebrates and unicellular organisms like bacteria and algae, it would be relevant to prolong the exposure time even further, by exposing the test organisms for their entire life span, so from the egg or juvenile phase till adulthood including their reproductive performance, or for several generations, assessing multigeneration effects (Figure 1) (see section on Multigeneration effects).

Figure 1. Consecutive steps of increasing ecological realism in single-species toxicity tests.

In contaminated environments, organisms are generally exposed to a wide variety of toxicants under variable and sub-optimal conditions. To further gain ecological realism, mixture toxicity and multistress scenarios should thus be considered (figure 1) (see sections on Mixture toxicity and Multistress). The highest ecological relevance of laboratory toxicity tests may be achieved by addressing the above mentioned issues all together in one type of experiment, chronic mixture toxicity tests assessing sublethal endpoints. Yet, even nowadays such studies remain scarce.

Another way of increasing ecological realism of toxicity testing is by moving towards multispecies test systems that allow for assessing the impacts of chemicals and other stressors on species interactions within communities (see chapter 5 on Population, community and ecosystem ecotoxicology).

4.4.1. Mixture Toxicity

Introduction

In contaminated environments, organisms are generally exposed to complex mixtures of toxicants. Hence, there is an urgent need for assessing their joint toxic effects. In theory, there are four classes of joint effects of compounds in a mixture as depicted in Figure 1.

Four classes of joint effects	No interaction (additive)	Interaction (non-additive)
Similar action	Simple similar action/ Concentration Addition	Complex similar action
Dissimilar action	Independent action/ Response Addition	Dependent action

Figure 1. The four classes of joint effects of compounds in a mixture, as proposed by Hewlett and Plackett (1959).

Simple Similar Action & Concentration Addition

The most simple case concerns compounds that share the same mode of action and do not interact (Figure 1 upper left panel: simple similar action). This holds for compounds acting on the same biological pathway, affecting strictly the same molecular target. Hence, the only difference is the relative potency of the compounds. In this case Concentration Addition is taken as the starting point, following the Toxic Unit (TU) approach. This approach expresses the toxic potency of a chemical as TU, which is calculated for each compound in the mixture as:

$$Toxic\,Unit = \frac{C}{EC_x}$$

with c = the concentration of the compound in the mixture, and EC_x = the concentration of the compound where the measured endpoint is affected by X % compared to the non-exposed control. Next, the toxic potency of the mixture is calculated as the sum of the TUs of the individual compounds:

$$TU\,Mixture = sumTU(compounds) = \sum \frac{ci}{EC_{x,1}}$$

Imagine that the EC_{50} of compound A is 300 µg.L^{-1} and that of compound B 60 µg.L^{-1}. In a mixture of A+B 30 µg.L^{-1} A and 30 µg.L^{-1} B are added. These concentrations represent 30/300 = 0.1 TU of A and 30/60 = 0.5 TU of B. Hence, the mixture consists of 0.1 + 0.5 = 0.6 TU. Yet, the two compounds in this mixture are not represented at equal toxic strength, since this specific mixture is dominated by compound B. To compose mixtures in which the compounds are represented at equal toxic strength, the equitoxicity concept is applied:

- 1 Equitoxic TU A+B = 0.5 TU A + 0.5 TU B
- 1 Equitoxic TU A+B = 150 µg.L^{-1} A + 30 µg.L^{-1} B

As in traditional concentration-response relationships, survival or a sublethal endpoint is plotted against the mixture concentration from which the EC_{50} value and the corresponding 95% confidence limits can be derived (see section on Concentration-response relationships). If the upper and lower 95% confidence limits of the EC_{50} value of the mixture include 1 TU, the EC_{50} of the mixture does not differ from 1 TU and the toxicity of the compounds in the mixture is indeed concentration additive (Figure 2).

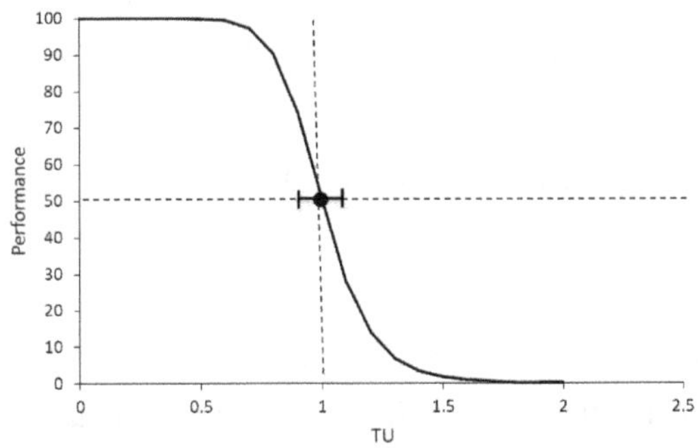

Figure 2. Concentration-response relationship for a mixture in which the toxicants have a concentration additive effect. The Y-axis shows the performance of the test organisms, e.g. their survival, reproduction or other endpoint measured. The horizontal dotted line represents the 50% effect level, the vertical dotted line represents 1 Toxic Unit (TU). The black dot represents the experimental EC50 value of the mixture with the 95% confidence limits.

An experiment appealing to the imagination was performed by Deneer et al. (1988), who tested a mixture of 50 narcotic compounds (see section on Toxicodynamics and Molecular interactions) and observed perfect concentration addition, even when the individual compounds were present at only 0.25% (0.0025 TU) of their EC50. This showed in particular that narcotic compounds present at concentrations way below their no effect level still contribute to the joint toxicity of a mixture (Deneer et al., 1988).

This was also shown for metals (Kraak et al., 1999). This is alarming, since even nowadays environmental legislation is still based on a compound-by-compound approach. The study by Deneer et al. (1988) also clearly demonstrated the logistical challenges of mixture toxicity testing. Since for composing equitoxic mixtures the EC_{50} values of the individual compounds need to be known, testing an equitoxic mixture of 50 compounds requires 51 toxicity tests: 50 individual compounds and 1 mixture.

Independent Action & Response Addition

When chemicals have a different mode of action, act on different targets, but still contribute to the same biological endpoint, the mixture is expected to behave according to Response Addition (also termed Independent Action; Figure 1, lower left panel). Such a situation would occur, for example, if one compound inhibits photosynthesis, and a second one inhibits DNA-replication, but both inhibit the growth of an exposed algal population.

To calculate the effect of a mixture of compounds with different modes of action, Response Addition is applied as follows: The probability that a compound, at the concentration at which it is present in the mixture, exerts a toxic effect (scaled from 0 to 1), differs per compound and the cumulative effect of the mixture is the result of combining these probabilities, according to:

$$E_{(mix)} = E_{(A)} + E_{(B)} - E_{(A)}E_{(B)}$$

Where $E_{(mix)}$ is the fraction affected by the mixture, and $E_{(A)}$ and $E_{(B)}$ are the fractions affected by the individual compounds A and B at the concentrations at which they occur in the mixture.

In fact, this equation sums the fraction affected by compound A and the fraction affected by compound B at the concentration at which they are present in the mixture, and then corrects for the fact that the fraction already affected by chemical A cannot be affected again by chemical B (or vice versa). The latter part of the equation is needed to account for the fact that the chemicals act independent from each other. This is visualised in Figure 3.

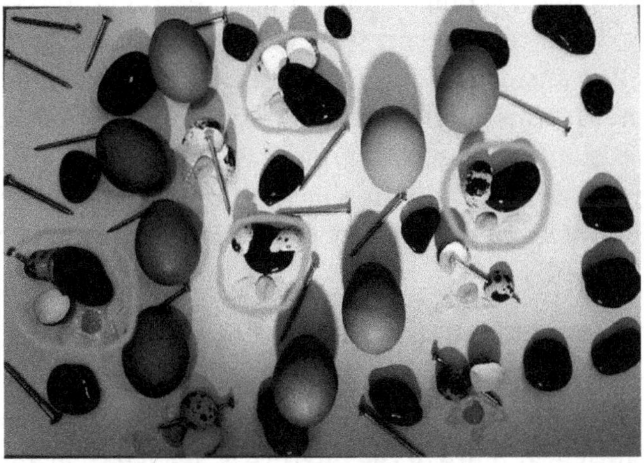

Figure 3. Illustration of stressors acting independent of each other, using the example given by Berenbaum (1981). Subsequently a handful of nails and a handful of pebbles are thrown to a collection of eggs. The nails break 5 eggs, and these 5 eggs broken by the nails cannot be broken again. The pebbles could break 4 eggs, but 1 egg was already broken by the nails. Hence, the pebbles break 3 additional eggs.

The equation: $E(mix) = E_{(A)} + E_{(B)} - E_{(A)}E_{(B)}$

can be rewritten as: $1\text{-}E_{(mix)} = (1\text{-}E_A)*(1\text{-}E_B)$

This means that the probability of not being affected by the mixture (1-E(mix)) is the product of the probabilities of not being affected by (the specific concentrations of) compound A and compound B. At the EC_{50}, both the affected and the unaffected fraction are 50%, hence $(1\text{-}E_A)*(1\text{-}E_B) = 0.5$. If both compounds equally contribute to the effect of the mixture, $(1\text{-}E_A) = (1\text{-}E_B)$ and thus $(1\text{-}E_{A \text{ or } B})^2 = 0.5$, so both $(1\text{-}E_A)$ and $(1\text{-}E_B)$ equal 0√.50.5 = 0.71. Since the probability of not being affected is 0.71 for compound A and compound B, the probability of being affected is 0.29. Thus at the EC50 of a mixture of two compounds acting according to Independent Action, both compounds should be present at a concentration equalling their EC29.

Interactions between the Compounds in a Mixture

Concentration Addition as well as Response Addition both assume that the compounds in a mixture do not interact (see Figure 1). However, in reality, such interactions can occur in all four steps of the toxic action of a mixture. The *first step* concerns chemical and physicochemical interactions. Compounds in the environment may interact, affecting each other's bioavailability. For instance, excess of Zn causes Cd to be more available in the soil solution as a result of competition for the same binding sites. The *second step* involves physiological interactions during uptake by an organism, influencing the toxicokinetics of the compounds, for example by competition for uptake sites at the cell membrane. The *third* step refers to the internal processing of the compounds, e.g. involving effects on each other's biotransformation or detoxification (toxicokinetics). The fourth step concerns interactions at the target site(s), i.e. the toxicodynamics during the actual intoxication process. The typical whole organism responses that are recorded in many ecotoxicity tests integrate the last three types of interactions, resulting in deviations from the toxicity predictions from Concentration Addition and Response Addition.

Deviations from Concentration Addition

If the EC_{50} of the mixture is higher than 1 TU and the lower 95% confidence limit is also above 1 TU, the toxicity of the compounds in the mixture is less than concentration additive, as more of the mixture is needed than anticipated to cause 50% effect (Figure 4, blue line; antagonism). Correspondingly, if the EC_{50} of the mixture is lower than 1 TU and the upper 95% confidence limit is also below 1 TU, the toxicity of the compounds in the mixture is more than concentration additive (Figure 4, red line; synergism).

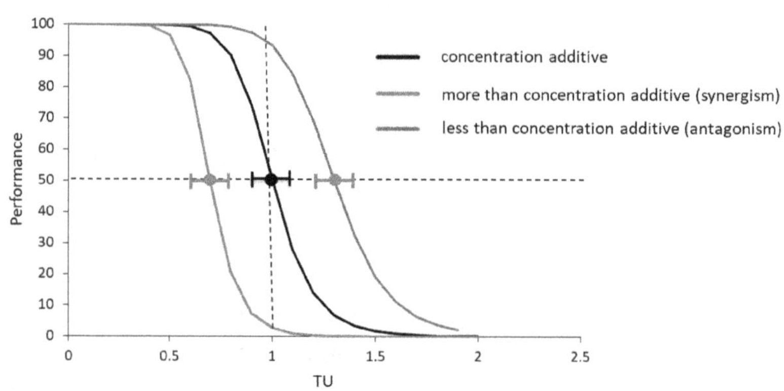

Figure 4. Concentration-response relationships for mixtures in which the toxicants have a less than concentration additive effect (blue line), a concentration additive effect (black line) and a more than concentration additive effect (red line). The Y-axis shows the performance of the test organisms, e.g. their survival, reproduction or other endpoint measured. The horizontal dotted line represents the 50% effect level, the vertical dotted line represents 1 Toxic Unit (TU). The coloured dots represent the EC50 values with the corresponding 95% confidence limits.

When the toxicity of a mixture is more than concentration additive, the compounds enhance each other's toxicity. When the toxicity of a mixture is less than concentration additive, the compounds reduce each other's toxicity. Both types of deviation from additivity can have two different reasons: 1. The compounds have the same mode of action, but do interact (Figure 1, upper right panel: complex similar action). 2. The compounds have different modes of actions (Independent action/Response Addition; Figure 1, lower left panel).

Concentration-response Surfaces and Isoboles

Elaborating on Figure 4, concentration-response relationships for mixtures can also be presented as multi-dimensional figures, with different axes for the concentration of each of the chemicals included in the mixture (Figure 5A). In case of a mixture of two chemicals, such a dose-response surface can be shown in a two-dimensional plane using isoboles. Figure 5B shows isoboles for a mixture of two chemicals, under different assumptions for interactions according to Concentration Addition. If the interaction between the two compounds decreases the toxicity of the mixture, this is referred to as antagonism (Figure 5B, blue line). If the interaction between the two compounds increases the toxicity of the mixture, this is referred to as synergism (Figure 5B, red line). Thus both antagonism and synergism are terms to describe deviations from Concentration Addition due to interaction between the compounds. Yet, antagonism in relation to Concentration Addition (less than concentration additive; Figure 5B blue line) can simply be caused by the compounds behaving according to Response Addition, and not behaving antagonistically.

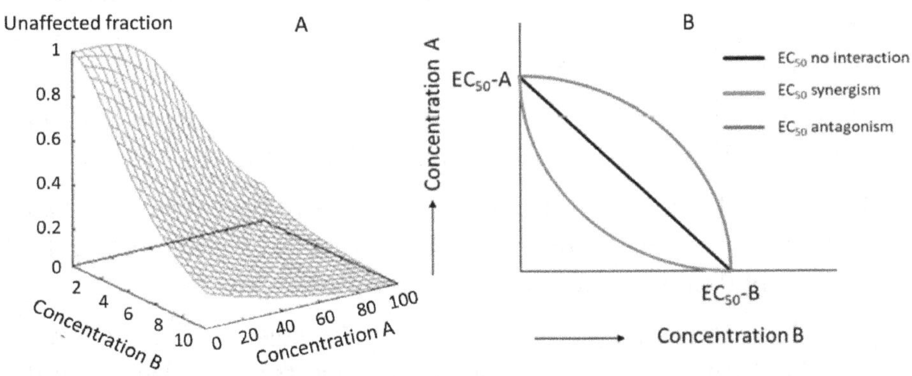

Figure 5. A. Dose-response surface showing the effect of chemicals A and B, single (sides of the surface), and in mixtures. B. Isoboles showing the toxicity of the same mixtures in a two-dimensional plane. The isoboles can be seen as a cross section through the dose-response surface. The isoboles show the 50% effect level according to Concentration Addition of mixtures of the two compounds in case they do not interact (black line), when they interact antagonistically (blue line) and when they interact synergistically (red line).

Synergism and Antagonism Evaluated by both Concepts

The use of the terms synergism and antagonism may be problematic, because antagonism in relation to Concentration Addition (less than concentration additive; Figure 5B blue line) can simply be caused by the compounds behaving according to Response Addition, and not behaving antagonistically. Similarly, deviations from Response Addition could also mean that chemicals in the mixture do have the same mode of action, so act additively according to Concentration Addition. One can therefore only conclude on synergism/antagonism if the experimental observations are higher/lower than the predictions by *both* concepts.

Suggested Further Reading

Rider, C.V., Simmons, J.E. (2018). Chemical Mixtures and Combined Chemical and Nonchemical Stressors: Exposure, Toxicity, Analysis, and Risk, Springer International Publishing AG. ISBN-13: 978-3319562322.

Bopp, S.K., Kienzler, A., Van der Linden, S., Lamon, L., Paini, A., Parissis, N., Richarz, A.N., Triebe, J., Worth, A. (2016). Review of case studies on the human and environmental risk assessment of chemical mixtures. JRC Technical Reports EUR 27968 EN, European Union, doi:10.2788/272583.

References

Berenbaum, M.C. (1981). Criteria for analysing interactions between biologically active agents. Advances in Cancer Research 35, 269-335.

Deneer, J.W., Sinnige, T.L., Seinen, W., Hermens, J.L.M. (1988). The joint acute toxicity to Daphnia magna of industrial organic chemicals at low concentrations. Aquatic Toxicology 12, 33–38.

Hewlett, P.S., Plackett, R.L. (1959). A unified theory for quantal responses to mixtures of drugs: non-interactive action. Biometrics 15, 691 610.

Kraak, M.H.S., Stuijfzand, S.C., Admiraal, W. (1999). Short-term ecotoxicity of a mixture of five metals to the zebra mussel Dreissena polymorpha. Bulletin of Environmental Contamination and Toxicology 63, 805-812.

Van Gestel, C.A.M., Jonker, M.J., Kammenga, J.E., Laskowski, R., Svendsen, C. (Eds.) (2011). Mixture toxicity. Linking approaches from ecological and human toxicology. SETAC Press, Society of Environmental Toxicology and Chemistry, Pensacola.

4.4.2. Multistress Introduction

Introduction

In contaminated environments, organisms are generally exposed to a wide variety of toxicants under variable and sub-optimal conditions. To gain ecological realism, multistress scenarios should thus be considered, but these are, however, understudied.

Definitions

Stress is defined as an environmental change that affects the fitness and ecological functioning of species (i.e., growth, reproduction, behaviour), ultimately leading to changes in community structure and ecosystem functioning. Multistress is subsequently defined as a situation in which an organism is exposed both to a toxicant and to stressful environmental conditions. This includes chemical-abiotic interactions, chemical-biotic interactions as well as combinations of these. Common abiotic stressors are for instance pH, drought, salinity and above all temperature, while common biotic stressors include predation, competition, population density and food shortage. Experiments on such stressors typically study, for instance, the effect of increasing temperature or the influence of food availability on the toxicity of compounds. The present definition of multistress thus excludes mixture toxicity (see section on Mixture toxicity) as well as situations in which organisms are confronted with several suboptimal (a)biotic environmental variables jointly without being exposed to toxicants. The next chapters deal with chemical-abiotic and chemical-biotic interactions and with practical issues related with the performance of multistress experiments, respectively.

4.4.3. Multistress - biotic

Introduction

Generally, organisms have to cope with the joint presence of chemical and natural stressors. Both biotic and abiotic stressors can affect the chemicals' bioavailability and toxicokinetics. Additionally, they can influence the behavior and physiology of organisms, which could result in higher or lower toxic effects. Vice versa, chemicals can alter the way organisms react to natural stressors.

By studying the effects of multiple stressors, we can identify potential synergistic, additive or antagonistic interactions, which are essential to adequately assess the risk of chemicals in nature. Relyea (2003), for instance, found that apparently safe concentrations of carbaryl can become deadly to some amphibian species when combined with predator cues. This section focuses on biotic stress, which can be defined as stress caused by living organisms and includes predation, competition, population density, food availability, pathogens and parasitism. It will describe how biotic stressors and chemicals act and interact.

Types of Biotic Stressors

Biotic stressors can have direct and indirect effects on organisms. For example, predators can change food web structures by consuming their prey and thus altering prey abundance and can indirectly affect prey growth and development as well, by inducing energetically-costly defense mechanisms. Also behaviors like (foraging) activity can be decreased and even morphological changes can be induced. For example, Daphnia pulex can develop neck spines when they are subject to predation. Similarly, parasites can alter host behavior or induce morphological changes, e.g., in coloration, but they usually do not kill their host. Yet, parasitism can compromise the immune system and alter the energy budget of the host.

High population density is a stressor that can affect energy budgets and intraspecific and interspecific competition for space, status or resources. By altering resource availability, changes in growth and size at maturity can be the result. Additionally, these competition-related stressors can affect behavior, for example by limiting the number of suitable mating partners. Also pathogens (e.g., viruses, bacteria and fungi) can lower fitness and fecundity.

It should be realized that the effects of different biotic stressors cannot be strictly separated from each other. For example, pathogens can spread more rapidly when population densities are high, while predation, on the other hand, can limit competition.

Effects of Biotic Stressors on Bioavailability and Toxicokinetics

Biotic stressors can alter the bioavailability of chemicals. For example in the aquatic environment, food level may determine the availability of chemicals to filter feeders, as they may adsorb to particulate organic matter, such as algae. As the exposure route (waterborne or via food) can influence the subsequent toxicokinetic processes, this may also change the chemicals' toxic effects.

Effects of Biotic Stressors on Behavior

Biotic stressors have been reported to cause behavioral effects in organisms that could change the toxic effects of chemicals. These effects include altered feeding rates and reduced activities. The presence of a predator, for example, reduces prey (foraging) activity to avoid being detected by the perceived predator and so decreases chemical uptake via food. On the other hand, the condition of the prey organisms will decrease due to the lower food consumption, which means less energy is available for other physiological processes (see below).

In addition to biotic stressors, also chemicals can disrupt essential behaviors by reduction of olfactory receptor sensitivity, cholinesterase inhibition, alterations in brain neurotransmitter levels, and impaired gonadal or thyroid hormone levels. This could lead to disruptive effects on communication, feeding rates and reproduction. An inability to find mating partners, for example, could then be worsened by a low population density. Furthermore, chemicals can alter predator-prey relationships, which might result in trophic cascades. Strong top-down effects will be observed when a predator or grazer is more sensitive to the contaminant than its prey. Alternatively, bottom-up effects are observed when the susceptibility of a prey species to predation is increased. For example, Cu exposure of fish and crustaceans can decrease their response to olfactory cues, making them unresponsive to predator stress and increasing the risk to be detected and consumed (Van Ginneken et al., 2018). Effects on the competition between species may also occur, when one species is more sensitive than the other. Thus, both chemical and biotic stressors can alter behavior and result in interactive effects that could change the entire ecosystem structure and function (Fleeger et al., 2003).

Physiology

Biotic stressors can cause elevated respiration rates of organisms, in aquatic organisms leading to a higher toxicant uptake through diffusion. On the other hand, they can also decrease respiration. For example, low food levels decrease metabolic activity and thus respiration. Additionally, a reduced metabolic rate could decrease the toxicity of chemicals which are metabolically activated. Also certain chemicals, such as metals, can cause a higher or lower oxygen consumption, which might counteract or reinforce the effects of biotic stressors.

Besides affecting respiration, both biotic and chemical stressors can induce physiological damage to organisms. For instance, predator stress and pesticides cause oxidative stress, leading to synergistic effects on the induction of antioxidant enzymes such as catalase and superoxide dismutase (Janssens and Stoks, 2013). Furthermore, the organism can eliminate or detoxify internal toxicant concentrations, e.g. by transformation via Mixed Function Oxidation enzymes (MFO) or by sequestration, i.e. binding to metallothioneins or storage in inert tissues such as granules. These defensive mechanisms for detoxification and damage control are energetically costly, leading to energy trade-offs. This means less energy can be used for other processes such as growth, locomotion or reproduction. Food availability and lipid reserves can then play an important role, as well-fed organisms that are exposed to toxicants can more easily pay the energy costs than food-deprived organisms.

Interactive Effects

The possible interactions, i.e. antagonism, synergism or additivity, between effects of stressors are difficult to predict and can differ depending on the stressor combination, chemical concentration, the endpoint and the species. For Ceriodaphnia dubia, Qin et al. (2011) demonstrated that predator stress influenced the toxic effects of several pesticides differently. While predator cues interacted antagonistically with bifenthrin and thiacloprid, they acted synergistically with fipronil.

It should also be noted that interactive effects in nature might be weaker than when observed in the laboratory as stress levels fluctuate more rapidly or animals can move away from areas with high predator risk or chemical exposure levels. On the other hand, because generally more than two stressors are present in ecosystems, which could interact in an additive or synergistic way as well, they might be even more important in nature. Understanding interactions among multiple stressors is thus essential to estimate the actual impact of chemicals in nature.

References

Fleeger, J.W., Carman, K.R., Nisbet, R.M. (2003). Indirect effects of contaminants in aquatic ecosystems. Science of the Total Environment 317, 207-233.

Janssens, L., Stoks, R. (2013). Synergistic effects between pesticide stress and predator cues: conflicting results from life history and physiology in the damselfly Enallagma cyathigerum. Aquatic Toxicology 132, 92-99.

Qin, G., Presley, S.M., Anderson, T.A., Gao, W., Maul, J.D. (2011). Effects of predator cues on pesticide toxicity: toward an understanding of the mechanism of the interaction. Environmental Toxicology and Chemistry 30, 1926-1934.

Relyea, R.A. (2003). Predator cues and pesticides: a double dose of danger for amphibians. Ecological Applications 13, 1515-1521.

Van Ginneken, M., Blust, R., Bervoets, L. (2018). Combined effects of metal mixtures and predator stress on the freshwater isopod Asellus aquaticus. Aquatic Toxicology 200, 148-157.

4.4.4. Multistress - abiotic

Introduction: Stress Related to the Ecological Niche Concept

The concept of stress can be defined at various levels of biological organization, from biochemistry to species fitness, ultimately leading to changes in community structure and ecosystem functioning. Yet, stress is most often studied in the context of individual organisms. The concept of stress is not absolute and can only be defined with reference to the normal range of ecological functioning. This is the case when organisms are within their range of tolerance (so-called ecological amplitude) or within their ecological niche, which describes the match of a species to specific environmental conditions. Applying this concept to stress allows it to be defined as a condition evoked in an organism by one or more environmental factors that bring the organism near or over the edges of its ecological niche (Van Straalen, 2003), see Figure 1.

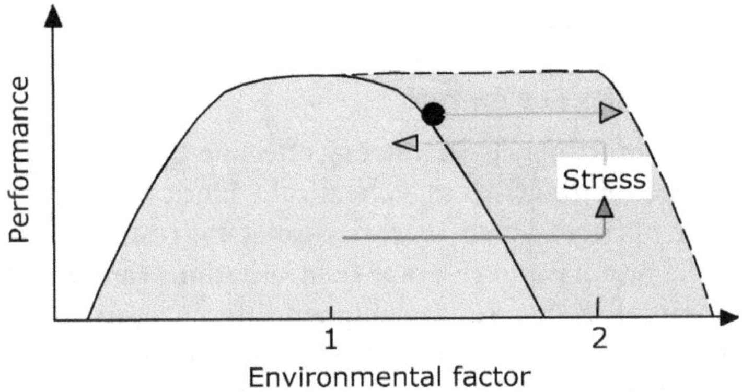

Figure 1. Schematic illustration of the ecological amplitude or niche-based (bell shaped curve) definition of stress. Stress arises when an environmental factor increases from point 1 to point 2 (red line) and the species is forced outside its ecological niche. By definition, the organism cannot grow and reproduce outside this niche, but it may survive there temporarily, if it can return in time to its niche (blue line). If the borders of the niche are extended through adaptation, then this specific state of the environmental factor does not result in stress anymore and the performance of the species falls within the normal operating range at that condition (green line). Redrawn from Van Straalen (2003) by Wilma IJzerman.

Multistress is subsequently defined as a situation in which an organism is exposed both to a toxicant and to stressful environmental conditions (see section Multistress – Introduction and definitions). This includes chemical-abiotic interactions, chemical-biotic interactions (see section Multistress – chemical – biotic interactions) as well as combinations of these. In general, organisms living under conditions close to their environmental tolerance limits appear to be more vulnerable to additional chemical stress. The opposite also holds: if organisms are stressed due to exposure to elevated levels of contaminants, their ability to cope with sub-optimal environmental conditions is reduced.

Chemical-abiotic Interactions

Temperature: One of the predominant environmental factors altering toxic effects obviously is temperature. For poikilothermic (cold-blooded) organisms, increases in temperature lead to an increase in activity, which may affect both the uptake and the effects of chemicals. In a review by Heugens et al. (2001), studies reporting the effect chemicals on aquatic organisms in combination with abiotic factors like temperature, nutritional state and salinity were discussed. Generally, toxic effects increased with increasing temperature. Dependent on the effect parameter studied, the differences in toxic effects between laboratory and relevant field temperatures ranged from a factor of 2 to 130.

Also freezing temperatures may interfere with chemical effects as was shown in another influential review of Holmstrup et al. (2010). Membrane damage is mentioned as an explanation for the synergistic interaction between combinations of metals and temperatures below zero.

Food: Food availability may have a strong effect on the sensitivity of organisms to chemicals (see section Multistress – chemical – biotic interactions). In general decreasing food or nutrient levels increased toxicity, resulting in differences in toxicity between laboratory and relevant field situations ranging from a factor of 1.2 to 10 (Heugens et al., 2001). Yet, way higher differences in toxic effects related to food levels have been reported as well: Experiments performed with daphnids in cages that were placed in outdoor mesocosm ditches (see sections on Cosm studies and In situ bioassays) showed stunning differences in sensitivity to the insecticide thiacloprid. Under conditions of low to ambient nutrient concentrations, the observed toxicity, expressed as the lowest observed effect concentration (LOEC) for growth and reproduction occurred at thiacloprid concentrations that were 2500-fold lower than laboratory-derived LOEC values. Contrary to the low nutrient treatment, such altered toxicity was often not observed under nutrient-enriched conditions (Barmentlo et al submitted). The difference was likely attributable to the increased primary production that allowed for compensatory feeding and perhaps also reduced the bioavailability of the insecticide. Similar results were observed for sub-lethal endpoints measured on the damselfly species Ischnura elegans, for which the response to thiacloprid exposure strongly depended on food availability and quality. Damselfies that were feeding on natural resources were significantly more affected than those that were offered high quality artificial food (Barmentlo et al submitted).

Salinity: The influence of salinity on toxicity is less clear (Heugens et al. 2001). If salinity pushes the organism towards its niche boundaries, it will worsen the toxic effects that it is experiencing. In case that a specific salinity fits in the ecological niche of the organism, processes affecting exposure will predominantly determine the stress it will experience. This for instance means that metal toxicity decreases with increasing salinity, as it is strongly affected by the competition of ions (see section on Metal speciation). The toxic effect induced by organophosphate insecticides however, increases with increasing salinity. For other chemicals, no clear relationship between toxicity and salinity was observed. A salinity increase from freshwater to marine water decreased toxicity by a factor of 2.1 (Heugens et al. 2001). However, as less extreme salinity changes are more relevant under field conditions, the change in toxicity is probably much smaller.

pH: Many organisms have a species-specific range of pH levels at which they function optimally. At pH values outside the optimal range, organisms may show reduced reproduction and growth, in extreme cases even reduced survival. In some cases, the effects of pH may be indirect, as pH may also have an important impact on exposure of organisms to toxicants. This is especially the case for metals and ionizable chemicals: metal speciation, but also the form in which ionizable chemicals occur in the environment and therefore their bioavailability, is highly dependent on pH (see sections on Metal speciation and Ionogenic organic chemicals). An example of the interaction between pH and metal effects was shown by Crommentuijn et al. (1997), who observed a reduced control reproduction of the springtail Folsomia candida, but also the lowest cadmium toxicity at a soil pHKCl 7.0 compared to pHKCl 3.1-5.7.

Drought: In soil, the moisture content (see section on Soil) is an important factor, since drought is often limiting the suitability of the soil as a habitat for organisms. Holmstrup et al. (2010), reviewing the literature, concluded that chemicals interfering with the drought tolerance of soil organisms, e.g. by affecting the functioning of membranes or the accumulation of sugars, may exacerbate the effects of drought. Earthworms are breathing through the skin and can only survive in moist soils, and the eggs of springtails can only survive at a relative air humidity close to 100%. This makes these organisms especially sensitive to drought, which may be enhanced by exposure to chemicals like metals, polycyclic aromatic hydrocarbon or surfactants (Holmstrup et al., 2010).

Many different abiotic conditions, such as oxygen levels, light, turbidity, and organic matter content, can push organisms towards the boundaries of their niche, but we will not discuss all stressors in this book.

Multistress in Environmental Risk Assessment

In environmental risk assessment, differences between stress-induced effects as determined in the laboratory under standardized optimal conditions with a single toxicant and the effects induced by multiple stressors are taken into account by applying an uncertainty factor. Yet, the choice for uncertainty factors is based on little ecological evidence. In 2001, Heugens already argued for obtaining uncertainty factors that sufficiently protect natural systems without being overprotective. Van Straalen (2003) echoed this and in current research the question is still raised if enough understanding has been gained to make accurate laboratory-to-field extrapolations.

It remains a challenge to predict toxicant-induced effects on species and even on communities while accounting for variable and suboptimal environmental conditions, even though these conditions are common aspects of natural ecosystems (see for instance the section on Eco-epidemiology).

References

Barmentlo, S.H., Vriend, L.M, van Grunsven, R.H.A., Vijver, M.G. (submitted). Evidence that neonicotinoids contribute to damselfly decline.

Crommentuijn, T., Doornekamp, A., Van Gestel, C.A.M. (1997). Bioavailability and ecological effects of cadmium on *Folsomia candida* (Willem) in an artificial soil substrate as influenced by pH and organic matter. Applied Soil Ecology 5, 261-271.

Heugens, E.H., Hendriks, A.J., Dekker, T., Van Straalen, N.M., Admiraal, W. (2001). A review of the effects of multiple stressors on aquatic organisms and analysis of uncertainty factors for use in risk assessment. Critical Reviews in Toxicology 31, 247-84.

Holmstrup, M., Bindesbøl, A.M., Oostingh, G.J., Duschl, A., Scheil, V., Köhler, H.R., Loureiro, S., Soares, A.M.V.M., Ferreira, A.L.G., Kienle, C., Gerhardt, A., Laskowski, R., Kramarz, P.E., Bayley, M., Svendsen, C., Spurgeon, D.J. (2010). Review Interactions between effects of environmental chemicals and natural stressors: A review. Science of the Total Environment 408, 3746–3762.

Van Straalen, N.M. (2003). Ecotoxicology becomes stress ecology Environmental Science and Technology 37, 324A-330A.

4.4.5. Chronic toxicity - Eco

Introduction

Most toxicity tests performed are short-term high-dose experiments, acute tests in which mortality is often the only endpoint. This is in sharp contrast with the field situation, where organisms are often exposed to relatively low levels of contaminants for their entire life span. The shorter the life cycle of the organism, the more realistic this scenario becomes. Hence, there is an urgent need for chronic toxicity testing. It should be realized though, that the terms acute and chronic have to be considered in relation to the length of the life cycle of the organism. A short-term exposure of four days is acute for fish, but chronic for algae, comprising already four generations.

From Acute to Chronic Toxicity Testing

The reason for the bias towards acute toxicity testing is obviously the higher costs involved in chronic toxicity testing, simply caused by the much longer duration of the test. Yet, chronic toxicity testing is challenging for several other reasons as well. First of all, during prolonged exposure organisms have to be fed. Although unavoidable, especially in aquatic toxicity testing, this will definitely influence the partitioning and the bioavailability of the test compound. Especially lipophilic compounds will strongly bind to the food, making toxicant uptake via the food more important than for hydrophilic compounds, thus causing compound specific changes in exposure routes. For chronic aquatic toxicity tests, especially for sediment testing, it may be challenging to maintain sufficiently high oxygen concentrations throughout the entire experiment (Figure 1).

Figure 1. Experimental design of a chronic sediment toxicity experiment, showing the experimental units and the aeration system.

Obvious choices to be made include the duration of the exposure and the endpoints of the test. Generally it is aimed at including at least one reproductive event or the completion of an entire life cycle of the organism within the test duration. To ensure this, validity criteria are set to the different test guidelines, such as:

- the mean number of living offspring produced per control parent daphnid surviving till the end of the test should be above 60 (OECD, 2012).
- 85% of the adult control chironomid midges from the control treatment should emerge between 12 and 23 days after the start of the experiment (OECD, 2010).
- the mean number of juveniles produced by 10 control collembolans should be at least 100 (OECD, 2016a).

Chronic Toxicity

Generally toxicity increases with increasing exposure time, often expressed as the acute-to-chronic ratio (ACR), which is defined as the LC_{50} from an acute test divided by the door NOEC of EC_{10} from the chronic test. Alternatively, as shown in Figure 2, the acute LC_{50} can be divided by the chronic LC_{50}. If compounds exhibit a strong direct lethal effect, the ACR will be low, but for compounds that slowly build up lethal body burdens (see section on Critical body concentrations) it can be very high. Hence, there is a relationship between the mode of action of a compound and the ACR. Yet, if chronic toxicity has to be extrapolated from acute toxicity data and the mode of action of the compound is unknown, an ACR of 10 is generally considered. It should be realized though that this number is chosen quite arbitrarily, potentially leading to under- as well as over estimation of the actual ACR.

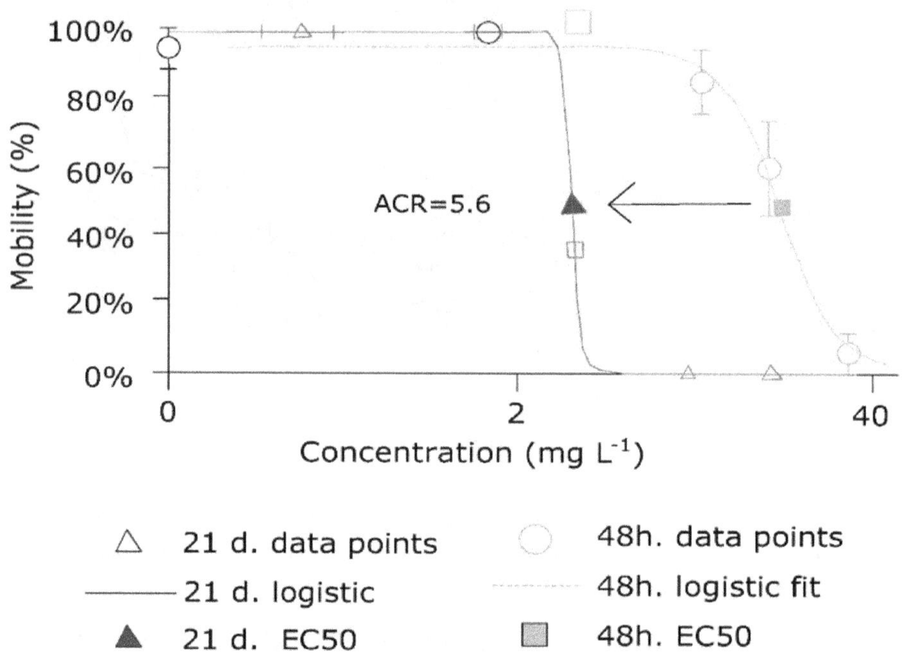

Figure 2. Average mobility (% of initial animals) of Daphnia magna (n = 15) exposed to a concentration range of the flame retardant ALPI (mg L–1) in Elendt medium after 48 h (± s.e. in x and y, n = 4 × 5 individuals per concentration) and after 21 days (± s.e. in x, n = 15 individuals per concentration). The toxicity increases with increasing exposure time with an Acute Chronic Ratio (ACR) of 5.6. Redrawn from Waaijers at al. (2013) by Wilma IJzerman.

Since reproductive events and the completion of life cycles are involved, chronic toxicity tests allow an array of sublethal endpoints to be assessed, including growth and reproduction, as well as species specific endpoints like emergence (time) of chironomids. Consequently, compounds with different modes of action may cause very diverse sublethal effects on the test organisms during chronic exposure (Figure 3). The polycyclic aromatic compound (PAC) phenanthrene did not affect the completion of the life cycle of the midges, but above a certain exposure concentration the larvae died and no emergence was observed at all, suggesting a non-specific mode of action (narcosis). In contrast, the PAC acridone caused no mortality but delayed adult emergence significantly over a wide range of test concentrations, suggesting a specific mode of action affecting life cycle parameters of the midges (Leon Paumen et al., 2008). This clearly demonstrates that specific effects on life cycle parameters of compounds with different modes of action need time to become expressed.

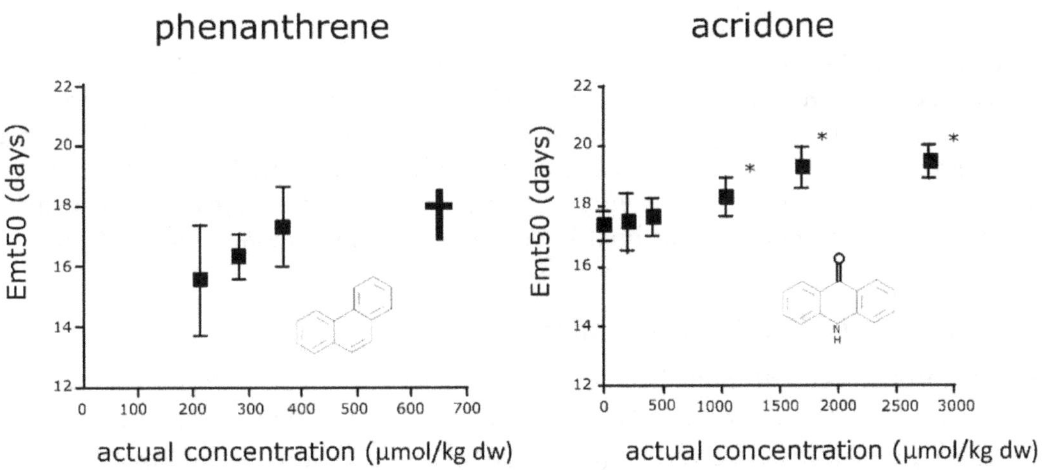

Figure 3. Effect of two polycyclic aromatic compounds (PACs) on the emergence time of Chironomus riparius males from spiked sediments. X-axis: actual concentrations of the compounds measured in the sediment. Y-axis: 50% male emergence time (EMt50, days, average plus standard deviations). †Concentrations with no emerging midges. *EMt50 value significantly different from control value ($p < 0.05$). Redrawn from Leon Paumen et al. (2008) by Wilma IJzerman.

Chronic toxicity tests are single species tests, but if the effects of toxicants are assessed on all relevant life-cycle parameters, these can be integrated into effects on population growth rate (r). For the 21-day daphnid test this is achieved by the integration of age-specific data on the probability of survival and fecundity. The population growth rates calculated from chronic toxicity data are obviously not related to natural population growth rates in the field, but they do allow to construct dose-response relationships for the effects of toxicants on r, the ultimate endpoint in chronic toxicity testing (Figure 4; Waaijers et al., 2013).

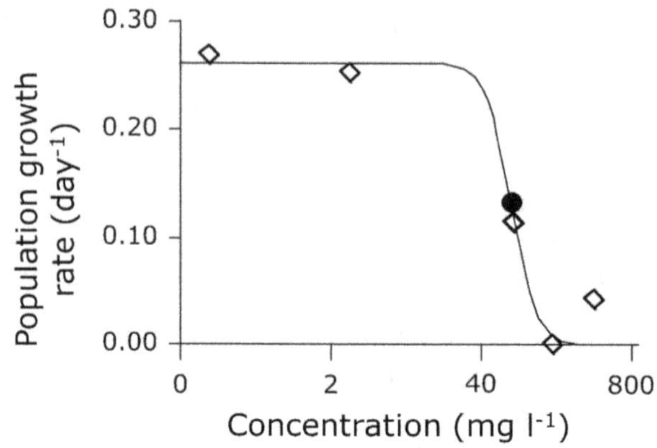

Figure 4: Population growth rate (d-1) of Daphnia magna (n = 15) exposed to a concentration range of the flame retardant DOPO (mg L-1) in Elendt medium after 21 days. The average population growth rate rate (◊) is shown (s.e. in x and y are smaller than the data points and therefore omitted). The EC50 is plotted as • (s.e. smaller than data point) and the logistic curve represents the fitted concentration-response relationship. Redrawn from Waaijers et al. (2013) by Wilma IJzerman.

Chronic Toxicity Testing in Practice

Several protocols for standardized chronic toxicity tests are available, although less numerous than for acute toxicity testing. For water, the most common test is the 21 day Daphnia reproduction test (OECD, 2012), for sediment 28-day test guidelines are available for the midge Chironomus riparius (OECD, 2010) and for the worm Lumbriculus variegatus (OECD, 2007). For terrestrial soil, the springtail Folsomia candida (OECD, 2016a) and the earthworm Eisenia fetida (OECD, 2016b) are the most common test species, but also for enchytraeids a reproduction toxicity test guideline is available (OECD, 2016c). For a complete overview see (https://www.oecd-ilibrary.org/environment/oecd-guidelines-for-the-testing-of-chemicals-section-2-effects-on-biotic-systems_20745761/datedesc#collectionsort).

References

Leon Paumen, M., Borgman, E., Kraak, M.H.S., Van Gestel, C.A.M., Admiraal, W. (2008). Life cycle responses of the midge Chironomus riparius to polycyclic aromatic compound exposure. Environmental Pollution 152, 225-232.

OECD (2007). OECD Guideline for Testing of Chemicals. Test No. 225: Sediment-Water Lumbriculus Toxicity Test Using Spiked Sediment. Section 2: Effects on Biotic Systems; Organization for Economic Co-operation and Development: Paris, 2007.

OECD (2010). OECD Guideline for Testing of Chemicals. Test No. 233: Sediment-Water Chironomid Life-Cycle Toxicity Test Using Spiked Water or Spiked Sediment. Section 2: Effects on Biotic Systems; Organization for Economic Co-operation and Development: Paris, 2010.

OECD (2012). OECD Guideline for Testing of Chemicals. Test No. 211. Daphnia magna Reproduction Test No. 211. Section 2: Effects on Biotic Systems; Organization for Economic Co-operation and Development: Paris, 2012.

OECD (2016a). OECD Guideline for Testing of Chemicals. Test No. 232. Collembolan Reproduction Test in Soil. Section 2: Effects on Biotic Systems; Organization for Economic Co-operation and Development: Paris, 2016.

OECD (2016b). OECD Guideline for Testing of Chemicals. Test No. 222. Earthworm Reproduction Test (Eisenia fetida/Eisenia andrei). Section 2: Effects on Biotic Systems; Organization for Economic Co-operation and Development: Paris, 2016.

OECD (2016c). OECD Guideline for Testing of Chemicals. Test No. 220: Enchytraeid Reproduction Test. Section 2: Effects on Biotic Systems; Organization for Economic Co-operation and Development: Paris, 2016.

Waaijers, S.L., Bleyenberg, T.E., Dits, A., Schoorl, M., Schütt, J., Kools, S.A.E., De Voogt, P., Admiraal, W., Parsons, J.R., Kraak, M.H.S. (2013). Daphnid life cycle responses to new generation flame retardants. Environmental Science and Technology 47, 13798-13803.

4.4.6. Multigeneration Toxicity Testing - Eco

Introduction

It is generally assumed that chronic life cycle toxicity tests are indicative of the actual risk that populations suffer from long-term exposure. Yet, at contaminated sites, organisms may be exposed during multiple generations and the shorter the life cycle of the organism, the more realistic this scenario becomes. There are, however, only few multigeneration studies performed, due to the obvious time and cost constraints. Since both aquatic and terrestrial life cycle toxicity tests generally last for 28 days (see section on Chronic toxicity), multigeneration testing will take approximately one month per generation. Moreover, the test compound often affects the life cycle of the test species in a dose-dependent manner. Consequently, the control population, for example, could already be in the 9th generation, while an exposed population could still be in the 8th generation due to chemical exposure related delay in growth and/or development. On top of these experimental challenges, multigeneration experiments are extremely error prone, simply because the chance that an experiment fails increases with increasing exposure time.

Experimental Considerations

Designing a multigeneration toxicity experiment is challenging. First of all, there is the choice of how many generations the experiment should last, which is most frequently, but completely arbitrarily, set at approximately 10. Test concentrations have to be chosen as well, mostly based on chronic life cycle EC_{50} and EC_{10} values (Leon Paumen et al. 2008). Yet, it cannot be anticipated if, and to what extent, toxicity increases (or decreases) during multigeneration exposure. Hence, testing only one or two exposure concentrations increases the risk that the observed effects are not dose related, but are simply due to stochasticity. If the test concentrations chosen are too high, many treatments may go extinct after few generations. In contrast, too low test concentrations may show no effect at all. The latter was observed by Marinkovic at al. (2012), who had to increase the exposure concentrations during the experiment (see Figure 1). Finally, since a single experimental treatment often consists of an entire population, treatment replication is also challenging.

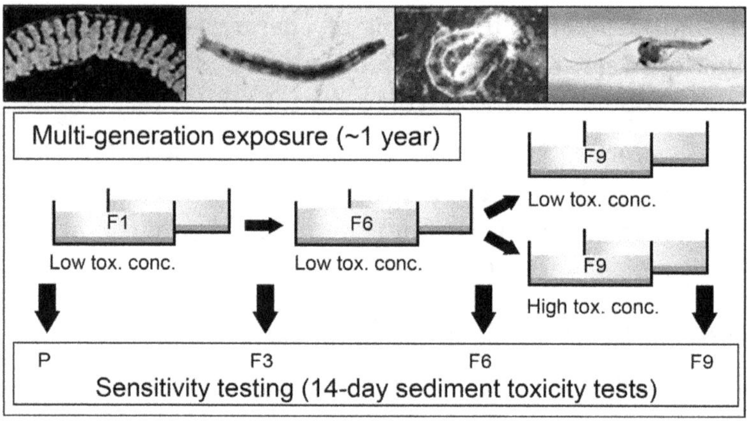

Figure 1. Experimental design of a multigeneration toxicity experiment with the non-biting midge Chironomus riparius. After six generations the exposure concentrations were increased due to lack of effect. To evaluate if multigeneration led to adaptation, after 3, 6 and 9 generations the sensitivity of the test organisms to the test compound was determined. Redrawn from Marinkovic et al. (2012) by Evelin Karsten-Meessen.

Once the experiment is running, choices have to be made on the transition from generation to generation. If a replicate is maintained in a single jar, vessel or aquarium, generations may overlap and exposure concentrations may decrease with time. Therefore, most often a new generation is started by exposing offspring from the previous exposed parental generation in a freshly spiked experimental unit.

If the aim is to determine how a population recovers when the concentration of the toxicant decreases with time, exposure to a single spiked medium also is an option, which seems most applicable to soils (Ernst et al., 2016; van Gestel et al., 2017). To assess recovery after several generations of (continuous) exposure to contaminated media, offspring from previous exposed generations may be maintained under control conditions.

A wide variety of endpoints can be selected in multigeneration experiments. In case of aquatic insects like the non-biting midge Chironomus riparius these include survival, larval development time, emergence, emergence time, adult life span and reproduction. For terrestrial invertebrates survival, growth and reproduction can be selected. Only a very limited number of studies evaluated actual population endpoints like population growth rate (Postma and Davids, 1995).

To Persist or to Perish

If organisms are exposed for multiple generations the effects tend to worsen, ultimately leading to extinction, first of the population exposed to the highest concentration, followed by populations exposed to lower concentrations in later generations (Leon Paumen et al. 2008). Yet, it cannot be excluded that extinction occurs due to the relatively small population sizes in multigeneration experiments, while larger populations may pass a bottleneck and recover during later generations.

Thresholds have also been reported, as shown in Figure 2 (Leon Paumen et al. 2008). Below certain exposure concentrations the exposed populations perform equally well as the controls, generation after generation. Hence, these concentrations may be considered as the 'infinite no effect concentration'. A mechanistic explanation may be that the metabolic machinery of the organism is capable of detoxifying or excreting the toxicants and that this takes so little energy that there is no trade off regarding growth and reproduction.

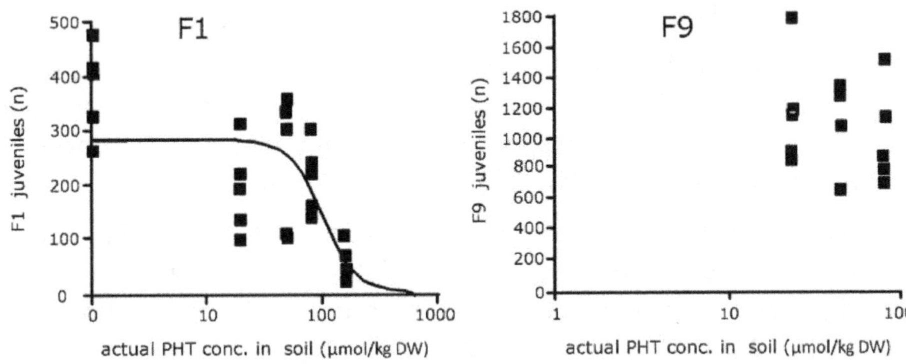

Figure 2. Transition from dose-response relationships to threshold concentrations during a multigeneration toxicity experiment with the collembolan Folsomia candida. Redrawn from Leon Paumen et al. (2008) by Wilma IJzerman.

It is concluded that the frequently reported worsening of effects during multigeneration toxicant exposure raises concerns about the use of single-generation studies in risk assessment to tackle long-term population effects of environmental toxicants.

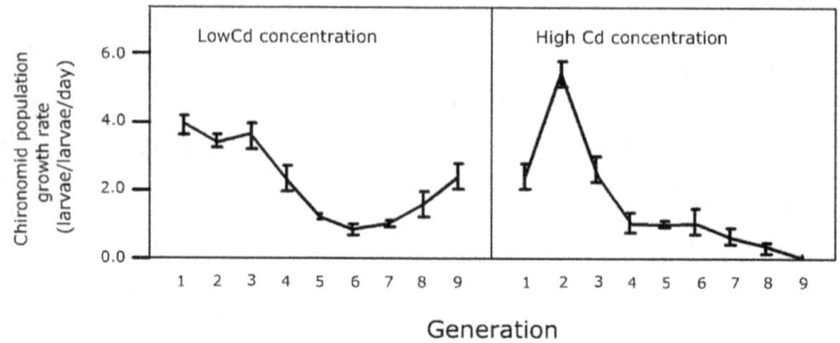

Figure 3. Extinction at a relatively high exposure concentration and adaptation at a relatively low exposure concentration during a multigeneration toxicity experiment with the non-biting midge Chironomus riparius. Redrawn from Postma & Davids (1995) by Wilma IJzerman.

If populations exposed for multiple generations do not get extinct and persist, they may have developed resistance or adaptation (Figure 3). Regular sensitivity testing can therefore be included in multigeneration experiments, as depicted in Figure 1. Yet, it is still under debate whether this lower sensitivity is due to genetic adaptation, epigenetics or phenotypic plasticity (Marinkovic et al., 2012).

References

Ernst, G., Kabouw, P., Barth, M., Marx, M.T., Frommholz, U., Royer, S., Friedrich, S. (2016). Assessing the potential for intrinsic recovery in a Collembola two-generation study: possible implementation in a tiered soil risk assessment approach for plant protection products. Ecotoxicology 25, 1–14.

Leon Paumen, M., Steenbergen, E., Kraak, M.H.S., Van Straalen, N. M., Van Gestel, C.A.M. (2008). Multigeneration exposure of the springtail Folsomia candida to phenanthrene: from dose-response relationships to threshold concentrations. Environmental Science and Technology 42, 6985-6990.

Marinkovic, M., De Bruijn, K., Asselman, M., Bogaert, M., Jonker, M.J., Kraak, M.H.S., Admiraal, W. (2012). Response of the nonbiting midge Chironomus riparius to multigeneration toxicant exposure. Environmental Science and Technology 46, 12105–12111.

Postma. J.F., Davids, C. (1995). Tolerance induction and life-cycle changes in cadmium-exposed Chironomus riparius (Diptera) during consecutive generations. Ecotoxicology and Environmental Safety 30, 195-202.

Van Gestel, C.A.M., De Lima e Silva, C., Lam, T., Koekkoek, J.C. Lamoree, M.H., Verwei, R.A. (2017). Multigeneration toxicity of imidacloprid and thiacloprid to Folsomia candida. Ecotoxicology 26, 320–328.

4.4.7. Tropical Ecotoxicology

Introduction

The tropics cover the area of the world (approx. 40%) that lies between the Tropic of Cancer, 23½° north of the equator and the Tropic of Capricorn, 23½° south of the equator. It is characterized by, on average, higher temperatures and sunlight levels than in temperate regions.

Based on precipitation patterns, three main tropical climates may be distinguished: Tropical rainforest, monsoon and savanna climates. Due to the intrinsic differences between tropical and temperate regions, differences in the risks of chemicals are also likely to occur. These differences are briefly exemplified by taking pesticides as an example, addressing the following subjects: 1) Climate-related factors; 2) Species sensitivities; 3) Testing methods; 4) Agricultural practices and legislation.

Climate-related Factors

Three basic climate factors are essential for pesticide risks when comparing temperate and tropical aquatic agroecosystems: rainfall, temperature and sunlight.

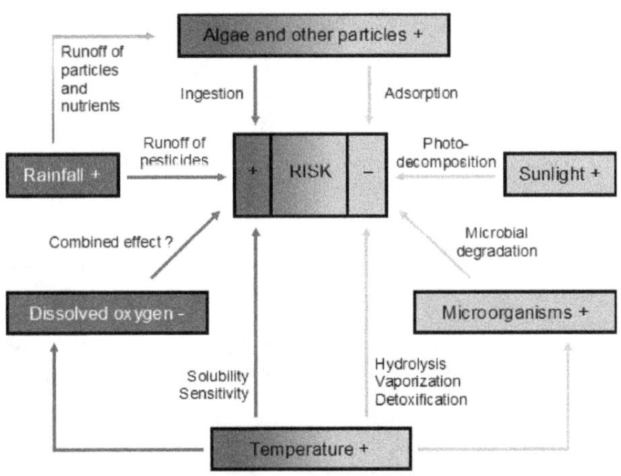

Figure 1. Schematic overview of the climatic related factors that have a possible influence on the risks of pesticides to aquatic ecosystems. The "+" and "-" in the parameter textboxes indicate relatively higher and lower levels of these parameters in tropical compared to temperate regions, respectively. Similarly, the "+" and "-" in the textbox "RISK" indicate a higher and lower risk in tropical compared to temperate freshwaters, respectively. Adapted from Daam and Van den Brink (2010).

For example, high tropical temperatures have been associated with higher microbial activities and hence enhanced microbial pesticide degradation, resulting in lower exposure levels. On the other hand, toxicity of pesticides to aquatic biota may be higher with increasing temperature. Regarding terrestrial ecosystems, other important abiotic factors to be considered are soil humidity, pH, clay and organic carbon content and ion exchange capacity (i.e. the capacity of a soil to adsorb certain compounds) (Daam et al., 2019). Although several differences in climatic factors may be distinguished between tropical and temperate areas, these do not lead to consistent greater or lesser pesticide risk (e.g. Figure 1).

Species Sensitivities

Tropical areas harbour the highest biodiversity in the world and generate nearly 60% of the primary production. This higher species richness, as compared to their temperate counterparts, dictates that the possible occurrence of more sensitive species cannot be ignored. However, studies comparing the sensitivity of species from the same taxonomic group did not demonstrate a consistent higher or lower sensitivity of tropical organisms compared to temperate organisms (e.g. Figure 2).

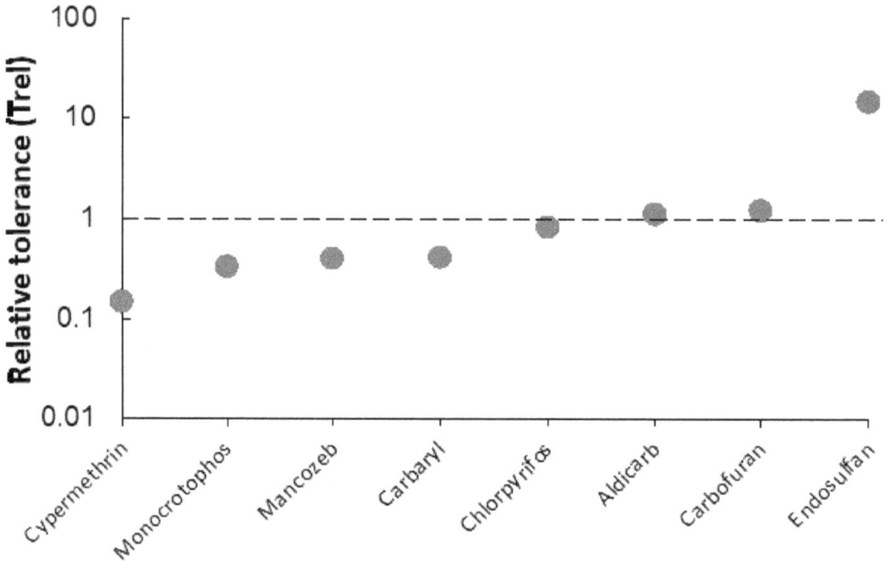

Figure 2. Comparison of the pesticide sensitivity of the tropical earthworm Perionyx excavatus with that of Eisenia fetida sensu lato using the relative tolerance (Trel) approach. The vertical dashed line at Trel = 1 indicates the sensitivity of E. fetida sensu lato. A Trel < 1 (red dots) and Trel > 1 (green dots) indicate a higher and lower sensitivity of P. excavatus relative to E. fetida sensu lato, respectively. PAF = potentially affected fraction. Modified from Daam et al. (2019).

Testing Methods

Given the vast differences in environmental conditions between tropical and temperate regions, the use of test procedures developed under temperate environments to assess pesticide risks in tropical areas has often been disputed. Subsequently, methods developed under temperate conditions need to be adapted to tropical environmental conditions, e.g. by using tropical test substrates and by testing at higher temperatures (Niva et al., 2016). As discussed above, tropical and temperate species from the same taxonomic group are not expected to demonstrate consistent differences in sensitivity. However, certain taxonomic groups may be more represented and/or ecologically or economically more important in tropical areas, such as freshwater shrimps (Daam and Rico, 2016) and (terrestrial) termites (Daam et al., 2019). Subsequently, the development of test procedures for such species and the incorporation in risk assessment procedures seems imperative.

Agricultural Practices and Legislation

Agricultural practices in tropical countries are likely to lead to a higher pesticide exposure and hence higher risks to aquatic and terrestrial ecosystems under tropical conditions. Some of the main reasons for this include i) unnecessary applications and overuse; ii) use of cheaper but more hazardous pesticides, and iii) dangerous transportation and storage conditions, all often a result of a lack in training of pesticide applicators in the tropics (Daam and Van den Brink, 2010; Daam et al., 2019). Finally, countries in tropical regions usually do not have strict laws and risk assessment regulations in place regarding the registration and use of pesticides, meaning that pesticides banned in temperate regions for environmental reasons are often regularly available and used in tropical countries such as Brazil (e.g. Waichman et al. 2002).

References and Recommended Further Reading

Daam, M.A., Van den Brink, P.J. (2010). Implications of differences between temperate and tropical freshwater ecosystems for the ecological risk assessment of pesticides. Ecotoxicology 19, 24–37.

Daam, M.A., Chelinho, S., Niemeyer, J., Owojori, O., de Silva, M., Sousa, J.P., van Gestel, C.A.M., Römbke, J. (2019). Environmental risk assessment of pesticides in tropical terrestrial ecosystems: current status and future perspectives. Ecotoxicology and Environmental Safety 181, 534-547.

Daam, M.A., Rico, A. (2016). Freshwater shrimps as sensitive test species for the risk assessment of pesticides in the tropics. Environmental Science and Pollution Research 25, 13235–13243.

Niemeyer, J.C., Moreira-Santos, M., Nogueira, M.A., Carvalho, G.M., Ribeiro, R., Da Silva, E.M., Sousa, J.P. (2010). Environmental risk assessment of a metal contaminated area in the Tropics. Tier I: screening phase. Journal of Soils and Sediments 10, 1557–1571.

Niva, C.C., Niemeyer, J.C., Rodrigues da Silva Júnior, F.M., Tenório Nunes, M.E., de Sousa, D.L., Silva Aragão, C.W., Sautter, K.D., Gaeta Espindola, E., Sousa, J.P., Römbke, J. (2016). Soil Ecotoxicology in Brazil is taking its course. Environmental Science and Pollution Research 23, 363-378.

Waichman, A.V., Römbke, J., Ribeiro, M.O.A., Nina, N.C.S. (2002). Use and fate of pesticides in the Amazon State, Brazil. Risk to human health and the environment. Environmental Science and Pollution Research 9, 423-428.

5 Chapter POPULATION, COMMUNITY AND ECOSYSTEM ECOTOXICOLOGY

If you want to re-use this chapter, for e.g. in your electronic learning environment, feel free to copy this url: maken.wikiwijs.nl/120181/5__Population__Community_and_Ecosystem_Ecotoxicology

5.1. INTRODUCTION: LINKING POPULATION, COMMUNITY AND ECOSYSTEM RESPONSES

In preparation

5.2. POPULATION ECOTOXICOLOGY IN LABORATORY SETTINGS

Introduction

The motivation to study ecotoxicological effects at the population level is that generally the targets of environmental protection are indeed populations, communities and ecosystems. Additionally, several phenomena are unique to this level, including age specific sensitivity and interaction between individuals. Studying the population level is distinguished from the individual level and lower by a less direct link between the chemical exposure and the observed effects, due to individual variability and several feedback loops, loosening the dose-response relationships. Research at the population level is thus characterized by an increasing level of uncertainty if these processes are not properly addressed and by increasing time and efforts. Hence, it is not surprising that effects at the population are understudied. This is even more the case for investigations on higher levels like meta-populations, communities and ecosystems (see sections on meta-populations, communities and ecosystems). It is thus highly important to obtain data and insights into mechanisms leading to effects at the population level, keeping in mind the relevant interactions with lower and higher levels of organisation.

Properties of populations are unique to this level of biological organization and include social structure (see section on invertebrate community ecotoxicology), genetic composition (see section on genetic variation), density and age structure. This gives room to age and developmental stage specific sensitivities to chemicals. For almost all species, young individuals like neonates or first instars are markedly more sensitive than adults or late instar larvae. This difference may run up to three orders of magnitude and consequently instar specific sensitivities may vary as much as species specific sensitivities (Figure 1). Population developmental stage specific sensitivities have also been reported. Exponentially growing daphnid populations exposed to the insecticide fenvalerate recovered much faster than populations that reached carrying capacity (Pieters and Liess, 2006). Given the age and developmental stage specific sensitivities, the timing of exposure to toxicants in relation to the critical life stage of the organism may seriously affect the extent of the adverse effects, especially in seasonally synchronised populations.

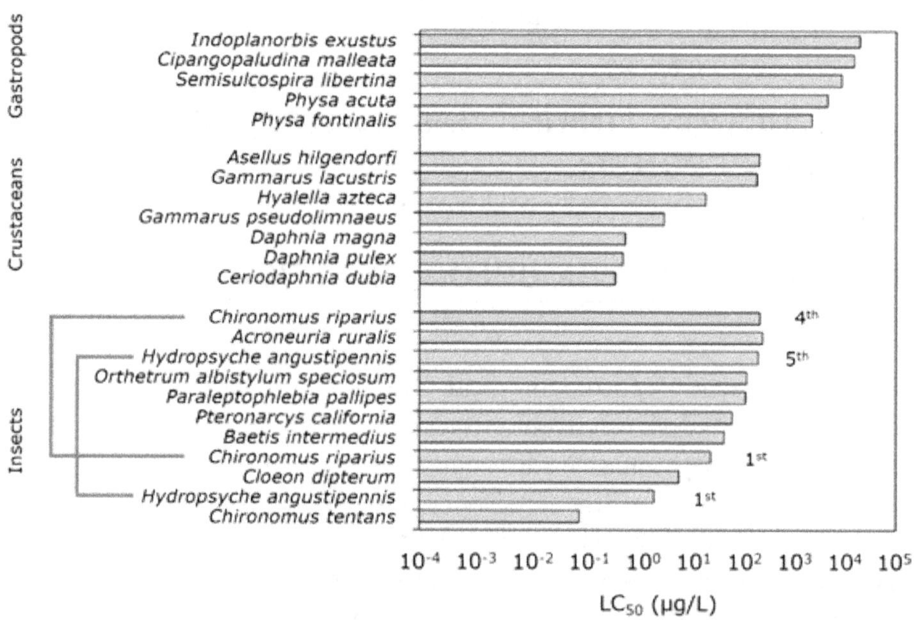

Figure 1. 48h LC50 values of the insecticide diazinon for insects, crustaceans, and gastropods ranked according to sensitivity (according to Stuijfzand et al., 2000), showing that instar specific sensitivities may vary as much as species specific sensitivities. Drawn by Wilma IJzerman.

A challenging question involved in population ecotoxicology is when a population is considered to be stable or in steady state. In spite of the various types of oscillation all populations depicted in Figure 2 can be considered to be stable. One could even argue that any population that does not go extinct can be considered stable. Hence, a single population could vary considerable in density over time, potentially strongly affecting the impact of exposure to toxicants.

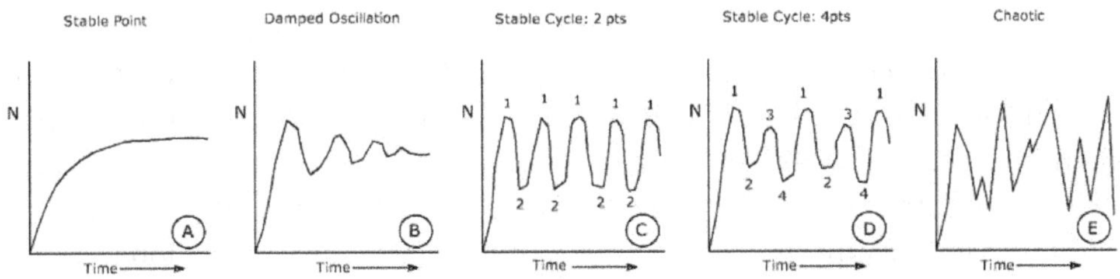

Figure 2. Different types of population development over time. Drawn by Wilma IJzerman.

When populations suffer from starvation and crowding due to high densities and intraspecific competition, they are markedly more sensitive to toxicants, sometimes even up to a factor of 100 (Liess et al., 2016). This may even lead to unforeseen, indirect effects. Relative population growth rate (individual/individual/day) of high density populations of chironomids actually increased upon exposure to Cd, because Cd induced mortality diminished the food shortage for the surviving larvae (Figure 3). Only at the highest Cd exposure population growth rate decreased again. For populations at low densities, the anticipated decrease in population growth rate with increasing Cd concentrations was observed. Yet, at all Cd exposure levels growth rate of low density populations was markedly higher than that of high density populations.

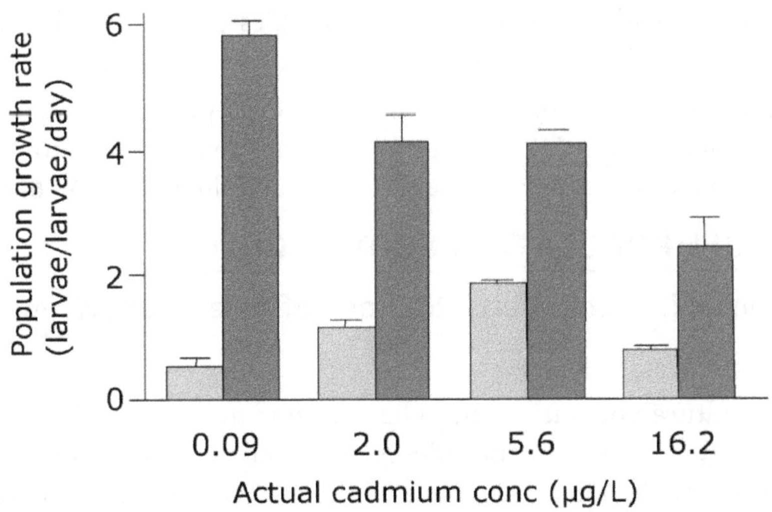

Figure 3. Effects of cadmium exposure and density on population growth rate of Chironomus riparius (according to Postma et al., 1994). Mean values with standard error. Blue bars represent the high larval density and purple bars the low larval density. Redrawn by Wilma IJzerman.

Population Ecotoxicity Tests

In chronic ecotoxicity studies, preferably cohorts of individuals of the same size and age are selected to minimize variation in the outcome of the test, whereas in population ecotoxicology the natural heterogenous population composition is taken into account. This does make it harder though to interpret the obtained experimental data. Especially when studying populations of higher organisms in the wild, the increasing time to complete the research due to the long life span of these organisms imposes practical limitations (see section on wildlife population ecotoxicology). In the laboratory, this can be circumvented by selecting test species with relatively short life cycles, like algae, bacteria and zooplankton. For algae, a three or four day test can be considered as a multigeneration experiment and during 21 d female daphnids may release up to three clutches of neonates. These population ecotoxicity tests offer the unique possibility to calculate the ultimate population parameter, the population growth rate (r). This is a demographic population parameter, integrating survival, maturity time and reproduction (see section on population modeling). Yet, such chronic experiments are typically performed with cohorts and not with natural populations, making these experiments rather an extension of chronic toxicity tests than true population ecotoxicity tests.

References

Knillmann, S., Stampfli, N.C., Beketov, M.A., Liess, M. (2012). Intraspecific competition increases toxicant effects in outdoor microcosms. Ecotoxicology 21, 1857–1866.

Liess, M., Foit, K., Knillmann, S., Schäfer, R.B., Liess, H.-D. (2016). Predicting the synergy of multiple stress effects. Scientific Reports 6, 32965.

Pieters, B.J., Liess, M. (2006). Population developmental stage determines the recovery potential of Daphnia magna populations after fenvalerate application. Environmental Science and Technology 40, 6157-6162.

5.3. WILDLIFE POPULATION ECOTOXICOLOGY

5.3.1. Forensic Investigation into Crash of Asian Vulture Populations

Introduction

Historically, vulture populations in India, Pakistan and Nepal were too numerous to be effectively counted. In the mid-1990s numbers in northern India started to decline catastrophically, which was evidenced in the Keoladeo National Park (figure 1, Prakash 1999). Further monitoring of population numbers indicated unprecedented declines of over 90-99% from the mid-1990s to the early 2000s for Oriental White-backed vultures (Gyps bengalensis), Long-billed vultures (Gyps indicus) and also Slender-billed vultures (Gyps tenuirostris) (Prakash 1999).

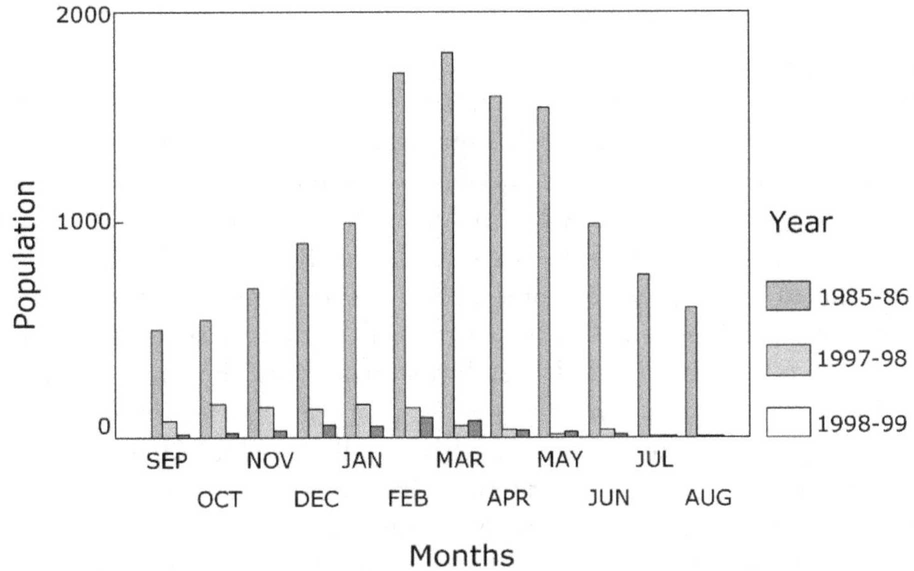

Figure 1. Populations of White-backed vultures in Keoladeo National park in different years. Redrawn from Prakash (1999) by Wilma IJzerman.

In the following years, similar declines were observed in Pakistan and Nepal, indicating that the causative factor was not restricted to a specific country or area. Total losses of vultures were estimated to be in the order of tens of millions. The first ideas about potential causes of those declines focussed on known infectious diseases or the possibility of new diseases to which the vulture population had not been previously exposed. However, no diseases were identified that had shown similar rates of mortalities in other bird species. Vultures are also considered to have a highly developed immune response given their diet of scavenging dead and often decaying animals. To obtain insights, initial interdisciplinary ecological studies were performed to provide a basic understanding of background mortality in the species affected. These studies started in large colonies in Pakistan, but were literally races against time, as some populations had already decreased by 50%, while others were already extirpated, (Gilbert et al., 2006). Despite those difficulties it was determined that mortalities were occurring principally in adult birds and not at the nestling phase. More in depth studies were performed to discriminate between natural mortality of for instance juvenile fledglings, which may be high in summer, just after fledging. After scrutinising the data no seasonality was observed in the abnormal, high mortality, indicating that this was not related to breeding activities. The investigations also revealed another important factor that these vultures were predominantly feeding on domestic livestock, while telemetric observations, using transmitters to assess flight and activity patterns f the birds, showed that the individual birds could range over very long distances to reach carcasses of livestock (up to over 100 km).

Since no apparent causes for mortality were obtained in the ecological studies, more diagnostic investigations were started, focussing on infectious diseases and carried out in Pakistan (Oaks, 2011). However, that was easier said than done. Since large numbers of birds died, it was deemed essential to establish the logistics necessary to perform the diagnostics, including post-mortems, on all birds found dead. Although high numbers of birds died, hardly any fresh carcasses were available, due to remoteness of some areas, the presence of other scavengers and often hot conditions which fostered rapid decay of carcasses. Post-mortems on a selection off birds revealed that birds suspect of abnormal mortality all suffered from visceral gout, which is a white pasty smear covering tissues in the body including liver and heart. In birds, this is indicative for kidney failure. Birds metabolise nitrogen into uric acid (mammals into urea) which is normally excreted with the faeces. However, in case of kidney failure the uric acid is not excreted but deposited in the body. Further inspections of more birds confirmed this, and the working hypothesis became that the increased mortality was caused by a factor inducing kidney failure in the birds.

Based on the establishment of kidney failure as the causative factor, histological and pathological studies were performed on several birds found dead which revealed that in birds with visceral gout, kidney lesions were severe with acute renal tubular necrosis (Oaks et al., 2004), confirming the kidney failure hypothesis. However, no indications of inflammatory cell infiltrations were apparent, ruling out the possibilities of infectious diseases. Those observation shifted the focus to potential toxic effects, although no previous case was known with a chemical causing such severe and extremely acute effects. First the usual suspects for kidney failure were addressed, like trace metals (cadmium, lead) but also other acute toxic chemicals like organophosphorus and carbamate pesticides and organochlorine chemicals None of those chemicals occurred at levels of concern and were ruled out. That left the researchers without leads to any clear causative factor, even after years of study!

Some essential pieces of information were available, however:

- acute renal failure seemed associated with the mortality,
- no infectious agent was likely to be causative pointing to chemical toxicity,
- since exposure was likely to be via the diet the chemical exposure needed to be related to livestock (the predominant diet for the vultures), pointing to compounds present in livestock such as veterinarian products,

- widespread use of veterinarian chemicals had started relatively recently.

After a survey of veterinarians in the affected areas of Pakistan, a single veterinarian pharmaceutical matched the criteria, diclofenac. This is a non-steroid anti-inflammatory drug (NSAID) since long used in human medicine but only introduced since the 1990s as a veterinarian pharmaceutical in India, Pakistan and surrounding countries. NSAIDs are known nephrotoxic compounds, although no cases were known with such acute and sever impacts. Chemical analyses of kidneys of vultures confirmed that kidneys of birds with visceral gout contained diclofenac, birds without signs of visceral gout did not. Also kidneys from birds that showed visceral gout and that died in captivity while being studied, were positive for diclofenac, as was the meat they were fed with. This all indicated diclofenac toxicity as the cause of the mortality, which was validated in exposure studies, dosing captive vultures with diclofenac. The species of *Gyps* vultures appeared extremely sensitive to diclofenac, showing toxic effects at 1% of the therapeutic dose for livestock mammalian species.

This underlying mechanism for that sensitivity has yet to be explained, but initially it was also unclear why the populations were impacted to such severe extent. That was found to be related to the feeding ecology of the vultures. They were shown to fly long ranges to search for carcasses, and as a result of that they show very aggregated feeding, i.e. a lot of birds on a single carcass (Green et al., 2004). Hence, a single contaminated carcass may expose an unexpected large part of the population to diclofenac. Hence, a combination of extreme sensitivity, foraging ecology and human chemical use caused the onset of extreme population declines of some Asian vulture species of the Gyps genus, or so called "Old World vultures".

This case demonstrated the challenges involved in attempting to disentangle the stressors causing very apparent population effects even on imperative species like vultures. It took several years of different groups of excellent researcher to perform the necessary research and forensic studies (under sometimes difficult conditions). Lessons learned are that even for compounds that have been used for a long time and thought to be well understood, unexpected effects may become evident. There is consensus that such effects may not be covered in current risk assessments of chemicals prior to their use and application, but this also draws attention to the need for continued post-market monitoring of organisms for potential exposure and effects. It should be noted that even nowadays, although the use of diclofenac is prohibited in larger parts of Asia, continued use still occurs due to its effectiveness in treating livestock and its low costs making it available to the farmers. Nevertheless, populations of Gyps vultures have shown to recover slowly.

References

Green, R.E., Newton, I.A., Shultz, S., Cunningham, A.A., Gilbert, M., Pain, D.J., Prakash, V. (2004). Diclofenac poisoning as a cause of vulture population declines across the Indian subcontinent. Journal of Applied Ecology 41, 793-800.

Gilbert, M., Watson, R.T., Virani, M.Z., Oaks, J.L., Ahmed, S., Chaudhry, M.J.I., Arshad, M., Mahmood, S., Ali, A., Khan, A.A. (2006). Rapid population declines and mortality clusters in three Oriental whitebacked vulture *Gyps bengalensis* colonies in Pakistan due to diclofenac poisoning. Oryx 40, 388-399.

Oaks, J.L., Gilbert, M., Virani, M.Z., Watson, R.T., Meteyer, C.U., Rideout, B.A., Shivaprasad, H.L., Ahmed, S., Chaudhry, M.J.I., Arshad, M., Mahmood, S., Ali, A., Khan, A.A. (2004). Diclofenac residues as the cause of vulture population decline in Pakistan. Nature 427, 630-633.

Oaks, J.L., Watson, R.T. (2011). South Asian vultures in crisis: Environmental contamination with a pharmaceutical. In: Elliott, J.E., Bishop, C.A., Morrissey, C.A. (Eds.) Wildlife Ecotoxicology. Springer, New York, NY. pp. 413-441.

Prakash, V. (1999). Status of vultures in Keoladeo National Park, Bharatpur, Rajasthan, with special reference to population crash in *Gyps* species. Journal of the Bombay Natural History Society 96, 365–378.

5.3.2. Otters, to PCB or not to PCB?

The European otter (Lutra lutra) is a lively species which historically ranges all over Europe. In the second half of last century populations declined in North-West Europe, and at the end of the 1980s the species was declared extinct in the Netherlands. Several factors contributed to these declines, exposure to polychlorinated biphenyls (PCBs) and other contaminants was considered a prominent cause. PCBs can have different effects on organisms, primarily Ah-receptor mediated (see section on Receptor interactions). In order to assess the actual contribution of chemical exposure to the extinction of the otters, and the potential for population recovery it is essential to gain insight in the ratios between exposure levels and risk thresholds. However, since otters are rare and endangered, limited toxicological data is available on such thresholds. Most toxicological data is therefore inferred from research on another mustelids species the mink (Mustela vison) (Basu et al., 2007) a high trophic level, piscivorous species often used in toxicological studies. Several studies show that mink is quite sensitive to PCBs, showing e.g. effects on the length of the baculum of juveniles (Harding et al., 1999) and induction of hepatic enzyme systems and jaw lesions (Folland et al., 2016). Based on such studies, several threshold levels for otters were derived, depending on the toxic endpoints addressed. Based on number of offspring size and kit survival, EC50 were derived of approximately 1.2 to 2.4 mg/kg *wet* weight (Leonards et al., 1995), while for decreases in vitamin A levels due to PCB exposure, a safety threshold of 4 mg/kg in blood was assessed (Murk et al., 1998).

To re-establish a viable population of otters in the Netherlands, a program was established in the mid-1990s to re-introduce otters in the Netherlands, including monitoring of PCBs and other organic contaminants in the otters. Otters were captured in e.g. Belarus, Sweden and Czech Republic. Initial results showed that these otters already contained < 1 mg/kg PCBs based on wet weight (van den Brink & Jansman, 2006), which was considered to be below the threshold limits mentioned before. Individual otters were radio-tagged, and most were recovered later as victims of car incidences. Over time, PCB concentrations had changed, although not in the same direction for all specimen. Females with high initial concentrations showed declining concentrations, due to lactation, while in male specimens most concentrations increased over time, as you would expect. Nevertheless, concentrations were in the range of the threshold levels, hence risks on effects could not be excluded. Since the re-introduction program was established in a relatively low contaminated area in the Netherlands, questions were raised for re-introduction plans in more contaminated areas, like the Biesbosch where contaminants may still affect otters .

To assess potential risks of PCB contamination in e.g. the Biesbosch for otter populations a modelling study was performed in which concentrations in fish from the Biesbosch were modelled into concentrations in otters. Concentrations of PCBs in the fish differed between species (lipid rich fish such as eel greater concentrations than lean white fish), size of the fish (larger fish greater concentrations than smaller fish) and between locations within the Biesbosch. Using Biomagnification Factors (BMFs) specific for each PCB-congener (see section on Complex mixtures), total PCB concentrations in lipids of otters were calculated based on fish concentrations and different compositions of fish diets of the otters (e.g. white fish versus eel, larger fish versus smaller fish, different locations). Different diets resulted in different modelled PCB concentrations in the otters, however all modelled concentrations were above the earlier mentioned threshold levels (van den Brink and Sluiter, 2015). This would indicate that risks of effects for otters could not be ruled out, and led to the notion that release of otters in the Biesbosch would not be the best option.

However, a major issue related to such risk assessment is whether the threshold levels derived from mink are applicable to otter. The resulting threshold levels for otter are rather low and exceedance of these concentrations has been noticed in several studies. For instance, in well-thriving Scottish otter populations PCBs levels have been recorded greater than 50 mg/kg lipid weight in livers (Kruuk

& Conroy, 1996). This is an order of magnitude higher than the threshold levels, which would indicate that even at higher concentrations, at which effects are to be expected based on mink studies, populations of free ranging otters do not seem to be affected adversely. Based on this, the applicability of mink-derived threshold levels for otters may be open to discussion.

The case on otters showed that the derivation of ecological relevant toxicological threshold levels may be difficult due to the fact that otters are not regularly used in toxicity tests. Application of data from a related species, in this case the American mink, however, may be limited due to differences in sensitivity. In this case, this could result in too conservative assessments of the risks, although it should be noted that this may be different in other combinations of species. Therefore, the read across of information of closely related species should therefore be performed with great care.

References

Basu, N., Scheuhammer, A.M., Bursian, S.J., Elliott, J., Rouvinen-Watt, K., Chan, H.M. (2007). Mink as a sentinel species in environmental health. Environmental Research 103, 130-144.

Harding, L.E., Harris, M.L., Stephen, C.R., Elliott, J.E. (1999). Reproductive and morphological condition of wild mink (Mustela vison) and river otters (Lutra canadensis) in relation to chlorinated hydrocarbon contamination. Environmental Health Perspectives 107, 141-147.

Folland, W.R., Newsted, J.L., Fitzgerald, S.D., Fuchsman, P.C., Bradley, P.W., Kern, J., Kannan, K., Zwiernik, M.J. (2016). Enzyme induction and histopathology elucidate aryl hydrocarbon receptor-mediated versus non-aryl receptor-mediated effects of Aroclor 1268 in American Mink (Neovison vison). Environmental Toxicology and Chemistry 35, 619-634.

Kruuk, H., Conroy, J.W.H. (1996). Concentrations of some organochlorines in otters (Lutra lutra L) in Scotland: Implications for populations. Environmental Pollution 92, 165-171.

Leonards, P.E.G., De Vries, T.H., Minnaard, W., Stuijfzand, S., Voogt, P.D., Cofino, W.P., Van Straalen, N.M., Van Hattum, B. (1995). Assessment of experimental data on PCB-induced reproduction inhibition in mink, based on an isomer- and congener-specific approach using 2,3,7,8-tetrachlorodibenzo-p-dioxin toxic equivalency. Environmental Toxicology and Chemistry 14, 639-652.

Murk, A.J., Leonards, P.E.G., Van Hattum, B., Luit, R., Van der Weiden, M.E.J., Smit, M. (1998). Application of biomarkers for exposure and effect of polyhalogenated aromatic hydrocarbons in naturally exposed European otters (Lutra lutra). Environmental Toxicology and Pharmacology 6, 91-102.

Van den Brink, N.W., Jansman, H.A.H. (2006). Applicability of spraints for monitoring organic contaminants in free-ranging otters (Lutra lutra). Environmental Toxicology & Chemistry 25, 2821-2826.

5.4. TRAIT-BASED APPROACHES

Introduction

It is impossible to assess the sensitivity of all species to all chemicals. Risk assessments therefore, need methods to extrapolate the sensitivity of a limited number of species to all species present in the environment is desired. Statistical approaches, like the species sensitivity distribution concept, perform this extrapolation by fitting a statistical distribution (e.g. log-normal distribution) to a selected set of sensitivity data (e.g. 96h-EC50 data) in order to obtain a distribution of the sensitivity of all species. From this distribution a threshold value associated with the lower end of the distribution can be chosen and used as a protective threshold value (Figure 1).

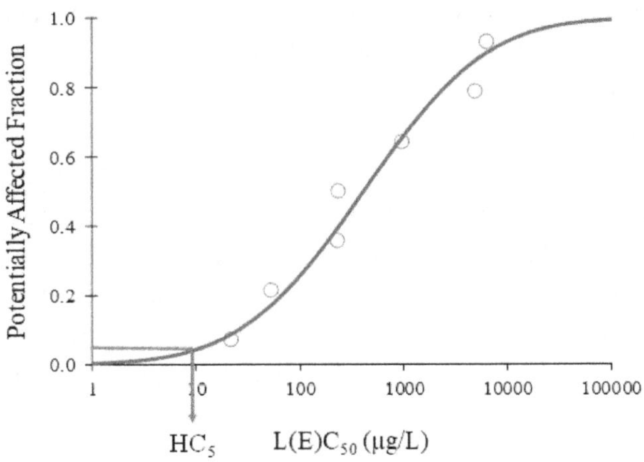

Figure 1. Species sensitivity distribution (line) fitted through a set of EC50 values (dots) and the threshold value protective for at least 95% of the species (Hazardous Concentration 5%, corresponding to the potentially affected fraction of species of 5%) derived from this distribution.

The disadvantage of this approach is that it does not include mechanistic knowledge on what determines species' sensitivity and uses species taxonomy rather than their characteristics. To overcome these and other problems associated with a taxonomy based approach (see Van den Brink et al., 2011 for a review) traits-based bioassessment approaches have been developed for assessing the effects of chemicals on aquatic ecosystem. In traits-based bioassessment approaches, species are not represented by their taxonomy but by their traits. A trait is a phenotypic or ecological characteristic of an organism, usually measured at the individual level but often applied as the average state/condition of a species. Examples of traits are body size, feeding habits, food preference, mode of respiration and lipid content. Traits describe the physical characteristics, ecological niche, and functional role of a species within the ecosystem. The recognized strengths of traits-based bioassessment approaches include: (1) traits add mechanistic and diagnostic knowledge, (2) traits are transferrable across geographies, (3) traits require no new sampling methodology as data that are currently collected can be used, (4) the use of traits has a long-standing tradition in ecology and can supplement taxonomic analysis.

When traits are used to study effects of chemical stressors on ecosystem structure (community composition) and function (e.g. nutrient cycling) it is important to make a distinction between response and effects traits (Figure 2). Response traits are traits that enable a response of the species to the exposure to a chemical. An example of a response trait may be size related surface area of an organism. Smaller organisms have relatively large surface areas, because their surface to volume ratio is higher than for larger animals Herewith, the uptake rate of the chemical stressor is generally higher in smaller animals compared to larger ones (Rubach et al., 2012). Effects traits of organisms influence the surrounding environment by the organisms, by altering the structure and functioning of the ecosystem. An example of an effect trait is the food preference of an organism. For instance, if the small (response trait) and herewith sensitive organisms happen to be herbivorous (effect trait) an increase in algal biomass may be expected when the organisms are affected (Van den Brink, 2008). So, to be able to predict ecosystem level responses it is important to know the (cor)relations between response and effect traits as traits are not independent from each other but can be linked phylogenetically or mechanistically and thus form trait syndromes (Van den Brink et al., 2011).

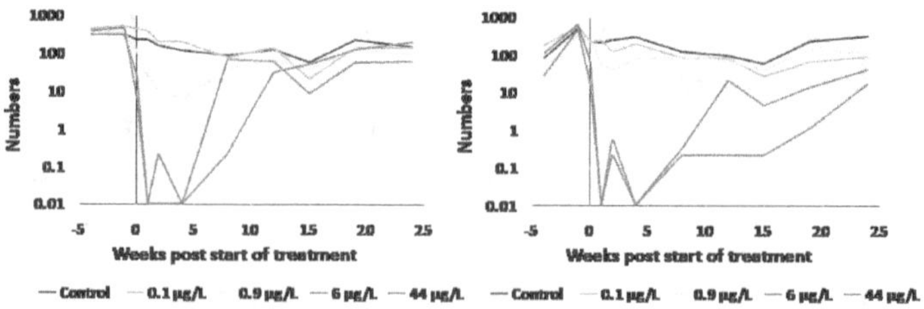

Figure 2. Conceptual diagram showing how species' traits determine both the sensitivity of species to chemicals and how the effects propagate to higher levels of biological organisation.

Predictive Models for Sensitivity using Response Traits

One of the holy grails of ecotoxicology is to find out which species traits make one species more sensitive to a chemical stressor than another one. In the past, two approaches have been used to assess the (cor)relationships between species traits and their sensitivity, one based on empirical correlations between species' traits and their sensitivity as represented by EC50's (Rico and Van den Brink, 2015) and one based on a more mechanistic approach using toxicokinetic/toxicodynamic experiments and models (Rubach et al., 2012). Toxicokinetic-toxicodynamic models (TKTD models) simulate the time-course of processes leading to toxic effects on organisms (Jager et al., 2011). Toxicokinetics describe what an individual does with the chemical and, in their simplest form, include the processes of uptake and elimination, thereby translating an external concentration of a toxicant to an internal body concentration over time (see Section on Toxicokinetics and Bioaccumulation).

Toxicodynamics describes what the chemical does with the organism, herewith linking the internal concentration to the effect at the level of the individual organism over time (e.g., mortality) (Jager et al., 2011) (see Sections on Toxicokinetics and Bioaccumulation and on Toxicodynamics and Molecular Interactions). Rubach et al. (2012) showed that almost 90% of the variation in uptake rates and 80% of the variation in elimination rates of an insecticide in a range of 15 freshwater arthropod species could be explained by 4 species traits. These traits were: i) surface area (without gills), ii) detritivorous feeding, iii) using atmospheric oxygen and iv) phylogeny in case of uptake, and i) thickness of exoskeleton, ii) complete sclerotization, iii) using dissolved oxygen and iv) % lipid of dry weight in case of elimination. For most of these traits, a mechanistic hypothesis between the traits and their influence on the uptake and elimination can be made (Rubach et al., 2012). For instance, a higher surface area to volume ratio increases the uptake of the chemical, so uptake is expected to be higher in small animals compared to larger animals. This shows that it is possible to construct mechanistic models that are able to predict the toxicokinetics of chemicals in species and herewith the sensitivity of species to chemicals based on their traits.

The use of Effect Traits to Model Recovery and Indirect Effects

Traits determining the way organisms within an ecosystem react to chemical stress are related to the intrinsic sensitivity of the organisms on the one hand (response traits) and their recovery potential and food web relations (effect traits) on the other hand (Van den Brink, 2008). Recovery of aquatic invertebrates is, for instance, determined by traits like number of life cycles per year, the presence of insensitive life stages like resting eggs, dispersal ability and having an aerial life stage (Gergs et al., 2011) (Figure 3).

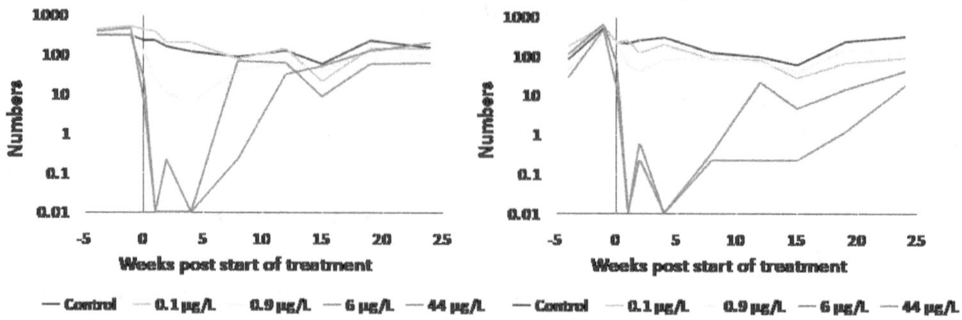

Figure 3. Effect and recovery patterns as observed for two mayfly species after a pulsed exposure to an insecticide. Both mayflies are equally susceptible to chlorpyrifos, but the species on the left (Cloeon dipterum) has many life cycles per year while the species to the right (Caenis horaria) has a full life cycle in spring (coinciding with the chlorpyrifos treatment) and a partial one in autumn.

Besides recovery, effect traits will also determine how individual level effects will propagate to higher levels of biological organisation like the community or ecosystem level. For instance, when Daphnia are affected by a chemical, their trait related to food preference (algae) will ensure that, under nutrient-rich conditions, the algae will not be subjected to top-down control and will increase in abundance. The latter effects are called indirect effects, which are not a direct result of the exposure to the toxicant but an indirect one through competition, food-web relationships, etc..

References

Gergs, A., Classen, S., Hommen, U. (2011). Identification of realistic worst case aquatic macroinvertebrate species for prospective risk assessment using the trait concept. Environmental Science and Pollution Research 18, 1316–1323.

Jager, T., Albert, C., Preuss, T.G., Ashauer, R. (2011). General unified threshold model of survival-A toxicokinetic-toxicodynamic framework for ecotoxicology. Environmental Science and Technology 45, 2529–2540

Rico, A., Van den Brink, P.J. (2015). Evaluating aquatic invertebrate vulnerability to insecticides based on intrinsic sensitivity, biological traits and toxic mode-of-action. Environmental Toxicology and Chemistry 34, 1907–1917.

Rubach, M.N., D.J. Baird, M-C. Boerwinkel, S.J. Maund, I. Roessink, Van den Brink P.J. (2012). Species traits as predictors for intrinsic sensitivity of aquatic invertebrates to the insecticide chlorpyrifos. Ecotoxicology 21, 2088-2101.

Van den Brink, P.J. (2008). Ecological risk assessment: from book-keeping to chemical stress ecology. Environmental Science and Technology 42, 8999–9004.

Van den Brink P.J., Alexander, A., Desrosiers, M., Goedkoop, W., Goethals, P., Liess, M., Dyer, S. (2011). Traits-based approaches in bioassessment and ecological risk assessment: strengths, weaknesses, opportunities and threats. Integrated Environmental Assessment and Management 7, 198-208.

5.5. POPULATION MODELS

Introduction

Ecological risk assessment of toxicants usually focuses on the risks run by individuals, by comparing exposures with no-effect levels. However, in many cases it is not the protection of individual plants or animals that is of interest but the protection of a viable population of a species in an ecological context. Risk assessment generally does not take into account the quantitative dynamics of populations and communities. Yet, understanding and predicting effects of chemicals at levels beyond that of individuals is urgently needed for several reasons. First, we need to know whether quality standards are sufficiently but not overly protective at the population level, when extrapolated from toxicity tests. Second, responses of isolated, homogenous cohorts in the laboratory may be different from those of interacting, heterogeneous populations in the field. Third, to set the right priorities in management, we need to know the relative and cumulative effect of chemicals in relation to other environmental pressures.

Ecological population models for algae, macrophytes, aquatic invertebrates, insects, birds and mammals have been widely used to address the risk of potentially toxic chemicals, however, until recently, these models were only rarely used in the regulatory risk assessment process due to a lack of connection between model output and risk assessment needs (Schmolke et al., 2010). Here, we will sketch the basic principles of population dynamics for environmental toxicology applications.

Exponential Growth

Ecological textbooks usually start their chapter on population ecology by introducing exponential and logistic growth. Consider a population of size N. If resources are unlimited, and the per capita birth (b) and death rates (d) are constant in a population closed to migration, the number of individuals added to the population per time unit (dN/dt) can be written as:

$$\frac{dN}{dt} = (b-d)N(t) \text{ or } \frac{dN}{dt} = rN(t)$$

where r is called the intrinsic rate of increase. This differential equation can be solved with boundary condition N(0) = N0 to yield

$$N(t) = N_0 e^{rt}$$

This is the well-known equation for *exponential growth* (Figure 1a). It applies for example, to animal populations early in the growing season or when they have colonized a new environment. The global human population has also seen exponential growth during most of its existence. The tendency for any population to grow exponentially was recognized by Malthus in his book "An Essay on Population", published in 1789, and it helped Darwin to formulate his theory of evolution by natural selection.

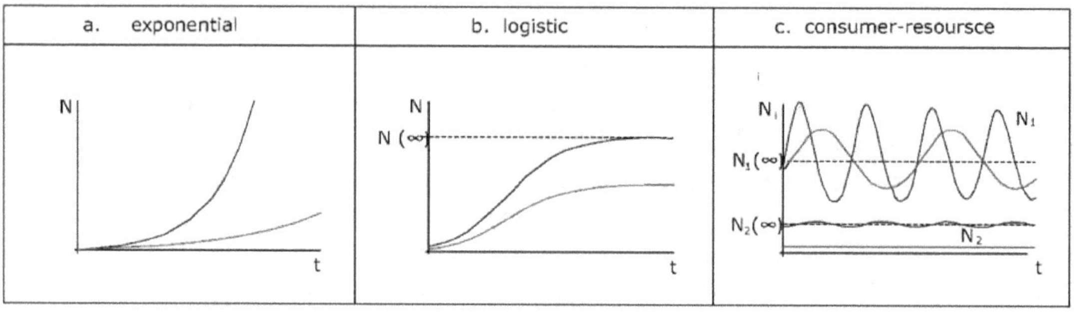

Figure 1. Population density as a function of time for exponential (a), logistic (b) and oscillating (c) populations without (blue) and with (red) exposure to toxicants. N = density, N(∞) equilibrium density, N1 = resource density, N_2 = consumer density. Drawn by Wilma IJzerman.

Since toxicants will affect either reproduction or survival, or both, they will also affect the exponential growth rate (Figure 1a). This suggests that r can be considered a measure of population performance under toxic stress. But rather than from observed population trajectories, r is usually estimated from life-history data. We know from basic demographic theory that any organism with "time-invariant" vital rates (that is, fertility and survival may depend on age, but not on time), will be growing exponentially at rate r. The intrinsic rate of increase can be derived from age-specific survival and fertility rates using the so-called Euler-Lotka equation, which reads:

$$\int_0^{x_m} l(x)e^{-rx}dx = 1$$

in which x is age, xm maximal age, l(x) survivorship from age zero to age x and m(x) the number of offspring produced per time unit at age x. Unfortunately this equation does not allow for a simple derivation of r; r must be obtained by iteration and the correct value is the one that, when combined with the l(x) and m(x) data, makes the integral equal to 1. Due to this complication approximate approaches are often applied. For example, in many cases a reasonably good estimate for r can be obtained from the age at first reproduction α, survival to first reproduction, S, and reproductive output, m, according to the following formula:

$$r = \frac{\ln(S\,m)}{a}$$

This is due to the fact that for many animals in the environment, especially those with high reproductive output and low juvenile survivorship, age at first reproduction is the dominant variable determining population growth (Forbes and Calow, 1999).

The classical demographic modelling approach, including the Euler-Lotka equation, considers time as a continuous variable and solves the equations by calculus. However, there is an equivalent formalism based on discrete time, in which population events are assumed to take place only at equidistant moments. The vital rates are then summarized in a so-called Leslie matrix, a table of survival and fertility scores for each age class, organized in such a way that when multiplied by the age distribution at any moment, the age distribution at the following time point is obtained.

This type of modelling lends itself more easily to computer simulation. The outcome is much the same: if the Leslie matrix is time-invariant the population will grow each time step by a factor λ, which is related to r as ln λ = r (λ = 1 corresponds to r = 0). Mathematically speaking λ is the dominant eigenvalue of the Leslie matrix. The advantage of the discrete-time version is that λ can be more easily decomposed into its component parts, that is, the life-history traits that are affected by toxicants (Caswell, 1996).

The demographic approach to exponential growth has been applied numerous times in environmental toxicology, most often in studies of water fleas (Suhett et al., 2015), and insects (Stark and Banks, 2003). The tests are called "life-table response experiments" (see section on Population ecotoxicology in a laboratory setting). The investigator observes the effects of toxicants on age-specific survival and fertility, and calculates r as a measure of population performance for each exposure concentration.

An example is given in Figure 2, derived from a study by Barata et al. (2000). Forbes and Calow (1999) concluded that the use of r in ecotoxicology adds ecological relevance to the analysis, but it does not necessarily provide a more sensitive or less sensitive endpoint: r is as sensitive as the vital rates underlying its estimation.

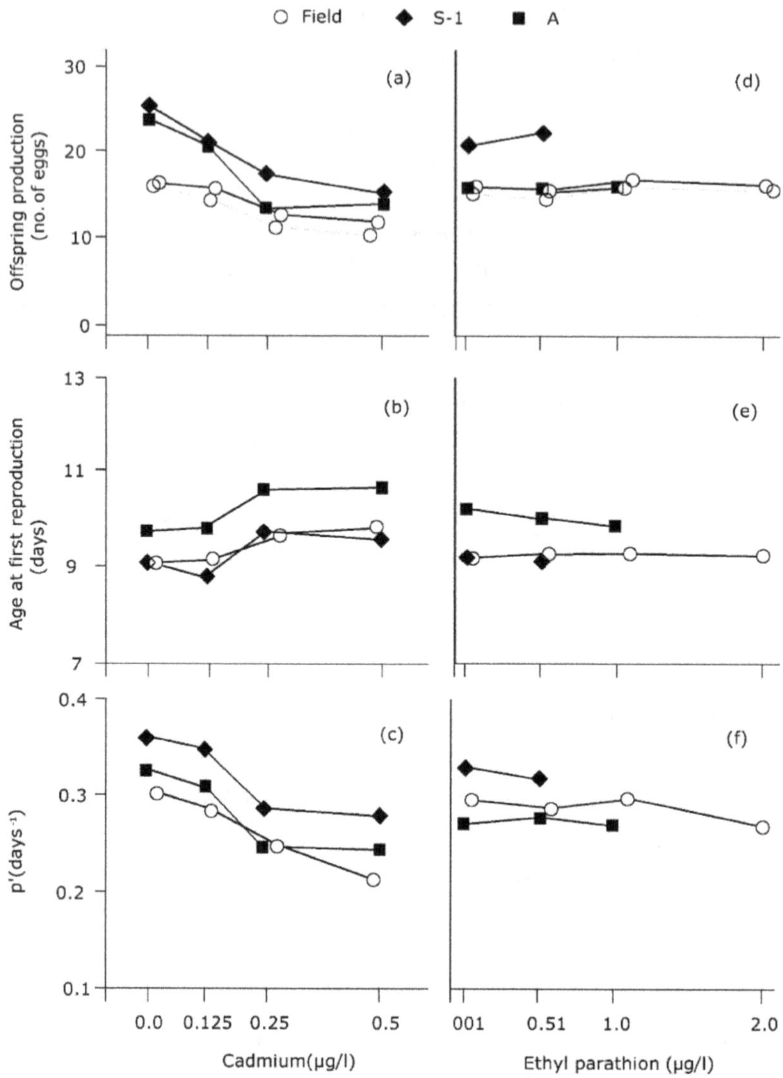

Figure 2. Example of results obtained in life-table response experiments with the water flea Daphnia magna. Cohorts of water fleas were cultured from field-collected resting eggs (ephippia) and two clonal lines under continuous culture in the laboratory (S-1 and A). Responses are shown for offspring production, age at first reproduction and the calculated intrinsic rate of population increase, as a function of cadmium and ethyl-parathion. Redrawn from Barata et al (2000) by Wilma IJzerman.

Hendriks et al. (2005) postulated that r should show a near-linear decrease with the concentration of a chemical, scaled to the LC_{50} (Figure 3). This relationship was confirmed in a meta-analysis of 200 laboratory experiments, mostly concerning invertebrate species (Figure 3). Anecdotal underpinning for large vertebrates comes from field cases where pollution limits population development.

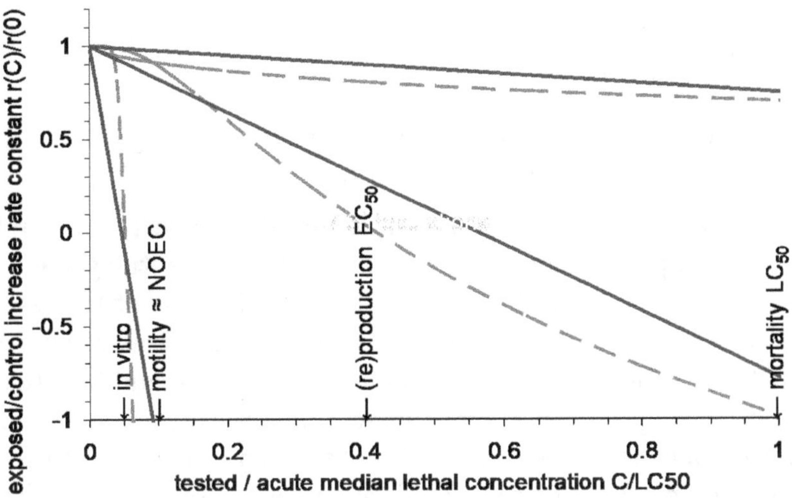

Figure 3. Modelled (red, μ ± 95% CI) and measured (blue, 5%-, 50%- and 95%-percentiles) population increase rates of organisms as a function of toxicant exposure concentrations C. Population growth r is expressed as the value observed in exposed organisms (r(C)) relative to the value in the control situation (r(0)) The exposure concentrations C, are scaled relative to the median lethal concentration LC50 of the toxicant. Characteristic (average) sub-lethal endpoints are given as a reference. The regression lines (individual data points not shown) are derived from a meta-analysis comprising 200 published laboratory experiments (adapted after Hendriks et al., 2005).

Logistic Growth

As exponentially growing populations are obviously rare, models that include some form of density-dependence are more realistic. One common approach is to assume that the birth rate b decreases with density due to increasing scarcity of resources. The simplest assumption is a linear decrease with N, expressed as follows:

$$\frac{dN}{dt} = rN(t)\left(1 - \frac{N(t)}{K}\right)$$

In this equation r is the maximum value of the per capita growth rate (1/N(t) dN/dt), realized at low density and K is a new parameter, the density at which dN/dt becomes zero (Figure 1b). This is the density at which the population will reach equilibrium, also called carrying capacity: N(∞) = K. The resulting model is known as the logistic model, also called the Verhulst-Pearl equation, described in 1844 by François Verhulst and rediscovered in 1920 by Raymond Pearl. The solution of the logistic differential equation is a bit more complicated than in the case of simple exponential growth, but can be obtained by regular calculus to read:

$$N(t)=\frac{K}{1+\left(\frac{K-N_0}{N_0}\right)e^{-rt}}$$

This equation defines an S-shaped curve with density beginning on t = 0 with N0 and increasing asymptotically towards K (Figure 1b).

The question is, can the parameters of the logistic growth equation be used to measure population performance like in the case of exponential growth? Practical application is limited because the carrying capacity is difficult to measure under natural and contaminated conditions. Many field populations of arthropods, for example, fluctuate widely due to predator-prey dynamics, and hardly ever reach their carrying capacity within a growing season. An experimental study on the springtail Folsomia candida (Noël et al., 2006) showed that zinc in the diet did not affect the carrying capacity of contained laboratory populations, although there were several interactions below K that were influenced by zinc, including hormesis (growth stimulation by low doses of a toxicant), and Allee effects (loss of growth potential at low density due to lower encounter rate).

Density-dependence is expected to act as buffering mechanism at the population level because toxicity-induced population decline diminishes competition, however, the effects very much depend on the details of population regulation. This was demonstrated in a model for peregrine falcon exposed to DDE and PBDEs (Schipper et al., 2013). While the equilibrium size of the population declined by toxic exposure, the probability of individual birds finding a suitable territory increased. However, at the same time the number of non-breeding birds shifting to the breeding stage became limiting and this resulted in a strong decrease in the equilibrium number of breeders.

Mechanistic Effect Models

To enhance the potential for application of population models in risk assessment, more ecological details of the species under consideration must be included, e.g. effects of dispersal, abiotic factors, predators and parasites, dispersal, landscape structure and many more. A further step is to track the physiology and ecology of each individual in the population. This is done in the dynamic energy budget modelling approach (DEB) developed by (Kooijman et al., 2009). By including such details, a model will become more realistic, and more precise predictions can be made on the effects of toxic exposures. These types of models are generally called "mechanistic effect models' (MEMs). They allow a causal link between the protection goal, a scenario of exposure to toxicants and the adverse population effects generated by model output (Hommen et al., 2015). The European Food Safety

Authority (EFSA) in 2014 issued an opinion paper containing detailed guidelines on the development of such models and how to adjust them to be useful in the risk assessment of plant protection products.

References

Caswell, H. (1996). Demography meets ecotoxicology: untangling the population level effects of toxic substances. In: Newman, M.C., Jagoe, C.H. (Eds.). Ecotoxicology. A hierarchical treatment. Lewis Publishers, Boca Raton, pp. 255-292.

Barata, C., Baird, D.G., Amata, F., Soares, A.M.V.M. (2000). Comparing population response to contaminants between laboratory and field: an approach using Daphnia magna ephippial egg banks. Functional Ecology 14, 513-523.

EFSA (2014). Scientific Opinion on good modeling practice in the context of mechanistic effect models for risk assessment of plant protection products. EFSA Panel on Plant Protection and their Residues (PPR). EFSA Journal 12, 3589.

Forbes, V.E., Calow, P. (1999). Is the per capita rate of increase a good measure of population-level effects in ecotoxicology. Environmental Toxicology and Chemistry 18, 1544-1556.

Hendriks, A.J., Maas, J.L., Heugens, E.H.W., Van Straalen, N.M. (2005). Meta-analysis of intrinsic rates of increase and carrying capacity of populations affected by toxic and other stressors. Environmental Toxicology and Chemistry 24, 2267-2277

Hommen, U., Forbes, V., Grimm, V., Preuss, T.G., Thorbek, P., Ducrot, V. (2015). How to use mechanistic effect models in environmental risk assessment of pesticides: case studies and recommendations from the SETAC workshop Modelink. Integrated Environmental Assessment and Management 12, 21-31.

Kooijman, S.A.L.M., Baas, J., Bontje, D., Broerse, M., Van Gestel, C.A.M., Jager, T. (2009). Ecotoxicological Applications of Dynamic Energy Budget theory. In: Devillers, J. (Ed.). Ecotoxicology Modeling, Volume 2, Springer, Dordrecht, pp. 237-260.

Noël, H.L., Hopkin, S.P., Hutchinson, T.H., Williams, T.D., Sibly, R.M. (2006). Towards a population ecology of stressed environments: the effects of zinc on the springtail Folsomia candida. Journal of Applied Ecology 43, 325-332.

Schipper, A.M., Hendriks, H.W.M., Kaufmann, M.J., Hendriks, A.J., Huijbregts, M.A.J. (2013). Modelling interactions of toxicants and density dependence in wildlife populations. Journal of Applied Ecology 50, 1469–1478.

Schmolke, A., Thorbek, P., Chapman, P., Grimm, V. (2010) Ecological models and pesticide risk assessment: current modelling practice. Environmental Toxicology and Chemistry 29, 1006-1012.

Stark, J.D., Banks, J.E. (2003) Population effects of pesticides and other toxicants on arthropods. Annual Review of Entomology 48, 505-519.

Suhett, A.L. et al. (2015) An overview of the contribution of studies with cladocerans to environmental stress research. Acta Limnologica Brasiliensia 27, 145-159.

5.6. METAPOPULATIONS

Implications of Meta-population Dynamics on Risks of Environmental Chemicals

Populations can be defined as a group of organisms from the same species which live in a specific geographical area. These organisms interact and breed with each other. At a higher level, one can define meta-populations which can be described as a set of spatially separated populations which interact to a certain extent. The populations may function separately, but organisms can migrate between the populations. Generally the individual populations occur in more or less favourable habitat patches which may be separated by less favourable areas. However, in between populations, good habitats may also occur, where populations have not yet established , or the local populations may have gone extinct. Exchange between populations within a meta-population depends on i) the distances between the individual populations, ii) the quality of the habitat between the populations, e.g. the availability of so-called stepping stones, areas where organisms may survive for a while but which are too small or of too low habitat quality to support a local population and iii) the dispersal potential of the species. Due to the interactions between the different populations within a meta-population, chemicals may affect species at levels higher than the (local) population, also at non-contaminated sites.

An important effect of chemicals at meta-population scale is that local populations may act as a source or sink for other populations within the meta-population. When a chemical affects the survival of organisms in a local population, the local population densities decline. This may increase the immigration of organisms from neighbouring populations within the meta-population. Decrease of local densities would decrease emigration, resulting in a net-influx of organisms into the contaminated site. This is the case when organisms do not sense the contaminants, or that the contaminants do not alter the habitat quality for the organisms. In case the immigration rate at the delivering/source population to replace the populations is too high to replace the leaving organisms, population densities in neighbouring populations may decline, even at the non-contaminated source sites. Consequently, local populations at contaminated sites may act as a sink for other populations within the meta-population, so chemicals may have a much broader impact than just local.

On the contrary, when the local population is relatively small, or the chemical stress is not chronic, meta-population dynamics may also mitigate local chemical stress. Population level impacts of chemicals may be minimised by influx of organisms of neighbouring populations, potentially recovering the population densities prior to the chemical stress. Such recovery depends on the extent and duration of the chemical impact on the local populations and the capacity of the other populations to replenish the loss of the organisms in the affected population.

Meta-population dynamics may thus alter the extent to which contaminants may affect local populations _ through migration between populations. However, chemicals may affect the total carrying capacity of the meta-population as a whole. This can be illustrated by the modelling approach developed by Levins in the late 1960s (Levins 1969). A first assumption in this model is that not all patches that can potentially carry a local population are actually occupied, so let F be the fraction of occupied patches (1-F being the fraction not occupied). Populations have an average change of extinction being e (day-1 when calculating on a daily base), while non-occupied patches have a change of c of being populated (day-1) from the populated patches. The daily change in numbers of occupied patches is therefore:

$$\frac{dF}{dt} = c * F * (1 - F) - e * F$$

In this formula c*F*(1-F) equals the number of non-occupied patches that are being occupied from the occupied patches, while *e* * F equals the fraction of patches that go extinct during the day. This can be recalculated to a carrying capacity (CC) of

$$CC = 1 - \frac{e}{c}$$

while the growth rate (GR) of the meta-population can be calculated by

$$GR = c - e$$

In case chemicals increase extinction risk (e), or decrease the chance on establishment in a new patch (c) this will affect the CC (which will decrease because e/c will increase) as well as the GR (will decrease, may even go below 0). However, this model uses average coefficients, which may not be directly applicable to individual contaminated sites within a meta-population. More (complex) recent approaches include the possibility to use local-population specific parameters and even more, such model include stochasticity, increasing their environmental relevance.

Besides affecting populations directly in their habitats, chemicals may also affect the areas between habitat patches. This may affect the potential of organisms to migrate between patches. This may decrease the chances of organisms to repopulate non-occupied patches, i.e. decrease c, and as such both CC and GR. Hence, in a meta-population setting chemicals even in a non-preferred habitat may affect long term meta-population dynamics.

References

Levins, R. (1969). Some demographic and genetic consequences of environmental heterogeneity for biological control. Bulletin of the Entomological Society of America 15, 237–240

5.7 COMMUNITY ECOTOXICOLOGY

5.7.1. Community Ecotoxicology: Theory and Concepts

Introduction

The motivation to study ecotoxicological effects at the community level is that generally the targets of environmental protection are populations, communities and ecosystems. Consequently, when scaling up research from the molecular level, via cells, organs and individual organisms towards the population, community or even ecosystem level the ecological and societal relevance of the obtained data strongly increase (Figure 1). Yet, the difficulty of obtaining data increases, due to the increasing complexity, lower reproducibility and the increasing time needed to complete the research, which typically involves higher costs. Moreover, when effects are observed in the field it may be difficult to link these to specific chemicals and to identify the drivers of the observed effects. Not surprisingly, ecotoxicological effects at the community and ecosystem level are understudied.

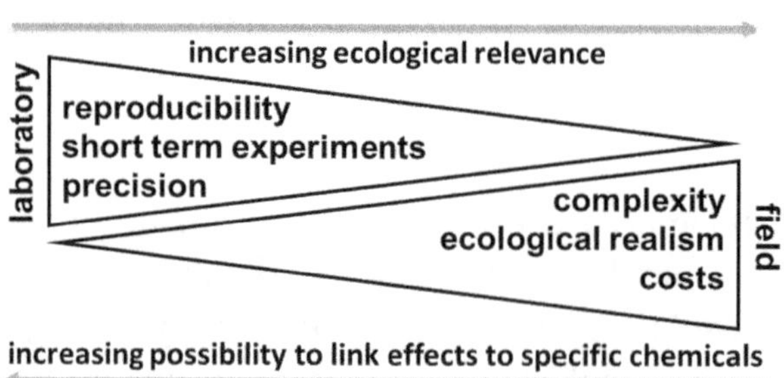

Figure 1. Characteristics of ecotoxicological research performed at different levels of biological organization.

Community Ecotoxicology: Definition and Indirect Effects

Community ecology is defined as the study of the organization and functioning of communities, which are assemblages of interacting populations of species living within a particular area or habitat. Building on this definition, community ecotoxicology is defined as the study of the effects of toxicants on patterns of species abundance, diversity, community composition and species interactions. These species interactions are unique to the community and ecosystem level and may cause direct effects of toxicants on specific species to exert indirect effects on other species. It has been estimated that the majority of effects at these levels of biological organization are indirect rather than direct. These indirect effects are exerted via:

- predator-prey relationships
- consumer-producer relationships
- competition between species
- parasite-host relationships
- symbiosis
- biotic environment

As an example, Roessink et al. (2006) studied the impact of the fungicide triphenyltin acetate (TPT) on benthic communities in outdoor mesocosms. For several species a dose-related decrease in abundance directly after application was observed, followed by a gradual recovery coinciding with decreasing exposure concentrations, all implying direct effects of the fungicide. For some species, however, opposite results were obtained and abundance increased shortly after application, followed by a gradual decline; see the example of the Culicidae in Figure 2. In this case, these typical indirect effects were explained by a higher sensitivity of the predators and competitors of the Culicidae.

Due to diminished predation and competition and higher food availability abundances of the Culicidae temporarily increased after toxicant exposure. With the decreasing exposure concentrations over time, the populations of the predators and competitors recovered, leading to a subsequent decline in numbers of the Culicidae.

The indirect effects described above are thus due to species-specific sensitivities to the compound of interest, which influence the interactions between species. Yet, at higher exposure concentrations also the less sensitive species will start to be affected by the chemical. This may lead to an "arch-shaped" relationship between the number of individuals of a certain species and the concentration of a toxicant.

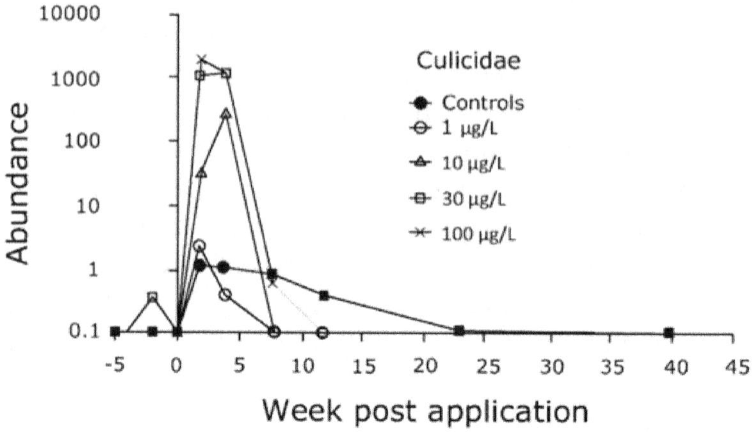

Figure 2. Dynamics of Culicidae in cosms treated with the fungicide triphenyltin acetate. Redrawn from Roessink et al. (2006) by Wilma IJzerman.

In a mesocosm study with the insecticide lambda-cyhalothin this was observed for Daphnia, which are prey for the more sensitive phantom midge Chaoborus (Roessink et al., 2005; Figure 3). At low exposure concentrations the indirect effects, such as release from predation by Chaoborus, led to an increase in abundance of the less sensitive Daphnia. At intermediate exposure concentrations there was a balance between the positive indirect effects and the adverse direct effects of the toxicant. At higher exposure concentrations the adverse direct effects overruled the positive indirect effects resulting in a decline in abundance of the Daphnia. These combined dose dependent direct and indirect effects are inherent to community-level experiments, but are compound and species-interaction specific.

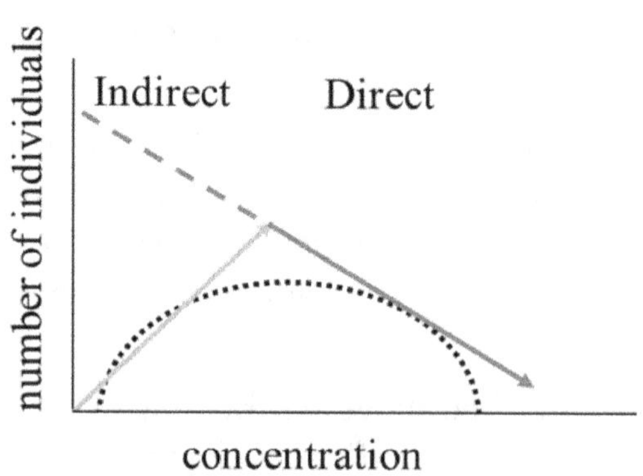

Figure 3. Arch-shaped relationship between the number of individuals of a certain species in a community-level experiment and the concentration of a toxicant, caused by the combination of indirect positive effects at low exposure concentrations and adverse direct effects at higher exposure concentrations.

Investigating Communities and Analysing and Interpreting Community Ecotoxicity Data

To study community ecotoxicology, experiments have to be scaled up and are therefore often performed in mesocosms, artificial ponds, ditches and streams, or even in the field, sometimes accompanied by the use of in- and exclosures. To assess the effects of toxicants on communities in such large systems requires meticulous sampling schemes, which often make use of artificial substrates and e.g. emergence traps for aquatic invertebrates with terrestrial adult life stages (see section on Community ecotoxicology in practice).

Alternatively to scaling up the experiments in community ecotoxicology, the size of the communities may be scaled down. Algae and bacteria grown on coin sized artificial substrates in the field or in experimental settings provide the unique advantage that the experimental unit is actually an entire community.

Given the large scale and complexity of experiments at the community level, such experiments generally generate overwhelming amounts of data, making appropriate analysis of the results challenging. Data analysis focusing on individual responses, so-called univariate analysis, that suffice in single species experiments, obviously falls short in community ecotoxicology, where cosm or (semi-)field communities sometimes consist of over a hundred different species. Hence, multivariate analysis is often more appropriate, similar to the approaches frequently applied in field studies to identify possible drivers of patterns in species abundances. Alternative approaches are also applied, like using ecological indices such as species richness or categorizing the responses of communities into effect classes (Figure 4). To determine if species under semi-field conditions respond equally sensitive to toxicant exposure as in the laboratory, the construction and subsequent comparison of species sensitivity distributions (SSD) (see section on SSDs) may be helpful.

Class 1 : No treatment-related effects demonstrated (NOEC most sensitive endpoint)

Class 2 : Slight short-term effects, usually on individual samplings directly after application

Class 3 : Pronounced short-term effects of sensitive endpoints; recovery within 8 weeks post last application

Class 4 : Pronounced effects, recovery of sensitive endpoint not studied (duration study too short)

Class 5 : Pronounced longer-term effects, full recovery not within 8 weeks post last application

Figure 4. Effect classes to categorize the responses of communities in community-level experiments.

The analysis and interpretation of community ecotoxicity data is also challenged by the dynamic development of each individual replicate cosm, artificial pond, ditch or stream, including those from the control. From the start of the experiment, each control replicate develops independently, matures, and at the end of the experiments that generally last for several months control replicates may differ not only from the treatments, but also among each other. The challenge is then to separate the toxic signal from the natural variability in the data.

In experiments that include a recovery phase, it is frequently observed that previously exposed communities do recover, but develop in another direction than the controls, which actually challenges the definition of recovery. Moreover, recovery can be decelerated or accelerated depending on the dispersal capacity of the species potentially inhabiting the cosms and the distance to nearby populations within a metapopulation (see section on Metapopulations).

Other crucial factors that may affect the impact of a toxicant on communities, as well as their recovery from this toxicant exposure include habitat heterogeneity and the state of the community in combination with the moment of exposure. Habitat heterogeneity may affect the distribution of toxicants over the different environmental compartments and may provide shelter to organisms. Communities generally exhibit temporal dynamics in species composition and in their contribution to ecosystem processes (see section on Structure versus function), as well in the lifecycle stages of the individual species. Exponentially growing populations recover much faster than populations that reached carrying capacity and for almost all species, young individuals are up to several orders of magnitude more sensitive than adults or late instar larvae (see section on Population ecotoxicology). Hence, the timing of exposure to toxicants may seriously affect the extent of the adverse effects, as well as the recovery potential of the exposed communities.

From Community Ecotoxicology towards Ecosystems and Landscapes

When scaling up from the community to the ecosystem level, again unique characteristics emerge: structural characteristics like biodiversity, but also ecosystem processes, quantified by functional endpoints like primary production, ecosystem respiration, nutrient cycling and decomposition. Although a good environmental quality is based on both ecosystem structure and functioning, there is definitely a bias towards ecosystem structure, both in science and in policy (see section on Structure versus function). Levels of biological organisation higher than ecosystems are covered by the field of landscape ecotoxicology (see section on Landscape ecotoxicology) and in a more practical way by the concept of ecosystem services (see section on Ecosystem services).

References

Roessink, I., Crum, S.J.H., Bransen, F., Van Leeuwen, E., Van Kerkum, F., Koelmans, A.A., Brock, T.C.M. (2006). Impact of triphenyltin acetate in microcosms simulating floodplain lakes. I. Influence of sediment quality. Ecotoxicology 15, 267-293.

Roessink, I., Arts, G.H.P., Belgers, J.D.M., Bransen, F., Maund, S.J., Brock, T.C.M. (2005). Effects of lambda-cyhalothrin in two ditch mesocosm systems of different trophic status. Environmental Toxicology and Chemistry 24, 1684-1696.

Further Reading

Clements, W.H., Newman, M.C. (2002). Community Ecotoxicology. John Wiley & Sons, Ltd.

5.7.2. Community Ecotoxicology in Practice

Introduction: Linking Effects at Different Levels of Biological Organization

It is generally anticipated that ecotoxicological tests should provide data useful for making realistic predictions of the fate and effects of chemicals in natural ecosystems (Landner et al., 1989). The ecotoxicological test, if used in an appropriate way, should be able to identify the potential environmental impact of a chemical before it has caused any damage to the ecosystem. In spite of the considerable amount of work devoted to this problem and the plethora of test methods being published, there is still reason to question whether current procedures for testing and assessing the hazard of chemicals in the environment do answer the questions we have asked. Most biologists agree that at each succeeding level of biological organization new properties appear that would not have been evident even by the most intense and careful examination of lower levels of organization (Cairns Jr., 1983).

These levels of biological hierarchy might be crudely characterized as subcellular, cellular, organ, organism, population, multispecies, community, and ecosystem (Figure 1). At the lower biological level, responses are faster than those occurring at higher levels of organization.

Experiments executed at the lower biological level often are performed under standard laboratory conditions (see Section on Toxicity testing). The laboratory setting has advantages like allowing for replication, the use of relatively easy and simplified conditions that enable outcomes that are rather robust across different laboratories, the stressor of interest being more traceable under optimal stable conditions, and easy repetition of experiments. As a consequence, at the lower biological level the responses of organisms to chemical stressors tend to be more tractable, or more causal, than those identified when studying effects at higher tiered levels.

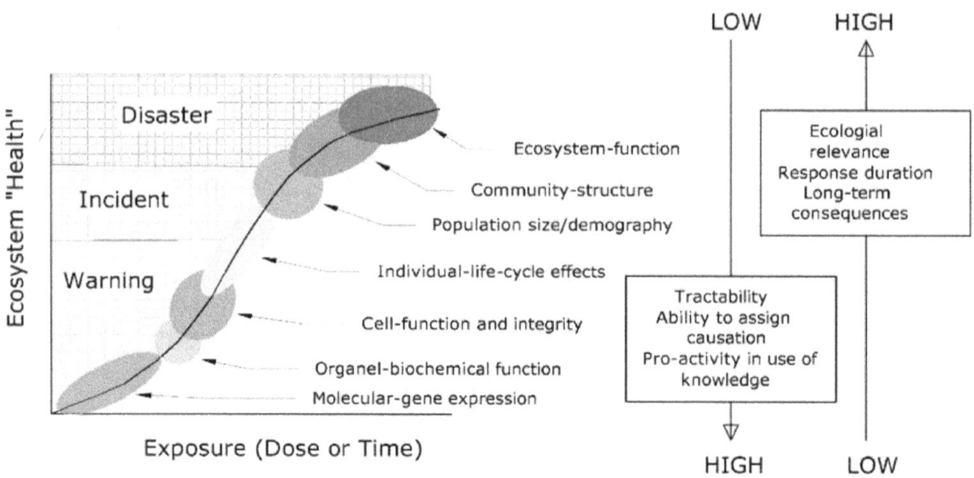

Figure 1: Tests at different biological levels, from molecule to ecosystem scale (modified from Newman, 2008). Each biological level is of equal importance in environmental toxicology, but has a different implication to the ecosystem health. If the fitness at the gene to the individual species level is affected due to exposure, this can be seen as a warning for the ecosystem health. If the impact of exposure can be detected at individual species (e.g. reproductive output) to population levels, this can be seen as an incident. If the impact of exposure is detectable at the community (structure and functioning) level, this is considered a disaster for ecosystem health. Measurements performed at the different biological level inform us differently: at the higher biological levels in general more ecological realism regarding exposure as well as species-interactions is gained, whereas at the lower biological levels causality and tractability of the link between the response and the dose of chemicals is achieved. Drawn by Wilma IJzerman.

The merit to perform cosm studies, so at the higher biological level (see Figure 1), is to investigate the impact of a stressor on a variety of species, all having interactions with each other. This enables detecting both direct and indirect effects on the structure of species assemblages due to the chemicals. Indirect effects can become manifest as disruptions of species interactions, e.g. competition, predator-prey interactions and the like. A second important reason for conducting cosm studies is that abiotic interactions at the level of the ecosystem can be accounted for, allowing for measurement of effects of chemicals under more environmentally realistic exposure conditions. Conditions that likely influence the fate and behavior of chemical are sorption to sediments and plants, photolysis, changes in pH (see section on Bioavailability for a more detailed description), and other natural fluctuations.

What are Cosm Studies?

Microcosm or mesocosm (or cosm) studies represent a bridge between the laboratory and the natural world (examples of aquatic cosms are given in Figure 2). The difference between micro- and mesocosms is mostly restricted to size (Cooper and Barmuta, 1993). Aquatic microcosms are 10-3 to 10 m^3 in size, while mesocosms are 10 to 104 m^3 or even larger equivalent to whole or natural systems. The originality of cosms is mainly based on the combination of ecological realism, achieved by the introduction of the basic components of natural ecosystems, and facilitated access to a number of physicochemical, biological, and toxicological parameters that can be controlled to some extent. The cosm approach also makes it possible to work with treatments that can be replicated, so enabling the study of multiple environmental factors which can be manipulated. The system allows the establishment of food webs, the assessment of direct and indirect effects, and the evaluation of effects of contamination on multiple trophic and taxonomic levels in an ecologically relevant context. Cosm studies make it possible to assess effects of contaminants by looking at the parts (individuals, populations, communities) and the whole (ecosystems) simultaneously.

Figure 2. Different aquatic ecotoxicological testing facilities: A) indoor microcosms (water-sediment interface, at Leiden University), B) in situ (or caged) outdoor enclosures (at Wageningen Environmental Research), and C) cosms or experimental ditches (at Living Lab, Leiden University).

As given in the OECD guidance document (OECD, 2004), the size to be selected for a meso- or microcosm study will depend on the objectives of the study and the type of ecosystem that is to be simulated. In general, studies in smaller systems are more suitable for short-term studies of up to three to six months and studies with smaller organisms (e.g. planktonic species). Larger systems are more appropriate for long-term studies (e.g. 6 months or longer). Numerous ecosystem-level manipulations have been conducted since the early 1970s (Hurlbert et al., 1972). The Experimental Lakes Area (ELA) situated in Ontario, Canada deserves special attention because of its significant contributions to the understanding of how natural communities respond to chemical stressors. This ELA consists of 46 natural, relatively undisturbed lakes, which were designated specially for ecosystem-level research. Many different questions have been tackled here, e.g. manipulations with nutrients (amongst others Levine and Schindler, 1999), synthetic estrogens (e.g. Kidd et al., 2014) and Wallace with pesticides in the Coweeta district (Wallace et al., 1999). It is nowadays realized that there is a need for testing more than just individual species and to take into account ecosystem elements such as fluctuations of abiotic conditions and biotic interactions when trying to understand the ecological effects of chemicals. Therefore a selection of study parameters is often considered as given by OECD (2004):

- Regarding treatment regime:
    - dosing regime, duration, frequency, loading rates, preparation of application
    - solutions, application of test substance, etc.;
    - meteorological records for outdoor cosms;
    - physicochemical water parameters (temperature, oxygen saturation, pH, etc.);
- Regarding *biological levels* it should be recorded what sampling methods and taxonomic identification methods are used;
    - **phytoplankton**: chlorophyll-a; total cell density; abundance of individual dominant taxa; taxa (preferably species) richness, biomass;
    - **periphyton**: chlorophyll-a; total cell density; density of dominant species; species richness, biomass;
    - **zooplankton**: total density per unit volume; total density of dominant orders (Cladocera, Rotifera and Copepoda); species abundance; taxa richness, biomass;

- **macrophytes**: biomass, species composition and % surface covering of individual plants;
- **emergent insects**: total number emerging per unit of time; abundance of individual dominant taxa; taxa richness; biomass; density; life stages;
- **benthic macroinvertebrates**: total density per unit area; species richness, abundance of individual dominant species; life stages;
- **fish**: total biomass at test termination; individual fish weights and lengths for adults or marked juveniles; condition index; general behaviour; gross pathology; fecundity, if necessary.

Two Typical Examples of Results Obtained in an Aquatic Cosm Study

A cosm approach assist in identifying and quantifying direct as well as indirect effects. Here two different types of responses are described, for more examples it is referred to the Section on Multistress.

Joint interactions: Barmentlo et al. (2018) used an outdoor mesocosm system consisting of 65 L ponds. Using a full factorial design, they investigated the population responses of macroinvertebrate species assemblages exposed for 35 days to environmentally relevant concentrations of three commonly used agrochemicals (imidacloprid, terbuthylazine, and NPK fertilizers). A detrivorous food chain as well as an algal-driven food chain were inoculated into the cosms. At environmentally realistic concentrations of binary mixtures, the species responses could be predicted based on concentration addition (see Section on Mixture toxicity). Overall, the effects of trinary mixtures were much more variable and counterintuitive. This was nicely illustrated by how the mayfly Cloeon dipterum reacted to the various combinations of the pesticides. Compared to single substance exposures and binary mixtures, extreme low recovery of C. dipterum (3.6% of control recovery for both mixtures) was seen. However, after exposure to the trinary mixture, recovery of C. dipterum no longer deviated from the control, and therefore was was higher than expected. Unexpected effects of the mixtures were also obtained for both zooplankton species (Daphnia magna and Cyclops sp.) As expected, the abundance of both zooplankton species was positively affected by nutrient applications, but pesticide addition did not lower their recovery. These type of unexpected results can only been identified when multiple species and multiple stressors are tested and cannot be detected in a lab-test with single species.

Indirect cascading effects: Van den Brink et al. (2009) studied the effects of chronic applications of a mixture of the herbicide atrazine and the insecticide lindane in indoor freshwater plankton-dominated microcosms. Both top-down and bottom-up regulation mechanisms of the species assemblage selected were affected by the pesticide mixture. Lindane exposure also caused a decrease in sensitive detritivorous macro-arthropods and herbivore arthropods. This allowed insensitive food competitors like worms, rotifers and snails to increase in abundance (although not always significantly). Atrazine inhibited algal growth and hence also affected the herbivores. A direct result of the inhibition of photosynthesis by atrazine exposure were lower dissolved oxygen and pH levels and an increase in alkalinity, nitrogen and electrical conductivity. See Figure 3 for a synthesis of all interactions observed in the study of Van den Brink et al. (2009).

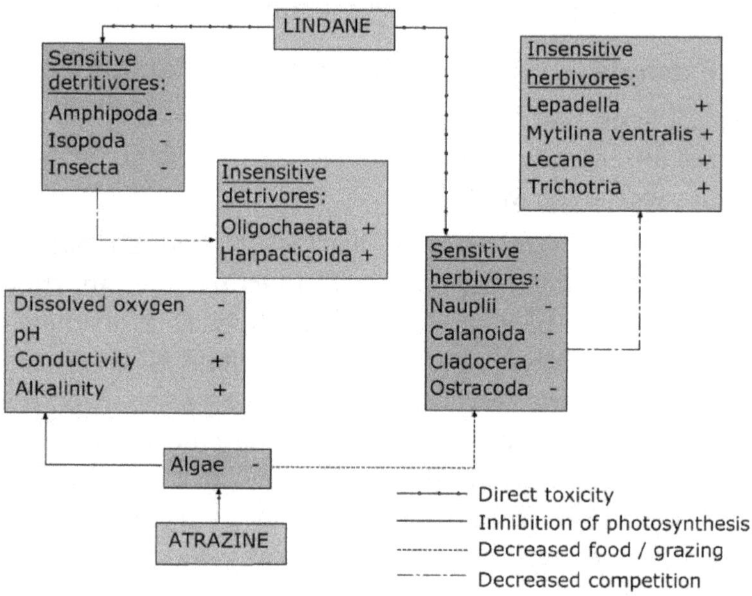

Figure 3. Ecological effect chain as observed in a microcosm experiments using atrazine and lindane as stressors. The arrows indicate the hypothesis-driven relationships between species. The red colors (with -) represent negative feedbacks, the green colors (with +) positive feedbacks. See the text for further explanation. Adapted from Van den Brink et al. (2009) by Wilma IJzerman.

Realism of Cosm Studies

There is a conceptual conflict between realism and replicability when applied to mesocosms. Replicability may be achieved, in part, by a relative simplification of the system. The crucial point in designing a model system may not be to maximize the realism, but rather to make sure that ecologically relevant information can be obtained. Reliability of information on ecotoxicological effects of chemicals tested in mesocosms closely depends on the representativeness of biological processes or structures that are likely to be affected. This means that within cosms key features at both structural and functional levels should be preserved as they ensure ecological representativeness. Extrapolation from small experimental systems to the real world seems generally more problematic than the use of larger systems in which more complex interactions can be studied experimentally as well. For that reason, Caquet et al. (2000) claim that testing chemicals using mesocosms refines the classical methods of ecotoxicological risk assessment because they provide conditions for a better understanding of environmentally relevant effects of chemicals.

References

Barmentlo S.H., Schrama M., Hunting E.R., Heutink R., Van Bodegom P.M., De Snoo G.R., Vijver M.G. (2018). Assessing combined impacts of agrochemicals: Aquatic macroinvertebrate population responses in outdoor mesocosms, Science of the Total Environment 631-632, 341-347.

Caquet, T., Lagadic, L., Sheffield, S.R. (2000) Mesocosm in ecotoxicology: outdoor aquatic systems. Reviews of Environmental Contamination and Toxicology 165, 1-38.

Cairns Jr. J. (1983). Are single species toxicity tests alone adequate for estimating environmental hazard? Hydrobiologica 100, 47-57.

Cooper, S.D., Barmuta, L.A. (1993) Field experiments in biomonitoring. In Rosenberg, D.M., Resh, V.H. (Eds.) Freshwater Biomonitoring and Benthic Macroinvertebrates. Chapman and Hall, New York, pp. 399–441.

OECD (2004). Draft Guidance Document on Simulated Freshwater Lentic Field Tests (Outdoor Microcosms and Mesocosms) (July 2004). Organization for Economic Cooperation and Development, Paris. http://www.oecd.org/fr/securitechimique/essais/32612239.pdf

Hurlbert, S.H., Mulla, M.S., Willson, H.R. (1972) Effects of an organophosphorus insecticide on the phytoplankton, zooplankton, and insect populations of fresh-water ponds. Ecological Monographs 42, 269-299.

Kidd, K.A., Paterson, M.J., Rennie, M.D., Podemski, C.L., Findlay, D.L., Blanchfield, P.J., Liber, K. (2014). Direct and indirect responses of a freshwater food web to a potent synthetic oestrogen. Philosophical Transactions of the Royal Society B Biological Sciences 369, Article AR 20130578, DOI:10.1098/rstb.2013.0578

Landner, L., Blanck, H., Heyman, U., Lundgren, A., Notini, M., Rosemarin, A., Sundelin, B. (1989) Community Testing, Microcosm and Mesocosm Experiments: Ecotoxicological Tools with High Ecological Realism. Chemicals in the Aquatic Environment. Springer, pp. 216-254.

Levine, S.N., Schindler, D.W. (1999). Influence of nitrogen to phosphorus supply ratios and physicochemical conditions on cyanobacteria and phytoplankton species composition in the Experimental Lakes Area, Canada. Canadian Journal of Fisheries and Aquatic Sciences 56, 451-466.

Newman, M.C. (2008). Ecotoxicology: The History and Present Directions. In Jørgensen, S.E., Fath, B.D. (Eds.), Ecotoxicology. Vol. 2 of Encyclopedia of Ecology, 5 vols. Oxford: Elsevier, pp.1195-1201.

Van den Brink, P.J., Crum, S.J.H., Gylstra, R., Bransen, F., Cuppen, J.G.M., Brock, (2009). Effects of a herbicide – insecticide mixture in freshwater microcosms: risk assessment and ecological effect chain. Environmental Pollution 157, 237-249.

Wallace, J.B., Grubaugh, J.W., Whiles, M.R. (1996). Biotic indices and stream ecosystem processes: Results from an experimental study. Ecological Applications 6, 140-151.

5.8. STRUCTURE VERSUS FUNCTION INCL. ECOSYSTEM SERVICES

Biodiversity at Three Different Levels

In ecology, biodiversity describes the richness of natural life at three levels: genetic diversity, species diversity (the most well-known) and landscape diversity. The most commonly used index, the Shannon Wiener index, expresses biodiversity in general terms as the number of species in relation to the number of individuals per species. Precisely, this index stands for the sum of the natural logarithm of the number of individuals per species present:

$$-\Sigma(p_i*\ln(p_i))$$

with $p_i = n_i/N$ in which ni is the number of individuals of species i and N the total number of individuals of all species combined. The index is higher for communities with more species, but also higher when the abundance is more equally distributed over species. A low index implies a community with a few very dominant species. Environmental pollution tends to increase dominance, i.e. a few species are favoured and many become rare (see section on Community ecotoxicology).

In environmental toxicology, most attention is paid to species diversity. Genetic diversity plays a role in the assessment of more or less sensitive or resistant subspecies or local populations of a species, like in various mining areas with persistent pollution. Landscape diversity is receiving attention only recently and aims primarily at the total load of e.g. pesticides applied in an agronomic landscape (see Section on Landscape ecotoxicology), although it should more logically focus on the interactions between the various ecosystems in a landscape, for instance a lake surrounded partly by a forest, partly by a grassland.

Structural and Functional Diversity

In general, the various types of interactions between species do not play a major role in the study of biodiversity neither within ecology nor in environmental toxicology. The diversity in interactions described in the food web or food chain is not expressed in a term like the Shannon-Wiener index. However, in aquatic as well as soil ecological research, extensive, quantitative descriptions have been made of various ecosystems. These model descriptions, like the one for arable soil below, are partly based on the taxonomic background of species groups and partly on their functional role in the food web, expressed as their way of feeding (see for instance the phytophagous nematodes feeding from plants, the fungivorous nematodes eating fungi and the predaceous nematodes eating other nematodes).

The scheme in Figure 1 shows a very general soil food web and the different trophic levels. Much more detailed soil food web descriptions also are available, that do not only link the different trophic groups but also describe the energy flows within the system and through these flows the intensity and thus strength of the interactions that together determine the stability of the system (see e.g. de Ruiter et al., 1998).

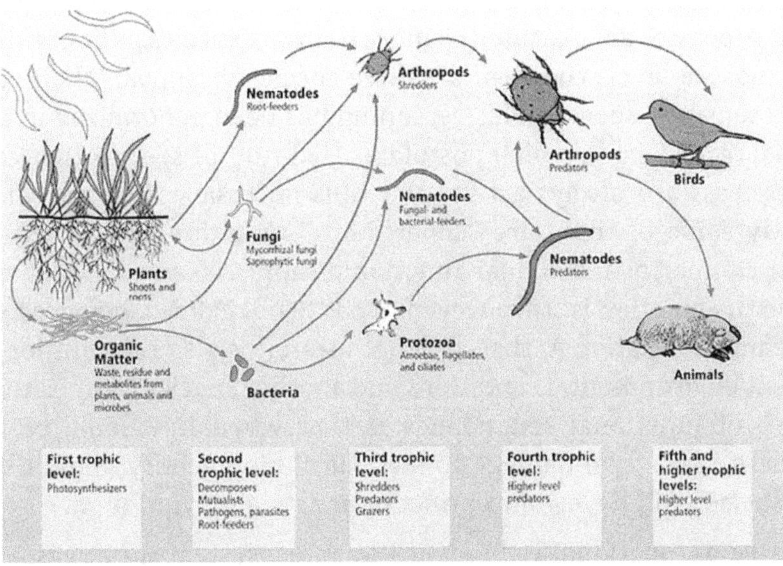

Figure 1. Simplified soil food web showing the relationships between different organisms living in and on the soils and their trophic relationships. See text for further explanation. Source: https://en.wikipedia.org/wiki/Soil_food_web#/media/File:Soil_food_webUSDA.jpg

This food web shown in Figure 1 illustrates that biodiversity not only has a structural side: the various types of species, but also a functional one: which species are involved in the execution of which process. Various functional aspects are indicated in Figure 1, e.g. photosynthesis, decomposition, predation, grazing, etc. Often functions are related to the nutritional ecology of species or to their dealings with specific nutrients (carbon, nitrogen, etc.), but sometimes also to their behaviour (e.g. litter decomposition). At the species level this functional aspect has been further elaborated in specific feeding leagues. At the ecosystem level this functional aspect has clearly been recognized in the last decades and resulted in the development of the concept of ecosystem services (see Section on Ecosystem services). However, these do not trace back to the individual species level and as such not to the functional aspect of biodiversity. Another development to be mentioned is that of trait-based approaches, which attempt to group species according to certain traits that are linked not only to exposure and sensitivity but also to their functioning. With that the trait-based approach may enable linking structural and functional biodiversity (see Section on Trait-based approaches).

Functional Redundancy

When effects of contaminants on species are compared to effects on processes, the species effects are mostly more distinct than the process effects. In other words: effects on structural diversity will be seen already at lower concentrations, and probably also sooner, than effects on functional diversity. This can be explained by the fact that processes are executed by more than one species. When with increasing chemical exposure levels the most sensitive species disappear, their role is taken over by less sensitive species. This reasoning has been generalized in the concept of "*functional redundancy*", which postulates that not all species that can perform a specific process are always active, and thus necessary, in a specific situation. Consequently some of them are "superfluous" or redundant. When a sensitive species that can perform a similar function disappears, a redundant species may take over, so the function is still covered. It has to be realized, however, that in case this is relevant in situation A, that does not mean it is also relevant for situation B with different environmental conditions and another species composition. Another consequence of functional redundancy is that when functional biodiversity is affected, there is (severe) damage to structural biodiversity: most likely several important species will have gone extinct or are strongly inhibited.

The degree of functional redundancy in an ecosystem is not easily measured. In general one may assume that the rate of ecosystem processes will increase with increasing species diversity. If such a relationship shows a curve levelling-off towards a ceiling, this represents a clear case of redundancy (Figure 2, top left graph). However, the relationship may take all kind of different forms. If ecosystem rates depend on keystone species, there may be discontinuities in the curve, related to the demise of these specific species. Figure 2 shows various theoretical shapes of the curves.

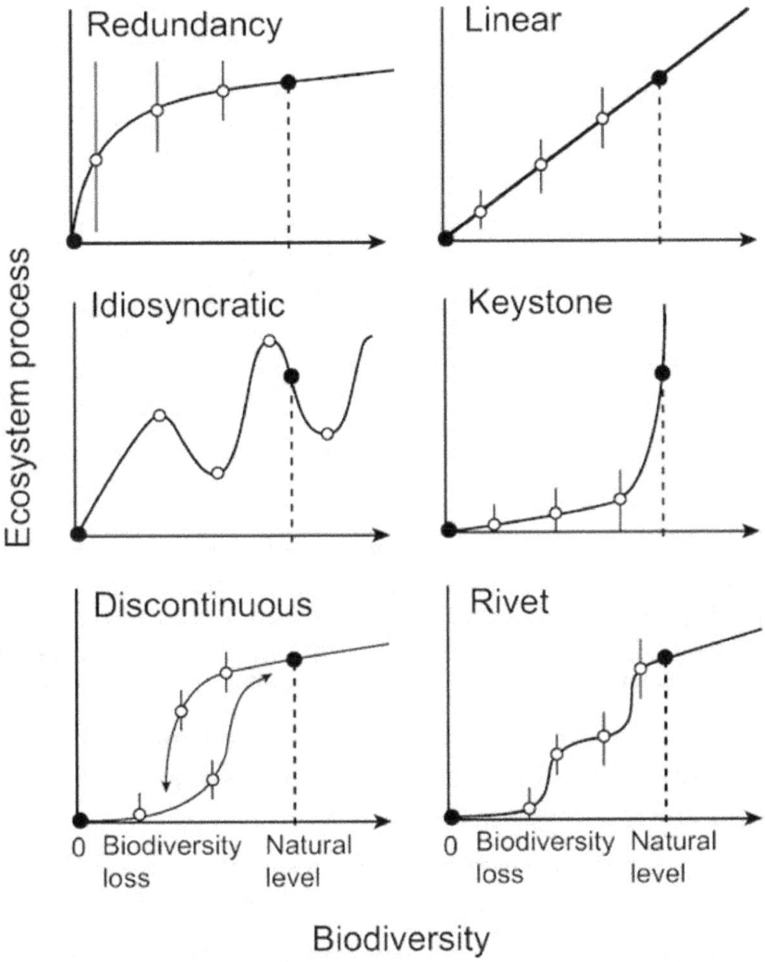

Figure 2. Six possible relationships between ecosystem process rate and species diversity in a community, slightly modified from Naeem et al. (2002).

Examples of the Relation between Structure and Functioning

Redundant species are often less efficient in performing a certain function. Tyler (1984) studied a gradient of soil copper contamination by a copper manufacturing plant in Gusum, Sweden. He observed that specific enzyme functions as well as general processes like mineralisation decreased faster than the total fungal biomass, with decreasing distance from the plant. (Figure 3b). The explanation was provided in subsequent experimental research (Rühling et al., 1984). A number of micro-fungi were isolated from the field and tested for their sensitivity to copper. The various species showed different concentration-effect relationships but all going to zero, except for two species which increased in abundance at the higher concentration so that the total biomass stayed more or less the same (Figure 3a).

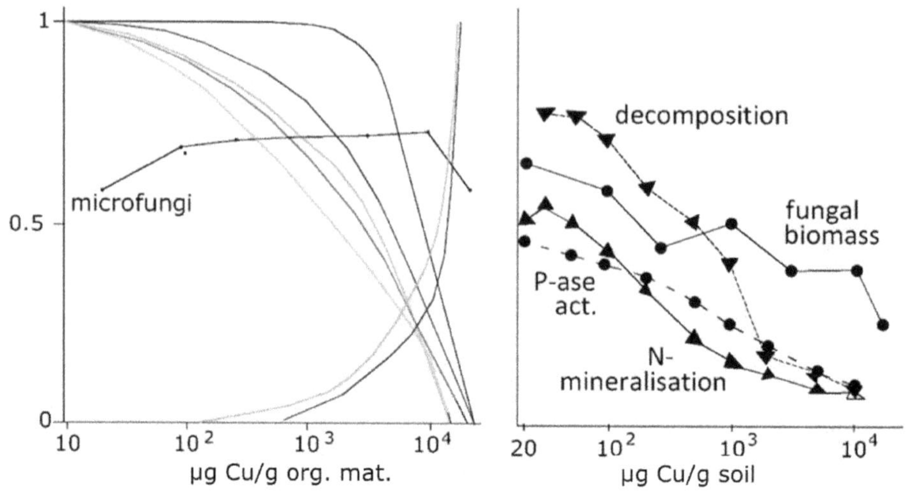

Figure 3. Left: Experimental dose responses of various microfungi species to increased levels of copper in the organic matter used as substrate and the total response for all tested species (Rühling et al., 1984). Right: Reduction of various breakdown processes and fungal biomass in the field with increasing copper levels in the soil (Tyler, 1984). Drawn by Wilma IJzerman.

Another example of the importance of a combined approach to structural and functional diversity are the different ecological types of earthworms. According to their behaviour and role they are classified as:

- anecics (deep burrowing earthworms moving up and down from deeper soil layers to the soil surface and consuming leaf litter),
- endogeics (active in the deeper laying mineral and humus soil layers and consuming fragmented litter material and humus), and
- epigeics (active in the upper soil litter layer and consuming leaf litter).

Adverse effects of contamination on the anecics will result in accumulation of litter at the soil surface, in reduced litter fragmentation by the epigeics and reduced humus-forming by the endogeics. In various studies it has been shown that these earthworms have different sensitivities for different types of pesticides. However, so far the ranking of more or less sensitive species is different for different groups of pesticides. So, there is no general relation between the function of a species e.g. surface active earthworms (epigeics) and their exposure to and sensitivity for pesticides. Nevertheless, pesticide effects on anecics generally lead to reduced litter removal, effects on endogeics result in slower fragmentation, reduced humification etc., and an effect on earthworm communities in general may hamper soil aeration and lead to soil compaction.

Another example of the impact of contaminants on functional diversity is from microbiological research on the impact of heavy metals by Doelman et al. (1994). They isolated fungi and bacteria from various heavy metal contaminated and clean areas, tested these species for their sensitivity to zinc and cadmium, and divided them accordingly in a sensitive and resistant group. As a next step they measured to what extent both groups were able to degrade and mineralize a series of organic compounds. Figure 4 shows that the sensitive group is much more effective in degrading a variety of organic compounds, whereas the heavy metal resistant microbes are far less effective. This would indicate that although functional redundancy may alleviate some of the effects that contaminants have on ecosystem functioning, the overall performance of the community generally decreases upon contaminant exposure.

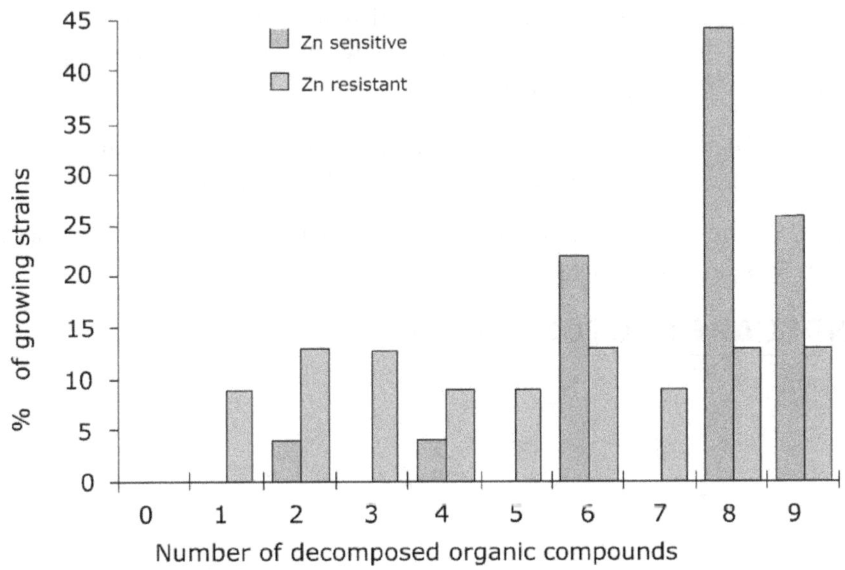

Figure 3. Decomposing capacity (measured as growth) of Zn resistant bacteria and Zn sensitive bacteria for increasing numbers of organic compounds. Redrawn from Doelman et al. (1994) by Wilma Ijzerman.

The latter example also shows that genetic diversity, expressed as the numbers of sensitive and resistant species, plays a role in the functional stability and sustainability of microbial degradation processes in the soil.

In conclusion, ecosystem services are worth studying in relation to contamination (Faber et al., 2019), but also more specific in relation to functional diversity at the species level. A promising field of research in this framework would include microorganisms in relation to the variety of degradation processes they are involved in.

References

De Ruiter, J.C., Neutel, A-M., Moore, J.C. 1995. Energetics, patterns of interaction strenghts and stability in real ecosystems. Science 269, 1257-60.

Doelman, P., Jansen, E., Michels, M., Van Til, M. (1994). Effects of heavy metals in soil on microbial diversity and activity as shown by the sensitivity-resistance index, an ecologically relevant parameter Biology and Fertility of Soils 17, 177-184.

Faber, J.H., Marshall, S., Van den Brink, P.J., Maltby, L. (2019). Priorities and opportunities in the application of the ecosystem services concept in risk assessment for chemicals in the environment. Science of the Total Environment 651, 1067-1077.

Naeem, S., Loreau, M., Inchausti, P. (2002). Biodiversity and ecosystem functioning: the emergence of a synthetic ecological framework. In: Loreau, M., Naeem, S., Inchausti, P. (Editors). Biodiversity and Ecosystem Functioning. Oxford University Press, Oxford, pp. 3-11.

Rühling, Å., Bååth, E., Nordergren, A., Söderström, B. (1984) Fungi in a metal-contaminated soil near the Gusum brass mill, Sweden. Ambio 13, 34-36.

Tyler, G. (1984) The impact of heavy metal pollution on forests: A case study of Gusum, Sweden. Ambio 13, 18-24.

5.9. LANDSCAPE ECOTOXICOLOGY

In preparation

6 Chapter RISK ASSESSMENT & REGULATION

If you want to re-use this chapter, for e.g. in your electronic learning environment, feel free to copy this url: http://maken.wikiwijs.nl/120183/6__Risk_Assessment___Regulation

6.1. INTRODUCTION: THE ESSENCE OF RISK ASSESSMENT

Introduction

We assess risks on a daily basis, although we may not always be aware of it. For example, when we cross the street, we – often implicitly – assess the benefits of crossing and weigh these against the risks of getting hit by a vehicle. If the risks are considered too high, we may decide not to cross the street, or to walk a bit further and cross at a safer spot with traffic lights.

Risk assessment is common practice for a wide range of activities in society, for example for building bridges, protection against floods, insurance against theft and accidents, and the construction of a new industrial plant. The principle is always the same: we use the available knowledge to assess the probability of potential adverse effects of an activity as good as we can. And if these risks are considered too high, we consider options to reduce or avoid the risk.

Terminology

Risk assessment of chemicals aims to describe the risks resulting from the use of chemicals in our society. In chemical risk assessment, risk is commonly defined as "the probability of an adverse effect after exposure to a chemical". This is a very practical definition that provides natural scientists and engineers the opportunity to quantify risk using "objective" scientific methods, e.g. by quantifying exposure and the likelihood of adverse effects. However, it should be noted that this definition ignores more subjective aspects of risk, typically studied by social scientists, e.g. the perceptions of people and (dealing with) knowledge gaps. This subjective dimension can be important for risk management. For example, risk managers may decide to take action if a risk is perceived as high by a substantial part of the population, even if the associated health risks have been assessed as negligible by natural scientists and engineers.

Next to the term "risk", the term "hazard" is often used. The difference between both terms is subtle, but important. A hazard is defined as the inherent capacity of a chemical (or agent/activity) to cause adverse effects. The labelling of a substance as "carcinogenic" is an example of a hazard-based action. The inherent capacity of the substance to trigger cancer, as for example demonstrated in an *in vitro* assay or an experiment with rats or mice, can be sufficient reason to label a substance as "carcinogenic". Hazard is thus independent of the actual exposure level of a chemical, whereas risk is not.

Risk assessment is closely related to risk management, i.e. the process of dealing with risks in society. Decisions to accept or reduce risks belong to the risk management domain and involve consideration of the socio-economic implications of the risks as well as the risk management options. Whereas risk assessment is typically performed by natural scientists and engineers, often referred to as "risk assessors", risk management is performed by policy makers, often referred to as "risk managers".

Risk assessment and risk management are often depicted as sequential processes, where assessment precedes management. However, strict separation of both processes is not always possible and management decisions may be needed before risks are assessed. For example, risk assessment requires political agreement on what should be protected and at what level, which is a risk management issue (see Section on Protection Goals). Similarly, the identification, description and assessment of uncertainties in the assessment is an activity that involves risk assessors as well as risk managers. Finally, it is often more efficient to define alternative management options before performing a risk assessment. This enables the assessment of the current situation and alternative management scenarios (i.e., potential solutions) in one round. The scenario with the maximum risk reduction that is also feasible in practice would then be the preferred management option. This mapping of solutions and concurrent assessment of the associated risks is also known as solution-focused risk assessment.

Risk Assessment Steps and Tiering

Chemical risk assessment is typically organized in a limited number of steps, which may vary depending on the regulatory context. Here, we distinguish four steps (Figure1):

- Problem definition (sometimes also called hazard identification), during which the scope of the assessment is defined;

- Exposure assessment, during which the extent of exposure is quantified;
- Effect assessment (sometimes also called hazard or dose-response assessment), during which the relationship between exposure and effects is established;
- Risk characterization, during which the results of the exposure and effect assessments are combined into an estimate of risk and the uncertainty of this estimate is described.

Figure 1. Risk assessment consists of four steps (problem definition, exposure assessment, effect assessment & risk characterization) and provides input for risk management.

The four risk assessment steps are explained in more detail below. The four steps are often repeated multiple times before a final conclusion on the acceptability of the risk is reached. This repetition is called tiering (Figure 2). It typically starts with a simple, conservative assessment and then, in subsequent tiers, more data are added to the assessment resulting in less conservative assumptions and risk estimates. Tiering is used to focus the available time and resources for assessing risks on those chemicals that potentially lead to unacceptable risks. Detailed data are gathered only for chemicals showing potential risk in the lower, more conservative tiers.

The order of the exposure and effect assessment steps has been a topic of debate among risk assessors and managers. Some argue that effect assessment should precede exposure assessment because effect information is independent of the exposure scenario and can be used to decide how exposure should be determined, e.g., information on toxicokinetics can be relevant to determine the exposure duration of interest. Others argue that exposure should precede effect assessment since assessing effects is expensive and unnecessary if exposure is negligible. The current consensus is that the preferred order should be determined on a case-by-case basis with parallel assessment of exposure and effects and exchange of information between the two steps as the preferred option.

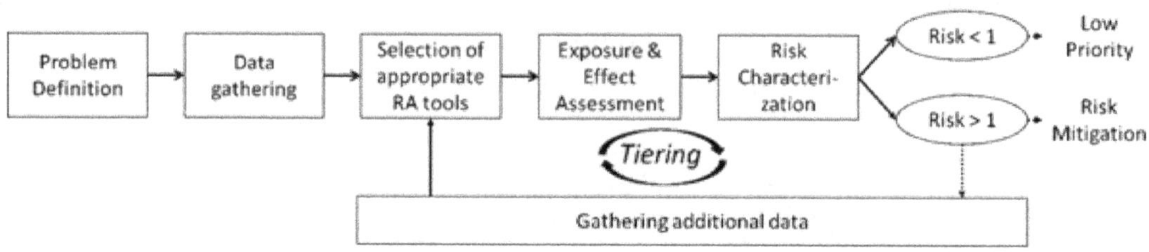

Figure 2: The principle of tiering in risk assessment. Initially risks are assessed using limited data and conservative assumptions and tools. When the predicted risk turns out unacceptable (Risk >1; see below), more data are gathered and less conservative tools are used.

Problem Definition

The scope of the assessment is determined during the problem definition phase. Questions typically answered in the problem definition include:

- What is the nature of the problem and which chemical(s) is/are involved?
- What should be protected, e.g. the general population, specific sensitive target groups, aquatic ecosystems, terrestrial ecosystems or particular species, and at what level?
- What information is already available, e.g. from previous assessments?
- What are the available resources for the assessment?
- What is the assessment order and will tiering be applied?
- What exposure routes will be considered?
- What is the timeframe of the assessment, e.g. are acute or (sub)chronic exposures considered?
- What risk metric will be used to express the risk?
- How will uncertainties be addressed?

Problem definition is not a task for risk assessors only, but should preferably be performed in a collaborative effort between risk managers, risk assessors and stakeholders. The problem definition should try to capture the worries of stakeholders as good as possible. This is not always an easy task as these worries may be very broad and sometimes also poorly articulated. Risk assessors need a clearly demarcated problem and they can only assess those aspects for which assessment methods are available. The dialogue should make transparent which aspects of the stakeholder concerns will be assessed and which not. Being transparent about this can avoid disappointments later in the process, e.g. if aspects considered important by stakeholders were not accounted for because suitable risk assessment methods were lacking. For example, if stakeholders are worried about the acute and chronic impacts of pesticide exposure, but only the acute impacts will be addressed, this should be made clear at the beginning of the assessment.

The problem definition phase results in a risk assessment plan detailing how the risks will be assessed given the available resources and within the available timeframe.

Exposure Assessment

An important aspect of exposure assessment is the determination of an exposure scenario. An exposure scenario describes the situation for which the exposure is being assessed. In some cases, this exposure situation may be evident, e.g. soil organisms living a contaminated site. However, especially when we want to assess potential risks of future substance applications, we have to come up a typical exposure scenario. Such scenarios are for example defined before a substance is allowed to be used as a food additive or before a new pesticide is allowed on the market. Exposure scenarios are often conservative, meaning that the resulting exposure estimate will be higher than the expected average exposure.

The exposure metric used to assess the risk depends on the protection target. For ecosystems, a medium concentration is often used such as the water concentration for aquatic systems, the sediment concentration for benthic systems and the soil concentration for terrestrial systems. These concentrations can either be measured or predicted using a fate model (see Section 3.8) and may or may not take into account bioavailability (see Section 3.6). For human risk assessment, the exposure metric depends on the exposure route. An air concentration is often used to cover inhalation, the average daily intake from food and water to cover oral exposure, and uptake through skin for dermal exposure. Uptake through multiple routes can also be combined in a dose metric for internal exposure, such as Area Under the Curve (AUC) in blood (see Section 6.3.1). Exposure metrics for specific wildlife species (e.g. top predators) and farm animals are often similar as those for humans. Measuring and modelling route-specific exposures is generally more complex than quantifying a simple medium concentration, because it does not only require the quantification of the substance concentration in the contact medium (e.g., concentration in drinking water), but also quantification of the contact intensity (e.g., how much water is consumed per day). Especially oral exposure can be difficult to quantify because it covers a wide range of different contact media (e.g. food products) and intensities varying from organism to organism.

Effect Assessment

The aim of the effect assessment is to estimate a reference exposure level, typically an exposure level which is expected to cause no or very limited adverse effects. There are many different types of reference levels in chemical risk assessment; each used in a different context. The most common reference level for ecological risk assessment is the Predicted No Effect Concentration (PNEC). This is the water, soil, sediment or air concentration at which no adverse effects at the ecosystem level are being expected. In human risk assessment, a myriad of different reference levels are being used, e.g. the Acceptable Daily Intake (ADI), the oral and inhalatory Reference Dose (RfD), the Derived No Effect Level (DNEL), the Point of Departure (PoD) and the Virtually Safe Dose (VSD). Each of these reference levels is used in a specific context, e.g. for addressing a specific exposure route (ADI is oral), regulatory domain (the DNEL is used in the EU for REACH, whereas the RfD is used in the USA), substance type (the VSD is typical for genotoxic carcinogens) or risk assessment method (the PoD is typical for the Margin of Safety approach).

What all reference levels have in common, is that they reflect a certain level of protection for a specific protection goal. In ecological risk assessment, the protection goal typically is the ecosystem, but it can also be a specific species or even an organism. In human risk assessment, the protection goal typically comprises all individuals of the human population. The definition of protection goals is a normative issue and it therefore is not a task of risk assessors, but of politicians. The protection levels defined by politicians typically involve a high level of abstraction, e.g. "the entire ecosystem and all individuals of the human population should be protected". Such abstract protection goals do not always match with the methods used to assess the risks.

For example, if one assumes that one molecule of a genotoxic carcinogen can trigger a deathly tumour, 100% protection for all individuals of the human population is feasible only by banning all genotoxic carcinogens (reference level = 0). Likewise, the safe concentration for an ecosystem is infinitely small if one assumes that the sensitivity of the species in the system follows a lognormal distribution which asymptotically approaches the x-axis. Hence, the abstract protection goals have to be operationalized, i.e. defined in more practical terms and matching the methods used for assessing effects. This is often done in a dialogue between scientific experts and risk managers. An example is the "one in a million lifetime risk estimated with a conservative dose response model" which is used by many (inter)national organizations as a basis for setting reference levels for genotoxic carcinogens. Likewise, the concentration at which the no observed effect concentration (NOEC) for only 5% of the species is being exceeded is often used as a basis for deriving a PNEC.

Once a protection goal has been operationalized, it must be translated into a corresponding exposure level, i.e. the reference level. This is typically done using the outcomes of (eco)toxicity tests, i.e. tests with laboratory animals such as rats, mice and dogs for human reference levels and with primary consumers, invertebrates and vertebrates for ecological reference levels. Often, the toxicity data are plotted in a graph with the exposure level on the x-axis and the effect or response level on the y-axis. A mathematical function is then fitted to the data; the so-called dose-response relationship. This dose-response relationship is subsequently used to derive an exposure level that corresponds to a predefined effect or response level. Finally, this exposure level is extrapolated to the ultimate protection goal, accounting for phenomena such as differences in sensitivity between laboratory and field conditions, between tested species and the species to be protected, and the (often very large) variability in sensitivity in the human population or the ecosystem. This extrapolation is done by dividing the exposure level that corresponds to a predefined effect or response level by one or more assessment or safety factors. These assessment factors do not have a pure scientific basis in the sense that they account for physiological differences which have actually been proven to exist. These factors also account for uncertainties in the assessment and should make sure that the derived reference level is a conservative estimate. The determination of reference levels is an art in itself and is further explained in sections 6.3.1 for human risk assessment and 6.3.2 for ecological risk assessment.

Risk Characterization

The aim of risk characterization is to come up with a risk estimate, including associated uncertainties. A comparison of the actual exposure level with the reference level provides an indication of the risk:

$$Risk\ Indicator = \frac{Exposure\ Level}{\text{Reference}\ Level}$$

If the reference level reflects the maximum safe exposure level, then the risk indicator should be below unity (1.0). A risk indicator higher than 1.0 indicates a potential risk. It is a "potential risk" because many conservative assumptions may have been made in the exposure and effect assessments. A risk indicator above 1.0 can thus lead to two different management actions: (1) if available resources (time, money) allow and the assessment was conservative, additional data may be gathered and a higher tier assessment may be performed, or (2) consideration of mitigation options to reduce the risk. Assessment of the uncertainties is very important in this phase, as it reveals how conservative the assessment was and how it can be improved by gathering additional data or applying more advanced risk assessment tools.

Risks can also be estimated using a margin-of-safety approach. In this approach, the reference level used has not yet been extrapolated from the tested species to the protection goal, e.g. by applying assessment factors for interspecies and interindividual differences in sensitivity. As such, the reference level is not a conservative estimate. In this case, the risk indicator reflects the "margin of safety" between actual exposure and the non-extrapolated reference level. Depending on the situation at hand, the margin-of-safety typically should be 100 or higher. The main difference between the traditional and the margin-of-safety approach in risk assessment is the timing for addressing the uncertainties in the effect assessment.

Reflection

Figure 3 illustrates the risk assessment paradigm using the DPSIR chain (Section 1.2). It illustrates how reference exposure levels are being derived from protection goals, i.e. the maximum level of impact that we consider acceptable. The actual exposure level is either measured or predicted using estimated emission levels and dispersion models. When measured exposure levels are used, this is called retrospective or diagnostic risk assessment: the environmental is already polluted and the assessor wants to know whether the risk is acceptable and which substances are contributing to it. When the environment is not yet polluted, predictive tools can be used. This is called prospective risk assessment: the assessor wants to know whether a projected activity will result in unacceptable risks. Even if the environment is already polluted, the risk assessor may still decide to prefer predicted over measured exposure levels, e.g. if measurements are too expensive. This is possible only if the pollution sources are well-characterized. Retrospective (diagnostic) and prospective risk assessments can differ substantially in terms of problem definitions and methods used, and are therefore discussed in separate sections in this online book.

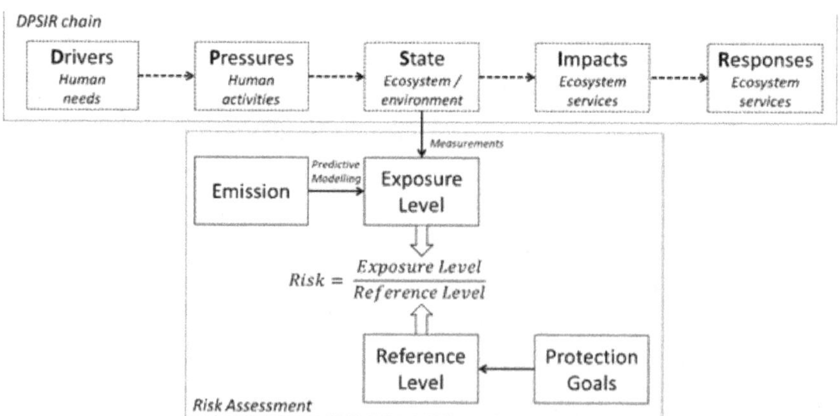

Figure 3: The risk assessment paradigm and the DPSIR chain.

Figure 3 can also be used to illustrate some important criticism on the current risk assessment paradigm, i.e. the comparison between the actual exposure level and a reference level. In current assessments, only one point of the dose-response relationship is being used to assess risk, i.e. the reference level. Critics argue that this is suboptimal and a waste of resources because the dose-response information is not used to assess the actual risk. A risk indicator with a value of 2.0 implies that the exposure is twice as high as the reference level but this does not give an indication of how many individuals or species are being affected or of the intensity of the effect. If the dose-response relationship would be used to determine the risk, this would result in a better-informed risk estimate.

A final critical remark that should be made, is the fact that risk assessment is often performed on a substance-by-substance basis. Dealing with mixtures of chemicals is difficult because each mixture has a unique composition in terms of compounds and concentration ratios between compounds. This makes it difficult to determine a reference level for mixtures. Mixture toxicology is slowly progressing and several methods are now available to address mixtures, i.e. whole mixture methods and compound-based approaches (Section 6.3.6). Another promising development are effect-based methods (Section 6.4.2). These methods do not assess risk based on chemical concentration, but on the toxicity measured in an environmental sample. In terms of DPSIR, these methods are assessing risks on the level of impacts rather than the level of state or pressures.

6.2. ECOSYSTEM SERVICES AND PROTECTION GOALS

In preparation

6.3. PREDICTIVE RISK ASSESSMENT APPROACHES AND TOOLS

6.3.1. Environmental Realistic Scenarios (PECs) – Human

Under review

6.3.2. Environmental Realistic Scenarios (PECs) – Eco

Role of Exposure Scenarios in Environmental Risk Assessment (ERA)

An exposure scenario describes the combination of circumstances needed to estimate exposure by means of models. For example, scenarios for modelling pesticides exposure can be defined as a combination of abiotic (e.g. properties and dimensions of the receiving environment and related soil, hydrological and climate characteristics) and agronomic (e.g. crops and related pesticide application) parameters that are thought to represent a realistic worst-case situation for the environmental context in which the exposure model is to be run. A scenario for exposure of aquatic organisms could be e.g. a ditch with a minimum water depth of 30 cm alongside a crop growing on a clay soil with annual applications of pesticide using a 20-year time series of weather data and including pesticide exposure via spray drift deposition and leaching from drainpipes. Such a scenario would require modelling of spray drift, leaching from drainpipes and exposure in surface water, ending up in a 20-year time series of the exposure concentration. In this chapter, we explain the use of exposure scenarios in prospective ERA by giving examples for the regulatory assessment of pesticides in particular.

Need for Defining Exposure Assessment Goals

Between about 1995 and 2001 groundwater and surface water scenarios were developed for EU pesticide registration; also referred to as the FOCUS scenarios. The European Commission indicated that these should represent 'realistic worst-cases', a political concept which leaves considerable room for scientific interpretation. Risk assessors and managers agreed that the intention was to generate 90th percentile exposure concentrations. The concept of a 90th percentile exposure concentration assumes a statistical population of concentrations and 90% of these concentrations are lower than this 90th percentile (and thus 10% are higher). This 90th percentile approach has since then been followed for most environmental exposure assessments for pesticides at EU level.

The selection of the FOCUS groundwater and surface water scenarios involved a considerable amount of expert judgement because this selection could not yet be based on well-defined GIS procedures and databases on properties of the receiving environment. The EFSA exposure assessment for soil organisms was the first environmental exposure assessment that could be based on a well-defined GIS procedure, using EU maps of parameters like soil organic matter, density of crops and weather. During the development of this exposure assessment, it became clear that the concept of a 90th percentile exposure concentration is too vague: it is essential to define also the statistical population of concentrations from which this 90th percentile is taken. Based on this insight, the EFSA Panel on Plant Protection Products and their Residues (PPR) developed the concept of the exposure assessment goals, which has become the standard within EFSA for developing regulatory exposure scenarios for pesticides.

Procedure for Defining Exposure Assessment Goals

Figure 1 shows how an exposure assessment goal for the risk assessment of aquatic organisms can be defined following this EFSA procedure. The left part specifies the temporal dimensions and the right part the spatial dimensions. In box E_1, the Ecotoxicologically Relevant type of Concentration (ERC) is defined, e.g. the freely dissolved pesticide concentration in water for pelagic organisms. In box E_2, the temporal dimension of this concentration is defined, e.g. annual peak or time-weighted average concentration for a pre-defined period. Based on these elements, the multi-year temporal population of concentrations can be generated for one single water body (E5) which would consist of e.g. 20 peak concentrations in case of a time series of 20 years. The spatial part requires definition of the type of water body (e.g. ditch, stream or pond; box E3) and the spatial dimension of this body (e.g. having a minimum water depth of 30 cm; box E4). Based on these, the spatial population of water bodies can be defined (box E6), e.g. all ditches with a minimum water depth of 30 cm alongside fields treated with the pesticide. Finally, then in box E7 the percentile combination to be taken from the spatial-temporal population of concentrations is defined. Specification of the exposure assessment goals does not only involve scientific information, but also political choices because this specification influences the strictness of the exposure assessment. For instance, in case of exposure via spray drift a minimum water depth of 30 cm in box E4 leads to about a three times lower peak concentration in the water than a minimum water depth of 10 cm.

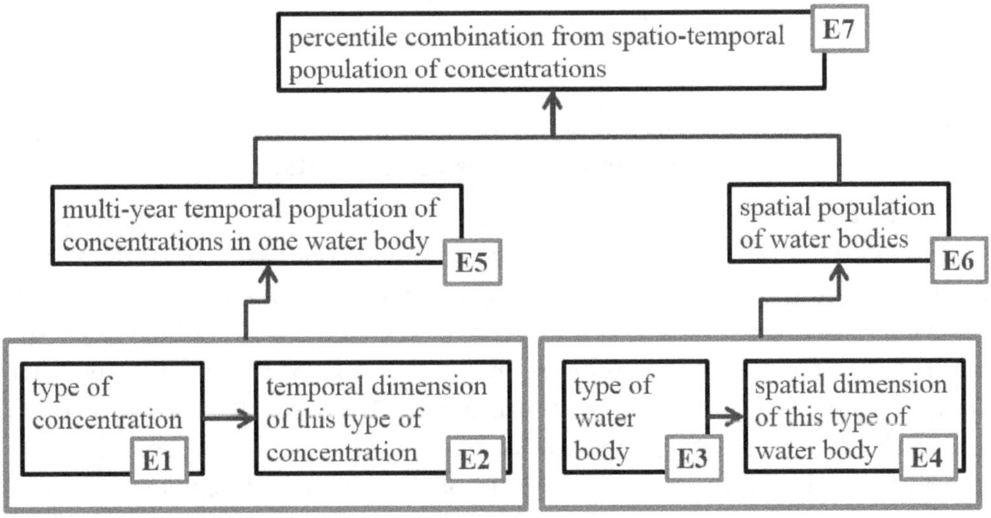

Figure 1. Scheme of the seven elements of the exposure assessment goal for aquatic organisms.

The schematic approach of Figure 1 can easily be adapted to other exposure assessment goals.

Interaction between Exposure and Effect Assessment for Organisms

Nearly all the environmental protection goals for pesticides involve assessment of risk for organisms; only groundwater and drinking water from surface water are based on a concentration of 0.1 µg/L which is not related to possible ecotoxicological effects. The risk assessment for organisms is a combination of an exposure assessment and an effect assessment as is illustrated by Figure 2.

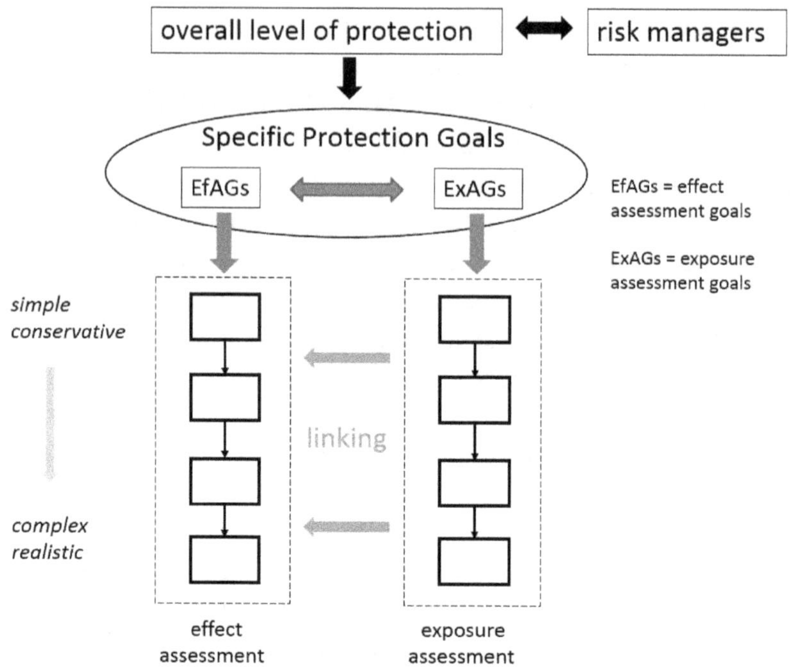

Figure 2. Overview of the risk assessment of organisms based on parallel tiered effect and exposure assessments.

Both the effect and the exposure assessment are tiered approaches with simple and conservative first tiers and less simple and more realistic higher tiers. A lower exposure tier may consist of a simple conservative scenario whereas a higher exposure tier may e.g. be based on a scenario selected using sophisticated spatial modelling. The top part of the scheme shows the link to the risk managers which are responsible for the overall level of protection. This overall level of protection is linked to the so-called Specific Protection Goals which will be explained in Section 6.5.3 and form the basis for the definition of the effect and exposure assessment goals. So the exposure assessment goals and resulting exposure scenarios should be consistent with the Specific Protection Goals (e.g. algae and fish may require different scenarios). When linking the two assessments, it has to be ensured that the type of concentration delivered by the exposure assessment is consistent with that required by the effect assessment (e.g. do not use time-weighted average concentrations in acute effect assessment). Figure 2 shows that in the assessment procedure information flows always from the exposure assessment to the effect assessment because the risk assessment conclusion is based on the effect assessment.

A relatively new development is to assess exposure and effects at the landscape level. This typically is a combination of higher-tier effect and exposure assessments. In such an approach, first the dynamics in exposure is assessed for the full landscape, and then combined with the dynamics of effects, for example based on spatially-explicit population models for species typical for that landscape. Such an approach makes a separate definition of the exposure and effect scenario redundant because this approach aims to deliver the exposure and effect assessment in an integrated way in space and time. Such an integrated approach requires the definition of "environmental scenarios". Environmental scenarios integrate both the parameters needed to define the exposure (exposure scenario) and those needed to calculate direct and indirect effects and recovery (ecological scenario) (see Figure 3). However, it will probably take at least a decade before landscape-level approaches, including agreed-upon environmental scenarios, will be implemented for regulatory use in prospective ERA.

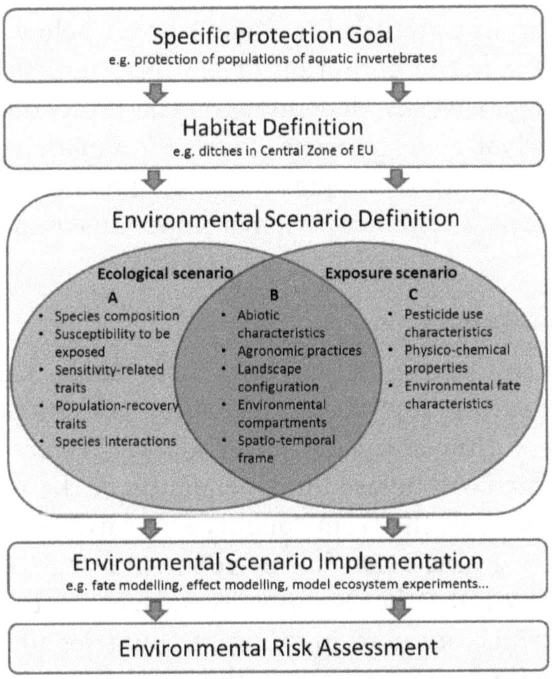

Figure 3. Conceptual framework of the role of an environmental scenario in prospective ERA (adapted after Rico et al. 2016).

References

Boesten, J.J.T.I. (2017). Conceptual considerations on exposure assessment goals for aquatic pesticide risks at EU level. Pest Management Science 74, 264-274.

Brock, T.C.M., Alix, A., Brown, C.D., et al. (2010). Linking aquatic exposure and effects: risk assessment of pesticides. SETAC Press & CRC Press, Taylor & Francis Group, Boca Raton, FL, 398 pp.

Rico, A., Van den Brink, P.J., Gylstra, R., Focks, A., Brock, T.C.M. (2016). Developing ecological scenarios for the prospective aquatic risk assessment of pesticides. Integrated Environmental Assessment and Management 12, 510-521.

6.3.3. Setting Reference Levels for Human Health Protection

In preparation

6.3.4. Setting Safe Standards

Introduction

The key question in environmental risk assessment is whether environmental exposure to chemicals leads to unacceptable risks for human and ecosystem health. This is done by comparing the measured or predicted concentrations in water, soil, sediment, or air, with a reference level. Reference levels represent a dose (intake rate) or concentration in water, soil, sediment or air below which unacceptable effects are not expected. The definition of 'no unacceptable effects' may differ between regulatory frameworks, depending on the protection goal. The focus of this section is the derivation of reference levels for aquatic ecosystems as well as for predators feeding on exposed aquatic species (secondary poisoning), but the derivation of reference values for other environmental compartments follows the same principles.

Terminology and Concepts

Various technical terms are in use as reference values, e.g. the Predicted No Effect Concentration (PNEC) for ecosystems or the Acceptable Daily Intake (ADI) for humans (Section on Human toxicology). The term "reference level" is a broad and generic term, which can be used independently of the regulatory context or protection goal. In contrast, the term "quality standard" is associated with some kind of legal status, e.g., inclusion in environmental legislation like the Water Framework Directive (WFD). Other terms exist, such as the terms 'guideline value' or 'screening level' which are used in different countries to indicate triggers for further action. While the scientific basis of these reference values may be similar, their implementation and the consequences of exceedance are not. It is therefore very important to clearly define the context of the derivation and the terminology used when deriving and publishing reference levels.

PNEC

A frequently used reference level for ecosystem protection is the Predicted No Effect Concentration (PNEC). The PNEC is the concentration below which adverse effects on the ecosystem are not expected to occur. PNECs are derived per compartment and apply to the organisms that are directly exposed. In addition, for chemicals that accumulate in prey, PNECs for secondary poisoning of predatory birds and mammals are derived. The PNEC for direct ecotoxicity is usually based on results from single species laboratory toxicity tests. In some case, data from field studies or mesocosms may be included.

A basic PNEC derivation for the aquatic compartment is based on data from single species tests with algae, water fleas and fish. Effects on the level of a complex ecosystem are not fully represented by effects on isolated individuals or populations in a laboratory set-up. However, data from laboratory tests can be used to extrapolate to the ecosystem level if it is assumed that protection of ecosystem structure ensures protection of ecosystem functioning, and that effects on ecosystem structure can be predicted from species sensitivity.

Accounting for Extrapolation Uncertainty: Assessment Factor (AF) Approach

To account for the uncertainty in the extrapolation from single species laboratory tests to effects on real life ecosystems, the lowest available test result is divided by an assessment factor (AF). In establishing the size of the AF, a number of uncertainties must be addressed to extrapolate from single-species laboratory data to a multi-species ecosystem under field conditions. These uncertainties relate to intra- and inter-laboratory variation in toxicity data, variation within and between species (biological variance), test duration and differences between the controlled laboratory set-up and the variable field situation. The value of the AF depends on the number of studies, the diversity of species for which data are available, the type and duration of the experiments, and the purpose of the reference level. Different AFs are needed for reference levels for e.g. intermittent release, short-term concentration peaks or long-term (chronic) exposure. In particular, reference levels for intermittent release and short-term exposure may be derived on the basis of acute studies, but short-term tests are less predictive for a reference level for long-term exposure and larger AFs are needed to cover this. Table 1 shows the generic AF scheme that is used to derive PNECs for long-term exposure of freshwater organisms in the context of European regulatory framework for industrial chemicals (REACH; see Section on REACH environment). This scheme is also applied for the authorisation of biocidal products, pharmaceuticals and for derivation of long-term water quality standards for freshwater under the EU Water Framework Directive. Further details on the application of this scheme, e.g., how to compare acute and chronic data and how to deal with irregular datasets, are presented in guidance documents (see suggested reading: EC, 2018; ECHA, 2008). Similar schemes exist for marine waters, sediment, and soil. However, for the latter two environmental compartments often too little experimental information is available and risk limits have to be calculated by extrapolation from aquatic data using the Equilibrium Partitioning concept. The derivation of Regulatory Acceptable Concentrations (RAC) for plant protection products (PPPs) is also based on the extrapolation of laboratory data, but follows a different approach focussing on generating data for specific taxonomic groups, taking account of the mode of action of the PPP (see suggested reading: EFSA, 2013).

Table 1. Basic assessment factor scheme used for the derivation of PNECs for freshwater ecosystems used in several European regulatory frameworks. Consult the original guidance documents for full schemes and additional information (see suggested reading: EC, 2018; ECHA, 2008).

Available data	Assessment factor
At least one short-term L(E)C50 from each of three trophic levels (fish, invertebrates (preferred *Daphnia*) and algae)	1000
One long-term EC10 or NOEC (either fish or *Daphnia*)	100
Two long-term results (e.g. EC10 or NOECs) from species representing two trophic levels (fish and/or *Daphnia* and/or algae)	50
Long-term results (e.g. EC10 or NOECs) from at least three species (normally fish, *Daphnia* and algae) representing three trophic levels	10

Application of Species Sensitivity Distribution (SSD) and Other Additional Data

The AF approach was developed to account for the uncertainty arising from extrapolation from (potentially limited) experimental datasets. If enough data are available for other species than algae, daphnids and fish, statistical methods can be applied to derive a PNEC. Within the concept of species sensitivity distribution (SSD), the distribution of the sensitivity of the tested species is used to estimate the concentration at which 5% of all species in the ecosystem is affected (HC5; see section on SSDs). When used for regulatory purposes in European regulatory frameworks, the dataset should meet certain requirements regarding the number of data points and the representation of taxa in the dataset, and an AF is applied to the HC5 to cover the remaining uncertainty from the extrapolation from lab to field.

Where available, results from semi-field experiments (mesocosms, see section on Community ecotoxicology) can also be used, either on its own or to underpin the PNEC derived from the AF or SSD approach. SSDs and mesocosm-studies are also used in the context of authorisation of PPP.

Reference Levels for Secondary Poisoning

Substances might be toxic to wildlife because of bioaccumulation in prey or a high intrinsic toxicity to birds and mammals. If this is the case, a reference level for secondary poisoning is derived for a simple food chain: water è fish or mussel è predatory bird or mammal. The toxicity data from bird or mammal tests are transformed into safe concentrations in prey. This can be done by simply recalculating concentrations in laboratory feed into concentrations in fish using default conversion factors (see e.g., ECHA, 2008). For the derivation of water quality standards under the WFD, a more sophisticated method was introduced that uses knowledge on the energy demand of predators and energy content in their food to convert laboratory data to a field situation. Also, the inclusion of other, more complex and sometimes longer food chains is possible, for which field bioaccumulation factors are used rather than laboratory derived values.

Suggested Additional Reading

EC (2018). Common Implementation Strategy for the Water Framework Directive (2000/60/EC). Guidance Document No. 27. Technical Guidance For Deriving Environmental Quality Standards. Updated version 2018. Brussels, Belgium. European Commission. https://circabc.europa.eu/ui/group/9ab5926d-bed4-4322-9aa7-9964bbe8312d/library/ba6810cd-e611-4f72-9902-f0d8867a2a6b/details

ECHA (2008). Guidance on information requirements and chemical safety assessment Chapter R.10: Characterisation of dose [concentration]-response for environment. Helsinki, Finland. European Chemicals Agency. May 2008. https://echa.europa.eu/documents/10162/13632/information_requirements_r10_en.pdf/bb902be7-a503-4ab7-9036-d866b8ddce69

EFSA (2013). Guidance on tiered risk assessment for plant protection products for aquatic organisms in edge-of-field surface waters. EFSA Journal 2013; 11(7): 3290 https://efsa.onlinelibrary.wiley.com/doi/epdf/10.2903/j.efsa.2013.3290

Traas, T.P., Van Leeuwen, C. (2007). Ecotoxicological effects. In: Van Leeuwen, C., Vermeire, T.C. (Eds.). Risk Assessment of Chemicals: an Introduction, Chapter 7. Springer.

6.3.5. Species Sensitivity Distributions (SSDs)

Introduction

The relationship between dose or concentration (X) and response (Y) is key in risk assessment of chemicals (see section on Concentration-response relationships). Such relationships are often determined in laboratory toxicity tests; a selected species is exposed under controlled conditions to a series of increasing concentrations to determine endpoints such as the No Observed Effect Concentration (NOEC), the EC50 (the Effect Concentration causing 50% effect on a studied endpoint such as growth or reproduction), or the LC50 (the Effect Concentration causing 50% lethal effects). For ecological risk assessment, multiple species are typically tested to characterise the (variation in) sensitivities across species or taxonomic groups within the ecosystem. In the mid-1980s it had been observed that–like many natural phenomena–a set of ecotoxicity endpoint data, representing effect concentrations for various species, follows a bell-shaped statistical distribution. The cumulative distribution of these data is a sigmoid (S-shaped) curve. It was recognized, that this distribution had particular utility for assessing, managing and protecting environmental quality regarding chemicals. The bell-shaped distribution was thereupon named a Species Sensitivity Distribution (SSD). Since then, the use of SSD models has grown steadily. Currently, the model is used for various purposes, providing important information for decision-making.

Below, the dual utility of SSD models for environmental protection, assessment and management are shown first. Thereupon, the derivation and use of SSD models are elaborated in a stepwise sequence.

The Dual Utility of SSD Models

A species sensitivity distribution (SSD) is a distribution describing the variance in sensitivity of multiple species exposed to a hazardous compound. The statistical distribution is often plotted using a log-scaled concentration axis (X), and a cumulative probability axis (Y, varying from 0 – 1; Figure 1).

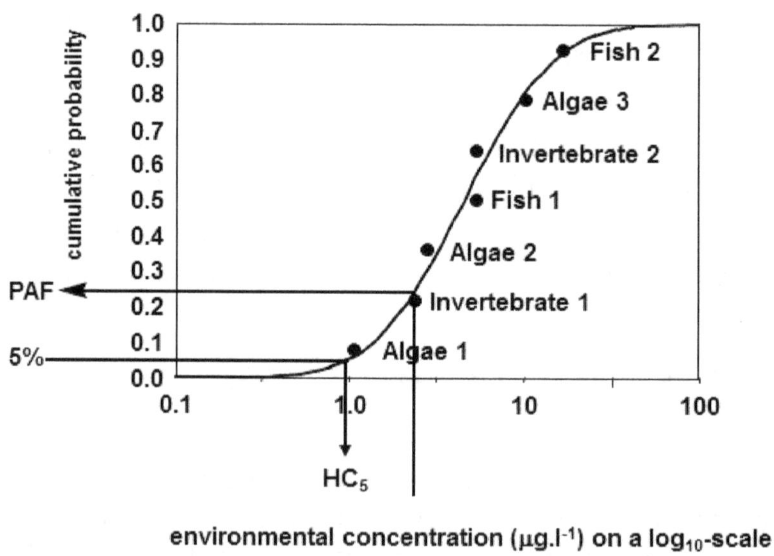

Figure 1. An species-sensitivity distribution (SSD) model, its data, and its dual use (from YàX, and from XàY). Dots represent the ecotoxicity endpoints (e.g., NOECs, EC50s, etc.) of different species.

Figure 1 shows that different species (here the dots represent 3 test data for algal species, 2 data for invertebrate species and 2 data fish species) have different sensitivities to the studied chemical. First, the ecotoxicity data are collected, and log10-transformed. Second, the data set can be visually inspected by plotting the bell-shaped distribution of the log-transformed data; deviations of the expected bell-shape can be visually identified in this step. They may originate from causes such as a low number of data points or be indicative for a selective mode of action of the toxicant, such as a high sensitivity of insects to insecticides. Third, common statistical software for deriving the two parameters of the log-normal model (the mean and the standard deviation of the ecotoxicity data) can be applied, or the SSD can be described with a dedicated software tool such as ETX (see below), including a formal evaluation of the 'goodness of fit' of the model to the data. With the estimated parameters, the fitted model can be plotted, and this is often done in the intuitively attractive form of the S-shaped cumulative distribution. This curve then serves two purposes. First, the curve can be used to derive a so-called Hazardous Concentration on the X-axis: a benchmark concentration that can be used as regulatory criterion to protect the environment (YàX). That is, chemicals with different toxicities have different SSDs, with the more hazardous compounds plotted to the left of the less hazardous compounds. By selecting a protection level on the Y-axis–representing a certain fraction of species affected, e.g. 5%–one derives the compound-specific concentration standards. Second, one can derive the fraction of tested species probably affected at an ambient concentration (XàY), which can be measured or modelled. Both uses are popular in contemporary environmental protection, risk assessment, and management.

Step 1: Ecotoxicity Data for the Derivation of an SSD Model

The SSD model for a chemical and an environmental compartment (e.g., surface water, soil or sediment) is derived based on pertinent ecotoxicity data. Those are typically extracted from scientific literature or ecotoxicity databases. Examples of such databases are the U.S. EPA's Ecotox database, the European REACH data sets and the EnviroTox database which contains quality-evaluated studies. The researcher selects the chemical and the compartment of interest, and subsequently extracts all test data for the appropriate endpoint (e.g., ECx-values). The set of test data is tabulated and ranked from most to least sensitive. Multiple data for the same species are assessed for quality and only the best data are used. If there is > 1 toxicity value for a species after the selection process, the geometric mean value is commonly derived and used. A species should only be represented once in the SSD. Data are often available for frequently tested species, representing different taxonomic and/or trophic levels. A well-known triplet of species frequently tested is "Algae, Daphnids and Fish", as this triplet is a requested minimum set for various regulations in the realm of chemical safety assessment (see section on Regulatory frameworks). For various compounds, the number of test data can be more than hundred, whilst for most compounds few data of acceptable quality may be available.

Step 2. The Derivation and Evaluation of an SSD Model

Standard statistical software (a spreadsheet program) or a dedicated software model such as ETX can be used to derive an SSD from available data. Commonly, the fit of the model to the data set is checked to avoid misinterpretation. Misfit may be shown using common statistical testing (Goodness of Fit tests) or by visual inspection and ecological interpretation of the data points. That is, when a chemical specifically affects one group of species (e.g., insects having a high sensitivity for insecticides), the user may decide the derive an SSD model for specific groups of species. In doing so, the outcome will consist of two or more SSDs for a single compound (e.g., an SSD_{Insect} and an SSD_{Other} when the compound is an insecticide, whilst the SSD_{Other} might be split further if appropriate). These may show a better goodness of fit of the model to the data, but – more importantly – they reflect the use of key knowledge of mode of action and biology prior to 'blindly' applying the model fit procedure.

Step 3a. The SSD Model used for Environmental Protection

The oldest use of the SSD model is the derivation of reference levels such as the PNEC (YàX). That is, given the policy goal to fully protect ecosystems against adverse effects of chemical exposures (see Section on Ecosystem services and protection goals), the protective use is as follows. First, the user defines which ecotoxicity data are used. In the context of environmental protection, these have often been NOECs or low-effect levels (ECx, with low x, such as EC_{10}) from chronic tests. This yields an SSD-NOEC or SSD-ECx. Then, the user selects a level of Y, that is: the maximum fraction of species for which the defined ecotoxicity endpoint (NOEC or ECx) may be exceeded, e.g., 0.05 (a fraction of 0.05 equals 5% of the species). Next, the user derives the Hazardous Concentration for 5% of the species (YàX). At the HC_5, 5% of the species are exposed to concentrations greater than their NOEC, but–which is the obverse–95% of the species are exposed to concentration less than their NOEC. It is often assumed that the structural and functional integrity of ecosystems is sufficiently protected at the HC_5 level if the SSD is based on NOECs. Therefore, many authorities use this level to derive regulatory PNECs (Predicted No Effect Concentration) or Environmental Quality Standards (EQS). The latter concepts are used as official reference levels in risk assessment, the first is the preferred abbreviation in the context of prospective chemical safety assessments, and the second is used in retrospective environmental quality assessment. Sometimes an extra assessment factor varying between 1 and 5 is applied to the HC_5 to account for remaining uncertainties. Using SSDs for a set of compounds yields a set of

HC_5 values, which–in fact–represent a relative ranking of the chemicals by their potential to cause harm.

Step 3b.The SSD Model used for Environmental Quality Assessment

The SSD model also can be used to explore how much damage is caused by environmental pollution. In this case, a predicted or measured ambient concentration is used to derive a Potentially Affected Fraction of species (PAF). The fraction ranges from 0–1 but, in practice, it is often expressed as a percentage (e.g., "24% of the species is likely affected"). According to this approach, users often have monitored or modelled exposure data from various water bodies, or soil or sediment samples, so that they can evaluate whether any of the studied samples contain a concentration higher than the regulatory reference level (previous section) and, if so how many species are affected. Evidently, the user must clearly express what type of damage is quantified, as damage estimates based on an SSD_{NOEC} or an SSD_{EC50} quantify the fractions of species affected beyond the no effect level and at the 50% effect level, respectively. This use of SSDs for a set of environmental samples yields a set of PAF values, which, in fact, represent a relative ranking of the pollution levels at the different sites in their potential to cause harm.

Practical uses of using SSD Model Outcomes

SSD model outcomes are used in various regulatory and practical contexts.

- The oldest use of the model, setting regulatory standards, has a global use. Organizations like the European Union and the OECD, as well as many countries, apply SSD models to set (regulatory) standards. Those standards are then used prospectively, to evaluate whether the planned production, use or release of a (novel) chemical is sufficiently safe. If the predicted concentration exceeds the criterion, this is interpreted as a warning. Dependent on the regulatory context, the compound may be regulated, e.g., prohibited from use, or its use limited. The data used to build SSD models for deriving regulatory standards are often chronic test data, and no or low effect endpoints. The resulting standards have been evaluated in validation studies regarding the question of sufficient protection. Note that some jurisdictions have both protective standards as well as trigger values for remediation, based on SSD modelling.
- The next use is in environmental quality assessment and management. In this case, the predicted or measured concentration of a chemical in an environmental compartment is often first compared to the reference level. This may already trigger management activities if the reference values have a regulatory status, such as a clean-up operation. The SSD may, however, be used to provide more detailed information on

expected magnitude of impact, so that environmental management can prioritize most-affected sites for earlier remediation. The use of SSDs needs be tailored to the situation. That is, if the exposure concentrations form an array close to the reference value, the use of $SSD_{NOEC\backslash s}$ is a logical step, as this ranks the site pollution levels (via the PAFs) regarding the potentially affected fraction of species experiencing slight exceedances of the no effect level. If the study area contains highly polluted sites, that approach may show that all measured concentrations are in the upper tail of the SSDNOEC sigmoid (horizontal part). In such cases, the SSD_{EC50} provides information on across-sites differences in expected impacts larger than the 50% effect level.

The third use is in Life Cycle Assessment of products. This use is comparative, so that consumers can select the most benign product, whilst producers can identify 'hot spots' of ecotoxicity in their production chains. A product often contains a suite of chemicals, so that the SSD model must be applied to all chemicals, by aggregating PAF-type outcomes over all chemicals. The model USEtox is the UN global consensus model for this application.

Today, these three forms of use of SSD models have an important role in the practice of environmental protection, assessment and management on the global scale, which relates to their intuitive meaning, their ease of use, and the availability of a vast number of ecotoxicity data in the global databases.

6.3.6. Mixtures

Under review

6.4. DIAGNOSTIC RISK ASSESSMENT APPROACHES AND TOOLS

To determine whether organisms are at risk when exposed to certain concentrations of hazardous compounds in the field, the toxicity of environmental samples can be analysed. To this purpose, several approaches and techniques have been developed, known as diagnostic tools. The tools described in Sections 6.5.1-6.5.8 have in common that they make use of living organisms to assess environmental quality. This is generally achieved by performing bioassays in which the selected test species are exposed to (concentrates or dilutions of) environmental samples after which their performance (survival, growth, reproduction etc) is measured. The species selected as test organisms for bioassays are generally the same as the ones selected for toxicity tests (see section on Selection of ecotoxicity test organisms).

Each biological organization level has its own battery of test methods. At the lowest level of biological organization, a wide variety of in vitro bioassays is available (see section Effect based monitoring: in vitro bioassays). These comprise tests based on cell lines, but also bacteria and zebra fish embryos are employed. If the response of a bioassay to an environmental sample exceeds the predefined effect-based trigger value, the response is considered to be indicative of ecological risks. Yet, the compounds causing the observed toxicity are initially unknown. However, these can subsequently be elucidated with Effect Directed Analysis (see section Effect Directed Analysis). The sample causing the effect is subjected to fractionation and the fractions are tested again. This procedure is repeated until the sample is reduced to a few individual compounds, which can then be identified allowing to confirm their contribution to the observed toxic effects.

At higher levels of biological organization, a wide variety of in vivo tests and test organisms are available, including terrestrial and aquatic plants and animals (see section Effect based monitoring: in vivo bioassays). Yet, different test species tend to respond very differently to specific toxicants and specific field collected samples. Hence, the results of a single species bioassay may not reliably reflect the risk of exposure to a specific environmental sample. To avoid over- and underestimation of environmental risks, it is therefore advisable to employ a battery of in vitro and in vivo bioassays. In a case study on effect-based water quality assessment, we showed the great potential of this approach, resulting in the ranking of sites based on ecological risks rather than on the absence or presence of compounds (see section Effect based water quality assessment).

At the higher levels of biological organization, effect-based monitoring tools include bioassays performed in mesocosms (see section Community Ecotoxicology in practice) and in the field itself, the so called in situ bioassays (see section Biomonitoring: in situ bioassays and contaminant concentrations in organisms). Cosm studies represent a bridge between the laboratory and the natural world. The originality of mesocosms is based on the combination of ecological realism, the ability to manipulate different environmental parameters and still having the opportunity to replicate treatments.

In the field, the aim of biomonitoring is the in situ assessment of environmental quality on a regular basis in time, using living organisms (see section Biomonitoring: in situ bioassays and contaminant concentrations in organisms). Organisms are collected from reference sites and exposed in cages or artificial substrates at the study sites, after which they are recollected and either their condition is analysed (in situ bioassay) or the internal concentrations of specific target compounds are measured, or both (see section Biomonitoring: in situ bioassays and contaminant concentrations in organisms).

Finally, two approaches will be introduced that aid to bridge policy goals and ecosystem responses to perturbation: the TRIAD approach and eco-epidemiology.

The TRIAD approach is a tool for site-specific ecological risk assessment, combining and integrating information on contaminant concentrations, bioassay results and ecological field inventories in a 'Weight of Evidence' approach (see section TRIAD approach). Eco-epidemiology is defined as the study of the distribution and causation of impacts of multiple stressor exposures in ecosystems, and the application of this study to reduce ecological impacts (see section Eco-epidemiology).

6.4.1. Effect-based Monitoring: In Vitro Bioassays

Effect-based Monitoring

Diagnosis of the chemical status of the environment is traditionally performed by the analytical detection of a limited number of chemical compounds. Environmental quality is then assessed by making a compound-by-compound comparison between the measured concentration of an individual contaminant and its environmental quality standard (EQS). Such a compound-by-compound approach, however, cannot cover the full spectrum of contaminants given the unknown identity of the vast majority of compounds released into the environment. It also ignores the presence of unknown breakdown products formed during degradation processes and the presence of compounds with concentration levels below the analytical limit of detection. Furthermore, it overlooks combined effects of contaminants present in the complex environmental mixture.

To overcome these shortcomings, effect-based monitoring has been proposed as a comprehensive and cost-effective, complementary strategy to chemical analysis for the diagnosis of environmental chemical quality. In effect-based monitoring the toxic potency of the complex mixture is determined as a whole by testing environmental samples in bioassays. Bioassays are defined as "biological test systems that consist of whole organisms or parts of organisms (e.g., tissues, cells, proteins), which show a measurable and potentially biologically relevant response when exposed to natural or xenobiotic compounds, or complex mixtures present in environmental samples" (Hamers et al. 2010).

Bioassays making use of whole organisms are further referred to as in vivo bioassays (in vivo means "while living"). In vivo bioassays have relatively high ecological relevance as they provide information on survival, reproduction, growth, or behaviour of the species tested. In vivo bioassays will be addressed in a separate section.

In vitro bioassays

Bioassays making use of tissues, cells, proteins are called *in vitro* bioassays (*in vitro* means "in glass"), as – in the past –they were typically performed in test tubes or petri dishes made from glass. Nowadays, *in vitro* bioassays are more often performed in microtiter wells-plates containing multiple (6, 12, 24, 48, 96, 384, or

1536) test containers (called "wells") per plate. Most *in vitro* bioassays show a very mechanism-specific response, which is for instance indicative of the inhibition of a specific enzyme or the activation of a specific molecular receptor.

In addition to the mechanism-specific information about the complex mixture present in the environment, *in vitro* bioassays have several other advantages. Small test volumes, for instance, make the *in vitro* assays suitable to test small samples. If sampling volumes are not restricted, however, the small volume of the *in vitro* bioassays allow that pre-concentrated samples (i.e. extracts) can be tested. Moreover, *in vitro* bioassays have short test durations (usually incubation periods range from 15 minutes to 48 hours) and can be performed in relatively high-throughput, i.e. multiple samples can be tested per microtiter plate experiment. Microtiter plate experiments require an easy read-out (e.g. luminescence, fluorescence, optical density), which is typically a direct measure for the toxic potency to which the bioassay was exposed. Finally, using cells or proteins for toxicity testing raises less ethical objections than the use of intact organisms as is done in *in vivo* bioassays.

Cell-based *in vitro* bioassays can make use of different types of cells. Cells can be isolated from animal tissue and be grown in medium in cell culture flasks. If a flask grows full, cells can be diluted in fresh medium and be distributed over several new flasks (i.e. "passaging"). For cells freshly isolated from animal tissue (called primary cells), however, the number of passages is limited, due to the fact that the cells have a limited number of cell doublings. Thus, the use of primary cells in environmental monitoring is not preferred, as preparation of cell cultures is time-consuming and requires the use of animals. Moreover, the composition and activity of the cells may change from batch to batch. Instead, environmental monitoring often makes use of cell lines. A cell line is a cell culture derived from a single cell which has been immortalized, allowing the cell to divide infinitely. Immortalization of cells is obtained by selecting either a (mutated) cancer cell from a donor animal or human being, or by causing a mutation in an a healthy cell after isolation using chemicals or viruses. The advantage of a cell line is that all cells are genetically identical and can be used for an indefinite number of experiments. The drawback of cell lines is that the cells are cancer cells that do not behave like a healthy cell in an intact organism. For instance, cancer cells have lost their differentiated properties and have a short cell cycle due to increased proliferation (see section on In vitro toxicity testing).

Examples of in vitro bioassays

Reporter gene bioassays are a type of *in vitro* bioassays that are frequently used in effect-based monitoring. Such bioassays make use of genetically modified cell lines or bacteria that contain an incorporated gene construct encoding for an easily measurable protein (i.e. the reporter protein). This gene construct is developed in such a way that its expression is triggered by a specific interaction between the toxic compound and a cellular receptor. If the receptor is activated by the toxic compound, transcription and translation of the reporter protein takes place, which can be easily measured as a change in colour, fluorescence, or luminescence.

The most well-known reporter gene bioassays are steroid hormone-sensitive bioassays. These bioassays are based on the principle by which steroid hormones act, i.e. activation of a receptor protein followed by translocation of the hormone-receptor complex to the nucleus where it binds to a hormone-responsive element of the DNA, thereby initiating transcription and translation of steroid hormone-dependent genes. In case of a hormone-responsive reporter gene bioassay, the reporter gene construct is also under transcriptional control of a hormone-responsive element. Activation of the steroid hormone receptor by an endocrine disrupting compound thus leads to expression of the reporter protein, which can easily be measured. Estrogenic activity, for instance, is typically measured in cell lines in which a plasmid is stably transfected into the cellular genome that encodes for the reporter protein luciferase (Figure 1).

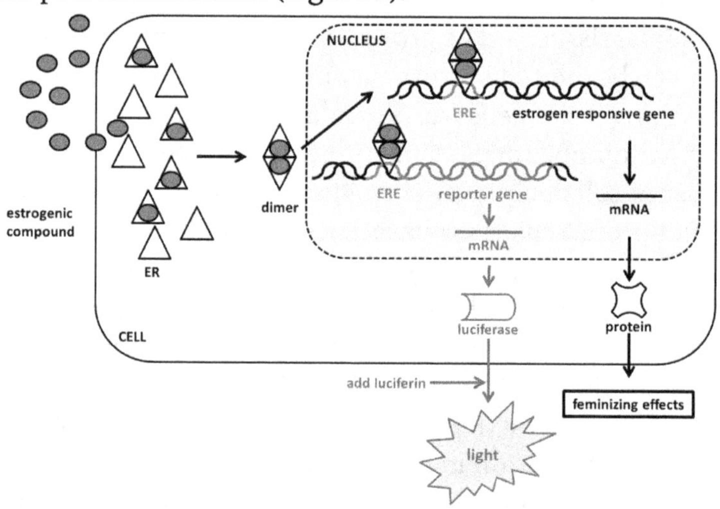

Figure 1: Principle of an estrogen responsive reporter gene assay: estrogenic compounds (red) enter the cell and activate the estrogen receptor (ER; triangle). Activated ERs form a dimer that is translocated to the nucleus where they bind to estrogen response elements (EREs). The regular subsequent pathway is indicated in black: estrogen responsive genes are transcribed into mRNA and translated into proteins that cause feminizing effects. The reporter gene pathway is indicated in blue: the reporter gene, which is also under transcriptional control of the ERE, is transcribed and translated into the reporter protein luciferase. Upon opening of the cell (lysis) and addition of the substrate luciferin and ATP as energy source, light is produced, which is a direct measure for the amount of luciferase produced, and thereby also for the estrogenic potency to which the cells were exposed.

Expression of this enzyme is under transcriptional control of an estrogen-responsive element (ERE). Upon exposure to an environmental sample, estrogenic compounds present in the sample may enter the cell and bind and activate the estrogen receptor (ER). The activated ER forms a dimer with another activated ER and is translocated to the nucleus where the dimer binds to the ERE, causing transcription and translation of the luciferase reporter gene. After 24 hours, the exposure is terminated and the amount of luciferase enzyme can be easily quantified by lysis of the cells and adding the energy source ATP and the substrate luciferin. Luciferin is hydrolysed by luciferase, which is associated with the emission of light (i.e. the same reaction as occurs in fireflies or in glowing worms). The amount of light produced by the cells is quantified in a luminometer and is a direct measure for the estrogenic potency of the complex mixture to which the cells were exposed.

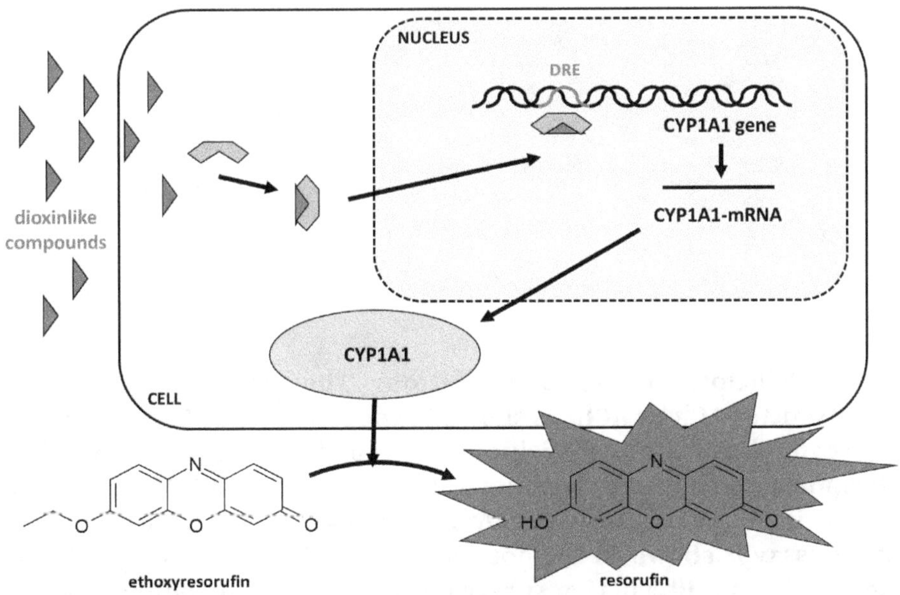

Figure 2: Simplified representation of the EROD (ethoxyresorufin-O-deethylase) assay. Dioxin-like compounds enter the hepatic cell and bind to the arylhydrocarbon receptor (AhR), which is translocated to the nucleus where it binds to dioxin-responsive elements (DREs) in the DNA. This causes transcription and translation of cytochrome P-4501A1 (CYP1A1). After 24h of incubation, the cells are lysed and the substrate ethoxyresorufin is added, which is oxidized by CYP1A1 into the fluorescent (pink) product resorufin

Another classic bioassay for the detection of dioxin-like compounds is the ethoxyresorufin-o-deethylase (EROD) bioassay (Figure 2). The EROD bioassay is an enzyme induction bioassay that makes use of a hepatic cell line (i.e. derived from liver cells). Similar as described for the estrogenic compounds, dioxin-like compounds can enter these cells upon exposure to an environmental sample, and bind and activate a receptor protein, i.c. the arylhydrocarbon receptor (AhR) (see section on Receptor interactions). The activated AhR is subsequently translocated to the nucleus where it forms a dimer with another transcription factor (ARNT) that binds to the dioxin responsive element (DRE), causing transcription and translation of dioxin-responsive genes. One of these genes encodes for CYP1A1, a typical Phase I biotransformation enzyme. Upon lysis of the cells and addition of the substrate ethoxyresorufin, CYP_1A_1 is capable of hydrolysing this substrate into ethanol and resorufin, which is a fluorescent reaction product that can be measured easily. As such, the amount of fluorescence is a direct measure for the dioxin-like potency to which the cells were exposed.

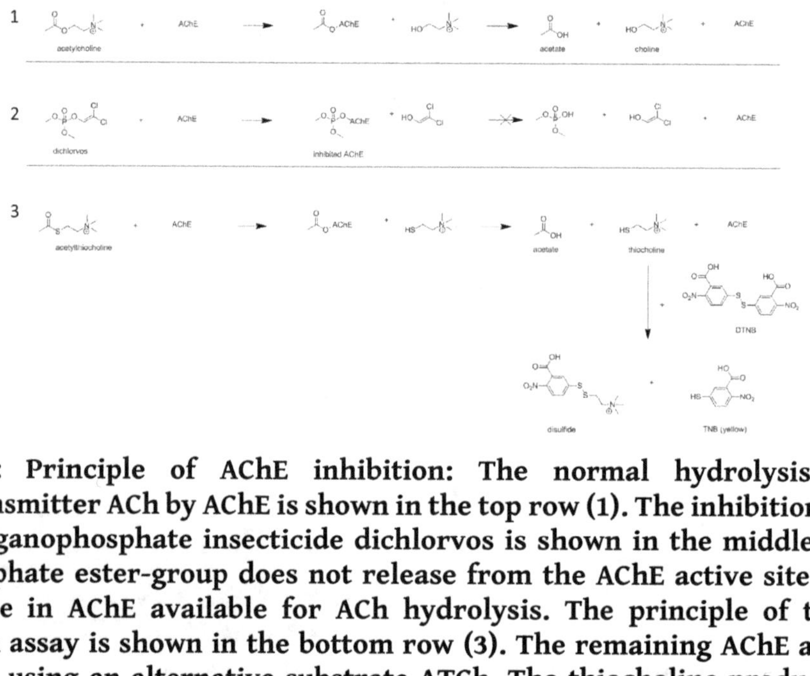

Figure 3: Principle of AChE inhibition: The normal hydrolysis of the neurotransmitter ACh by AChE is shown in the top row (1). The inhibition of AChE by the organophosphate insecticide dichlorvos is shown in the middle row (2). The phosphate ester-group does not release from the AChE active site, causing a decrease in AChE available for ACh hydrolysis. The principle of the AChE inhibition assay is shown in the bottom row (3). The remaining AChE activity is measured using an alternative substrate ATCh. The thiocholine product can be measured using the DTNB indicator (Ellman's reagent), which reacts with the thiol group, leading to a disulphide and a free TNB molecule. The yellow colour of the latter allows photometric quantification of the reaction.

Another classic bioassay is the acetylcholinesterase (AChE) inhibition assay for the detection of organophosphate and carbamate insecticides (Figure 3). By making a covalent bond to the active site of the AChE enzyme, these compounds are capable of inhibiting the hydrolysis of the neurotransmitter acetylcholine (ACh) (see section on Protein inactivation). The in vitro AChE inhibition assay makes use of the principle that AChE can also hydrolyse an alternative substrate called acetylthiocholine (ATCh) into acetic acid and thiocholine (TCh). AChE inhibition leads to a decreased rate of TCh formation, which can be measured using an indicator, called Ellman's reagent. This indicator reacts with the thiol (-SH) group of TCh, resulting in a yellow breakdown product that can easily be measured photometrically. In the bioassay, purified AChE (commercially available for instance from electric eel) is incubated with an environmental sample in the presence of ATCh and Ellman's reagent. A decrease in the rate by which the yellow reaction product is formed is a direct measure for the inhibition of the AChE activity.

Another bioassay that is used to detect mutagenic compounds in environmental samples is the Ames assay, which has been described in the section on Carcinogenicity and Genotoxicity.

Interpretation of the Toxicity Profile

In practice, multiple mechanism-specific *in vitro* bioassays are often combined into a test battery to cover the spectrum of toxicological endpoints in an (eco)system. As such, the battery can be considered as a safety net that signals the presence of toxic compounds at low concentrations. However, the question what combination of *in vitro* tests provides a sufficient level of coverage for the toxicological endpoints of concern still is an open one.

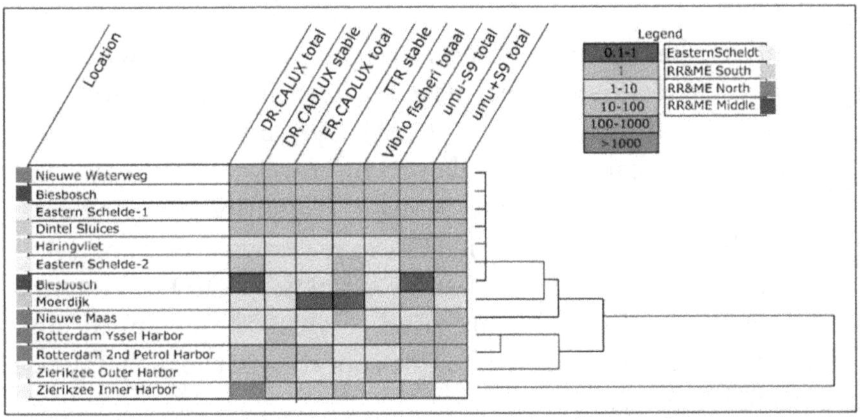

Figure 4: Example of a benchmark approach, in which toxicity profiles for sediment samples from different water systems (different colours blue) have been compared to their own reference profile (all green boxes). Colours green-yellow-orange-red indicate an increase in bioassay response. Different bioassays have been indicated at the top of the Figure. The tree-like structure (dendrogram) at the right indicates the relative distance between the different toxicity profiles. It clearly distinguishes between reference sites and clean sites on the one hand and harbour sites on the other hand. In between are samples from shipping lanes (Moerdijk and Nieuwe Maas). Zierikzee Inner Harbor is clearly a location with a deviating toxicity profile that is not similar to other harbour sites. Redrawn from Hamers et al. (2010) by Wilma IJzerman.

Still, testing an environmental sample in a battery of mechanism-specific *in vitro* bioassays yields a toxicity profile of the sample, indicating its toxic potency towards different endpoints. Two main strategies have been described to interpret *in vitro* toxicity profiles in terms of risk. In the "benchmark strategy", the toxicity profiles are compared to one or more reference profiles (Figure 4). A reference profile may be defined as the profile that is generally observed in environmental samples from locations with good chemical and/or ecological quality. The benchmark approach indicates to what extent the observed toxicity profile deviates from a toxicity profile corresponding to the desired environmental quality. It also indicates the endpoints that are most affected by the environmental sample.

In the "trigger value strategy" the response of each individual bioassay is compared to a bioassay response level at which chemicals are not expected to cause adverse effects at higher levels of biological organization. This endpoint-specific "safe" bioassay response level is called an effect-based trigger (EBT) value. The method for deriving EBT values is still under development. It can be based on different criteria, such as laboratory toxicity data, field concentrations, or EU environmental quality standards (EQS) of individual compounds, which are translated into bioassay-specific effect-levels (see section on Effect-based water quality assessment).

In addition to the benchmark and trigger value approaches focusing on environmental risk assessment, effect-based monitoring with in vitro bioassays can also be used for effect-directed analysis (EDA). EDA focuses on samples that cause bioassay responses that cannot be explained by the chemicals that were analyzed in these samples. The goal of EDA is to detect and identify emerging contaminants that are responsible for the unexplained bioassay response and are not chemically analyzed because their presence or identity is unknown. In EDA, in vitro bioassay responses to fractionated samples are used to steer the chemical identification process of unknown compounds with toxic properties in the bioassays (see section on Effect-Directed Analysis).

Further Reading:

Hamers, T., Leonards, P.E.G., Legler, J., Vethaak, A.D., Schipper, C.A. (2010). Toxicity profiling: an integrated effect-based tool for site-specific sediment quality assessment. Integrated Environmental Assessment and Management 6, 761-773

6.4.2. Effect Directed Analysis

In general, the quality of the environment may be monitored by two complementary approaches: i) quantitative chemical analysis of selected (priority) pollutants and ii) effect-based monitoring using in vitro/vivo bioassays. Compared to the more classical chemical analytical approach that has been used for decades, effect-based monitoring is currently applied in an explorative manner and has not yet matured into a routinely implemented monitoring tool that is anchored in legislation. However, in an international framework, developments to formalize the role of effect-based monitoring and to standardize the use of bioassay testing for environmental quality assessment are underway.

A weakness of the chemical approach is that because of the preselection of target compounds for quantitative analysis other compounds that are of relevance for the environmental quality may be missed. In comparison, inclusiveness is one of the advantages of effect-based monitoring: all compounds – and not only a few pre-defined ones – having a specific effect will contribute to the total, measured biological activity (see Section In vitro bioassays). In turn, the effect-based approach strongly benefits from chemical analytical support to pinpoint which compounds are responsible for the observed activity and to be able to take measures for environmental protection, e.g. the reduction of the emission or discharge of a specific toxic compound into the environment.

In Effect-Directed Analysis (EDA), the strengths of analytical chemical techniques and effect-based testing are combined with the aim to identify novel compounds that show activity in a biological analysis and that would have gone unnoticed using the chemical and the effect-based approach separately. A schematic representation of EDA is shown in Figure 1 and the various steps are described below in more detail. There is no limitation regarding the sample matrix: EDA has been applied to e.g. water, soil/sediment and biota samples. It is used for in-depth investigations at locations that are suspected to be contaminated but where the compounds responsible for the observed adverse effects are not known. In addition to environmental quality assessment, EDA is applied in the fields of food security analysis and drug discovery. In Table 1 examples of EDA studies are given.

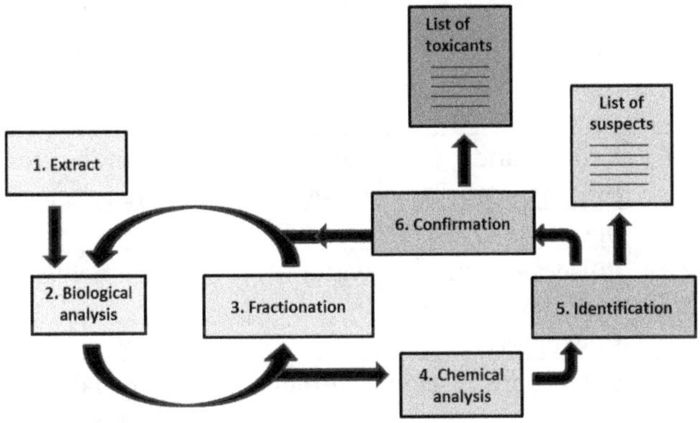

Figure 1. Schematic representation of Effect-Directed Analysis (EDA).

1. Extract

The first step is the preparation of an extract of the sample. For soil/sediment samples, a sieving step prior to the actual extraction may be necessary in order to remove large particles and obtain a sample that is well-defined in terms of particle size (e.g. <200 μm). Examples of biota samples are whole organism homogenates or parts of the organism, such as blood and liver. For the extraction of the samples, analytical techniques such as liquid/liquid or solid phase extraction are applied to concentrate the compounds of interest and to remove matrix constituents that may interfere with the later steps of the EDA.

2. Biological Analysis

The choice of endpoint to include in an EDA study is very important, as it dictates the nature of the toxicity of the compounds that may be identified (see Section on Toxicodynamics and Molecular Interaction). For application in EDA, typically in vitro bioassays that are carried out in multiwell (≥ 96 well) plates can be used, because of their low cost, high throughput and ease of use (see Section on In vitro bioassays), although sometimes in vivo assays (see Section on In vivo bioassays) are applied too.

Table 1. Examples of EDA studies, including endpoint, type of bioassay, sample matrix and compounds identified.

Endpoint	Type of bioassay	Sample matrix	Type of compounds identified
	In vitro		
Estrogenicity	Cell based reporter gene	Sediment	Endogenic hormones
Anti-androgenicity	Cell based reporter gene	Sediment	Plasticizers, organophosphorus flame retardants, synthetic fragrances
	idem	Water	Pharmaceuticals, pesticides, plasticizers, flame retardants, UV filters
Mutagenicity	Bacterial luminescence reporter strain	Water	Benzotriazoles
Thyroid hormone disruption	Radioligand binding	Polar bear plasma	Metabolites of PCBs, nonylphenols
	In vivo		
Photosystem II toxicity	Pulse Amplitude Modulation fluorometry	Water	Pesticides
Endocrine disruption	Snail reproduction	Sediment	Phthalates, synthetic fragrances, alkylphenols

3. Fractionation

Fractionation of the extract is achieved by the application of chromatography, resulting in the separation of the – in most cases – multitude of different compounds that are present in an extract of an environmental sample. Chromatographic separation is obtained after the migration of compounds through a sorbent bed. In most cases, the separation principle is based on the distribution of compounds between the liquid mobile phase and the solid stationary phase (liquid chromatography, or LC), but a chromatographic separation using the partitioning between the gas phase and a sorbent bed (gas chromatography, or GC) is also possible. At the end of the separation column, at specified time intervals fractions can be collected that are simpler in composition in comparison to the original extract: a reduction in the number of compounds per fraction is obtained. The collected fractions are tested in the bioassay and the responsive fractions are selected for further chemical analysis and identification (step 4). The time intervals for fraction collection vary between a few minutes in older applications and a few seconds in new applications of EDA, which enables fractionation directly into multiwell plates for high throughput bioassay testing. In cases where fractions are collected during time intervals in the order of minutes, the fractions are still so complex that a second round of fractionation to obtain fractions of reduced complexity is often necessary for the identification of compounds that are responsible for the observed effect (see Figure 2).

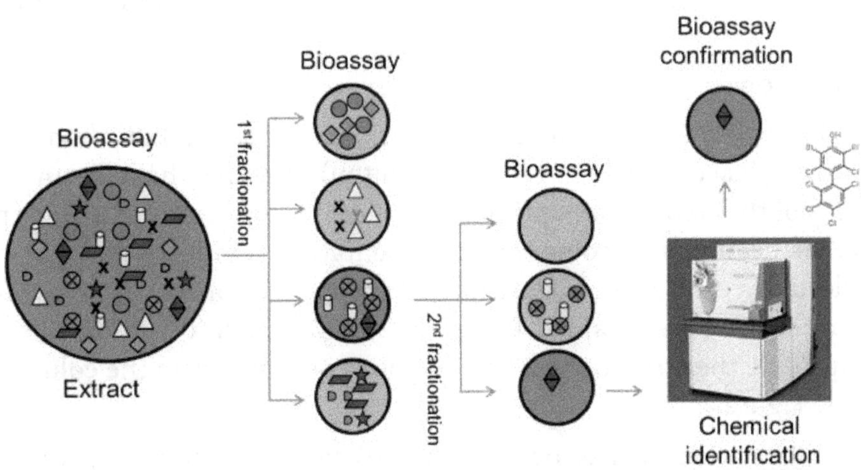

Figure 2. Schematic representation of extract fractionation and selection of fractions for further testing, identification and confirmation.

4. Chemical Analysis

Chemical analysis for the identification of the compounds that cause the effect in the bioassay is usually done by LC coupled to mass spectrometric (MS) detection. To obtain high mass accuracy that facilitates compound identification, high resolution mass spectrometry (HR-MS) is generally applied. Fractions obtained after one or two fractionation steps are injected into the LC-MS system. In studies where fractionation into multiwell plates is used (and thus small fractions in the order of microliters are collected), only one round of fractionation is applied. In these cases, identification and fraction collection can be done in parallel, using a splitter after the chromatographic column that directs part of the eluent from the column to the well plate and the other part to the MS (see Figure 3). This is called high throughput EDA (HT-EDA).

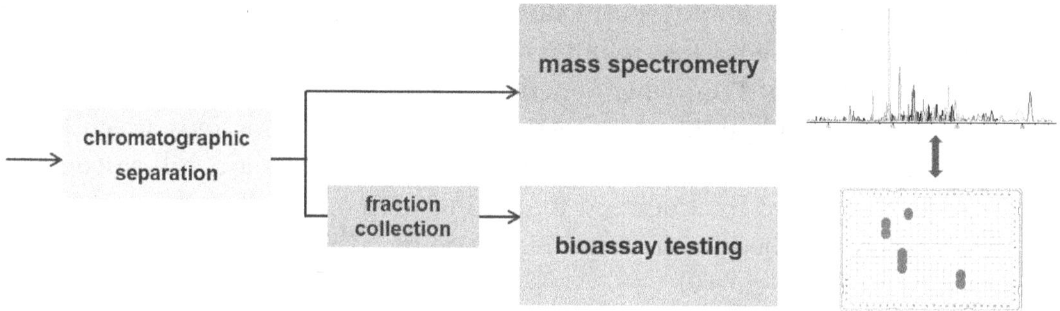

Figure 3. Schematic representation of high-throughput Effect-Directed Analysis (HT-EDA).

5. Identification

The use of HR-MS is necessary to obtain mass information to establish the molecular weight with high accuracy (e.g. 119.12423 Dalton) to derive the molecular formula (e.g. $C_6H_5N_3$) of the compound. Optimally, HR-MS instrumentation is equipped with an MS-MS mode, in which compound fragmentation is induced by collisions with other molecules, resulting in fragments that are specific for the original compound. Fragmentation spectra obtained using the MS-MS mode of HR-MS instruments help to elucidate the structure of the compounds eluting from the column, see for an example Figure 4.

Other information such as log Kow may be calculated using dedicated software packages that use elemental composition and structure as input. To aid the identification process, compound and mass spectral libraries are used as well as the more novel databases containing toxicity information (e.g. PubChem Bioassay, Toxcast). Mass spectrometry instrumentation vendor software, public/web-based databases and databases compiled in-house enable suspect screening to identify compounds that are known, e.g. because they are applied in consumer products or construction materials. When MS signals cannot be attributed to known compounds or their metabolites/transformation products, the identification approach is called non-target screening, where additional identification techniques such as Nuclear Magnetic Resonance (NMR) may aid the identification. The identification process is complicated and often time consuming, and results in a suspect list that needs to be evaluated for further confirmation of the identification.

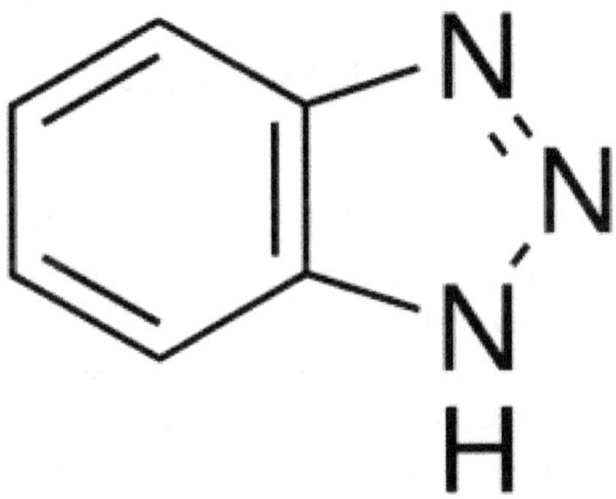

Figure 4. Example of a chemical structure corresponding to an accurate mass of 119.12423 Dalton and the corresponding molecular formula C6H5N3: 1,2,3-benzotriazole.

6. Confirmation

For an unequivocal confirmation of the identity of a tentatively identified compound, it is necessary to obtain a standard of the compound to investigate whether its analytical chemical behaviour corresponds to that of the tentatively identified compound in the environmental sample. In addition, the biological activity of the standard should be measured and compared with the earlier obtained data. In case both the chemical analysis and bioassay testing results support the identification, confirmation of compound identity is achieved.

In principle, the confirmation step of an EDA study is very straightforward, but in current practice the standards are mostly not commercially available. Dedicated synthesis is time consuming and costly, therefore the confirmation step often is a bottleneck in EDA studies.

The application of EDA is suitable for samples collected at specific locations where comprehensive chemical analysis of priority pollutants and other chemicals of relevance has been conducted already, and where ecological quality assessment has revealed that the local conditions are compromised (see other Sections on

Diagnostic risk assessment approaches and tools). Especially those samples that show a significant difference between the observed (in vitro) bioassay response and the activity that may be calculated according to the concept of Concentration Addition (see Section on Mixture Toxicity) by using the relative potencies and the concentrations of compounds active in that bioassay need a further in-depth investigation. EDA can be implemented at these 'hotspots' of environmental contamination to unravel the identity of compounds that have an effect, but that were not included in the chemical monitoring of the environmental quality. Knowledge with regard to the main drivers of toxicity at a specific location supports accurate decision making that is necessary for environmental quality protection.

6.4.3. Effect-based Monitoring: In vivo bioassays

Introduction

To determine whether organisms are at risk when exposed to hazardous compounds present at contaminated field sites, the toxicity of environmental samples can be analysed. To this purpose, several diagnostic tools have been developed, including a wide variety of in vitro, in vivo and in situ bioassays (see sections on In vitro bioassays and on In situ bioassays). In vivo bioassays make use of whole organisms (in vivo means "while living"). The species selected as test organisms for in vivo bioassays are generally the same as the ones selected for single species toxicity tests (see sections 4.3.4, 4.3.5, 4.3.6 and 4.3.7 on the Selection of ecotoxicity test organisms). Likewise, also the endpoints measured in in vivo bioassays are the same as those in single species ecotoxicity tests (see section on Endpoints). In vivo bioassays therefore have a relatively high ecological relevance, as they provide information on the survival, reproduction, growth, or behaviour of the species tested. A major difference between toxicity tests and bioassays is the selection of the controls. In laboratory toxicity experiments the controls consist of non-spiked 'clean' test medium (see section on Concentration response relationships). In bioassays the choice of the controls is more complicated though. Non-treated test medium may be incorporated as a control in bioassays to check for the health and quality of the test organisms.

But control media, like standard test water or artificial soil and sediment may differ in numerous aspects from natural environmental samples. Therefore, the control should preferably be a test medium that has exactly the same physicochemical properties as the contaminated sample, except for the chemical pollutants being present. This ideal situation, however, hardly ever exists. Hence, it is recommended to also incorporate environmental samples from less or non-contaminated reference sites into the bioassay and to compare the response of the organism to samples from contaminated sites with those from reference sites.

Alternatively, controls can be selected as the least contaminated environmental samples from a gradient of pollution or as the dilution required to obtain no effect. As dilution medium artificial control medium can be used or medium from a reference site.

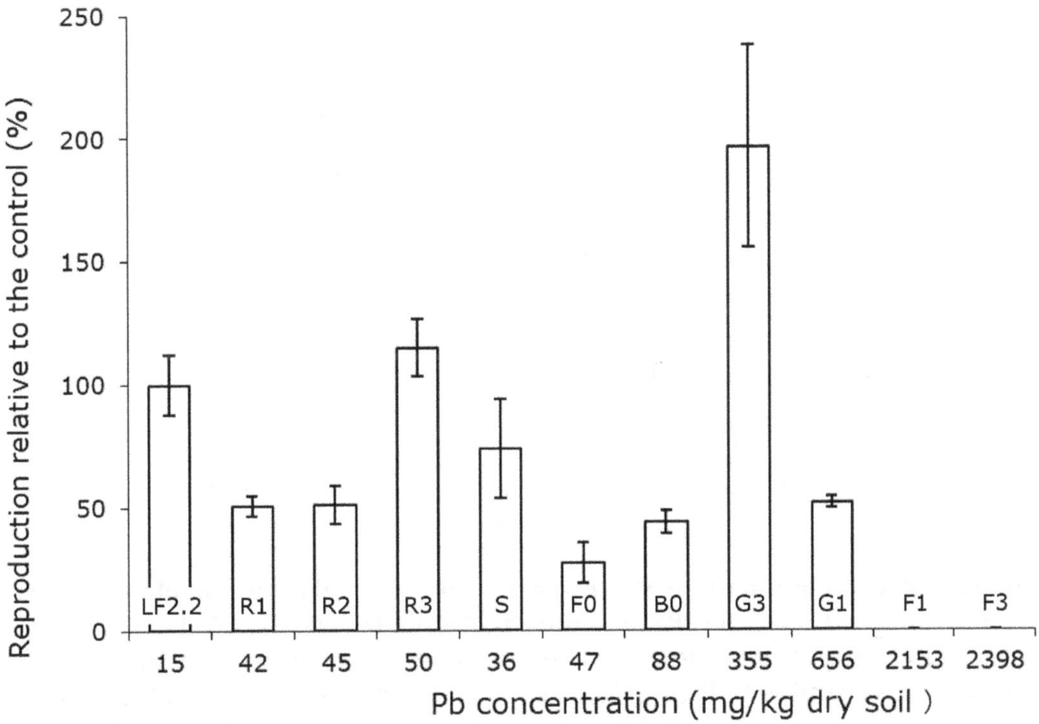

The Most Commonly used in vivo Bioassays

For the soil compartment, the earthworms *Eisenia fetida, E. andrei* and *Lumbricus rubellus*, the enchytraeid *Enchytraeus crypticus* and the collembolan *Folsomia candida* are most frequently selected as *in vivo* bioassay test organisms. An example of employing earthworms to assess the ecotoxicological effects of Pb contaminated soils is given in Figure 1. The figure shows the total Pb concentrations in different field soils taken from a soccer field (S), a bullet plot (B), grassland (G1, G3) and a forest (F1-F3) site near a shooting range. The pH of the grassland soils was near neutral (pHCaCl2 = 6.5-6.8), but the pH was rather low (3.2-3.7) for all other field sites. Earthworms exposed to these soils showed a significantly reduced reproductive output (Figure 1) at the most contaminated sites. At the less contaminated sites, earthworm responses were also affected by the difference in soil pH, leading to low juvenile numbers in the acid soil F0 but high numbers in the near neutral reference R3 and the field soil G3. In fact, earthworm reproduction was highest in the latter soil, even though it did contain an elevated concentration of 355 ± 54 mg Pb/kg dry soil. In soil G1, which contained almost twice as much Pb (656 ± 60 mg Pb/kg dry soil), reproduction was much lower and also reduced compared to the control, suggesting the presence of additional, unknown stressor (Luo et al., 2014).

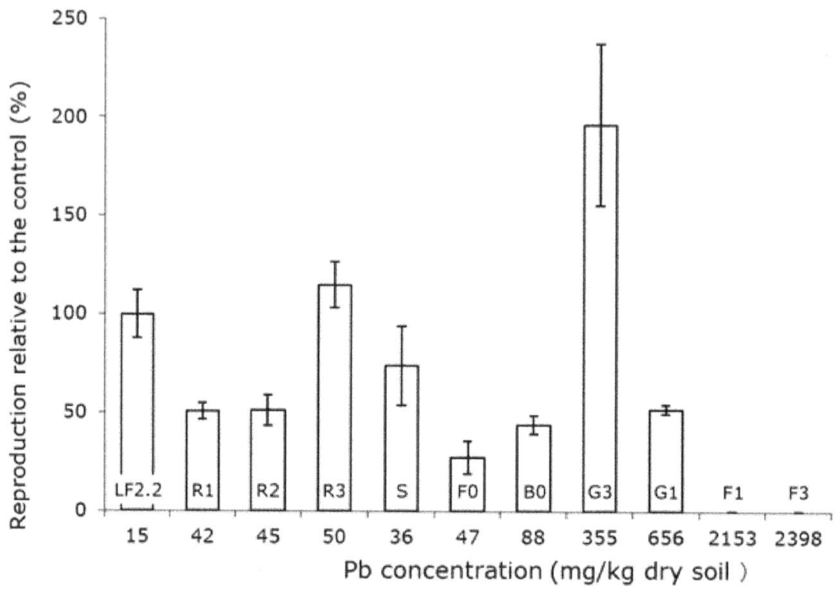

Figure 1. Reproduction of the earthworm Eisenia andrei after 4 weeks of exposure to control soils (LF2.2, R1, R2, R3) and field soils (S, B0, G1, G3, F0, F1, F3) from a Pb pollution gradient near a shooting range. Shown are the mean relative numbers of juveniles ± SD (n=4-5), compared to the control Lufa 2.2 (LF2.2) soil, as a function of average total Pb concentrations in the soils. Data from Luo et al. (2014).

For water, predominantly daphnids are employed, mainly Daphnia magna, but sometimes also other daphnid species or other aquatic invertebrates are selected. Also bioassays with several primary producers are available. An example of exposing daphnids (Chydorus sphaericus) to water samples is shown in Figure 2. The bars show the toxicity of the water samples and the diamonds the concentrations of cholinesterase inhibitors, as a proxy for the presence of insecticides. The toxicity of the water samples was higher when also the concentrations of insecticides were higher. Hence, in this case, the observed toxicity is well explained by the measured compounds. Yet, it has to be realized that this is an exception rather than a rule, since mostly a large portion of the toxic effects observed in surface waters cannot be attributed to compounds measured by water authorities and moreover, interactions are also not covered by such analytical data (see section on Effect based water quality assessment).

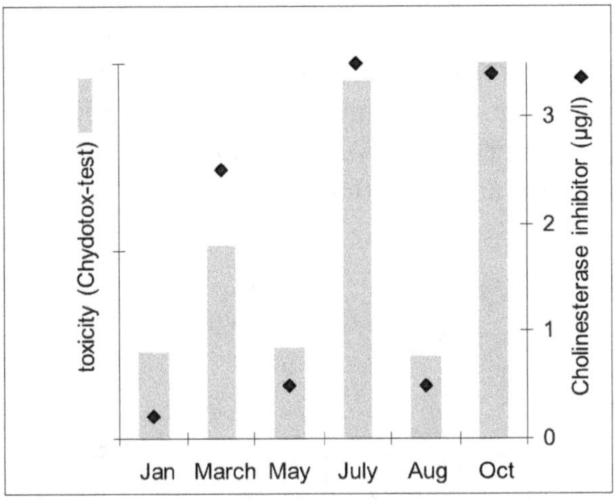

Figure 2. Toxicity of water samples to daphnids (Chydorus sphaericus)(bars) and the concentrations of cholinesterase inhibitors, as a proxy for the presence of insecticides (diamonds). Data from Pieters et al. (2008).

For sediments, oligochaetes and chironomids are selected as test organisms, but sometimes also rooting macrophytes and benthic diatoms. An example of exposing chironomids (Chironomus riparius) to contaminated sediments is shown in Figure 3. Whole sediment bioassays with chironomids allow the assessment of sensitive species-specific sublethal endpoints (see section on Chronic toxicity), in this case emergence. Figure 3 shows that more chironomids emerged on the reference sediment than on the contaminated sediment and that the chironomids on the reference sediment also emerged faster than on the contaminated sediment.

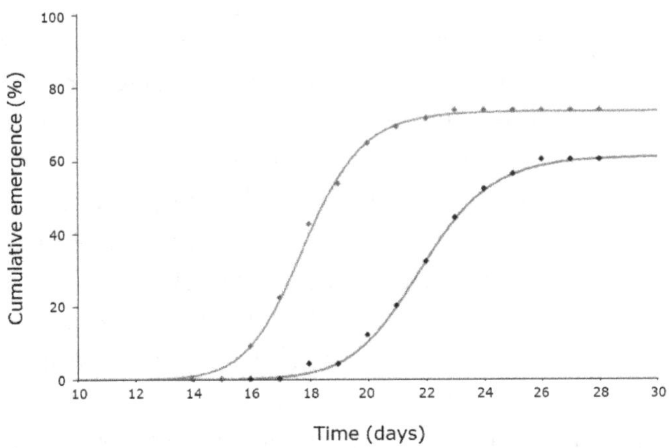

Figure 3. Emergence of chironomids (Chironomus riparius) on a reference (blue line) and a contaminated sediment (red line). Data from Nienke Wieringa.

For sediment, also benthic diatoms are selected as in vivo bioassay test organisms. Figure 4 shows the growth of the benthic diatom Nitzschia perminuta after 4 days of exposure to 160 sediment samples. The dotted line represents control growth. The growth of the diatoms ranged from higher than the controls to no growth at all, raising the question which deviation from the control should be considered a significant adverse effect.

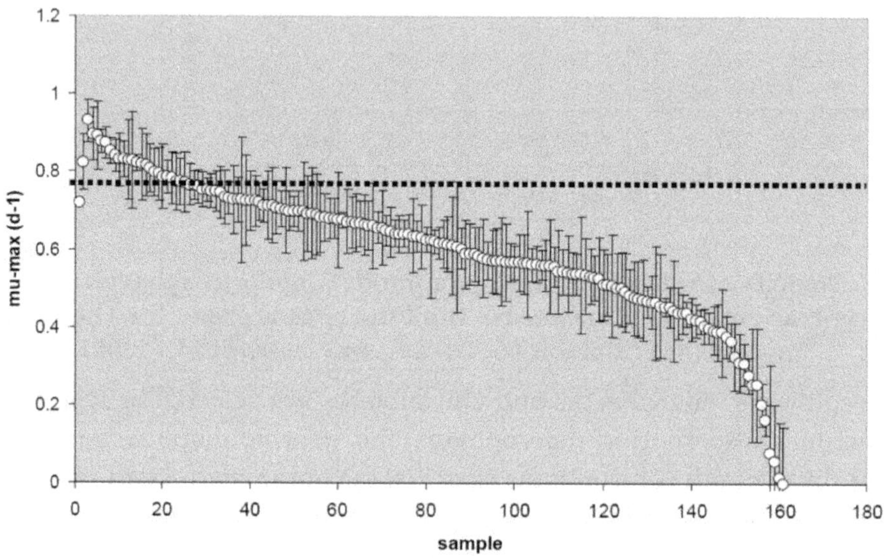

Figure 4. Growth of the benthic diatom Nitzschia perminuta after 4 days of exposure to 160 sediment samples. The dotted line represents control growth. Data from Harm van der Geest.

In vivo Bioassay Batteries

Environmental quality assessments are often performed with a single test species, like the four examples given above. Yet, toxicity is species and compound specific and this may therefore result in large margins of uncertainty in the environmental quality assessments, consequently leading to over- or underestimation of environmental risks. Obvious examples include the presence of herbicides that only would induce responses in bioassays with primary producers and the other way around, the presence of insecticides that induces strong effects on insects and to a lesser extent on other animals, but that would be completely overlooked in bioassays with primary producers. To reduce these uncertainties and to increase ecological relevance it is therefore advised to incorporate more test species belonging to different taxa in a bioassay battery (see section on Effect based water quality assessment).

References

Luo, W., Verweij, R.A., Van Gestel, C.A.M. (2014). Determining the bioavailability and toxicity of lead to earthworms in shooting range soils using a combination of physicochemical and biological assays. Environmental Pollution 185, 1-9.

Pieters, B.J., Bosman-Meijerman, D., Steenbergen, E., Van den Brandhof, E.-J., Van Beelen, P., Van der Grinten, E., Verweij, W., Kraak, M.H.S. (2008). Ecological quality assessment of Dutch surface waters using a new bioassay with the cladoceran *Chydorus sphaericus*. Proceedings Netherlands Entomological Society Meetings 19, 157-164.

6.4.4. Effect Based Water Quality Assessment

Introduction

Traditional chemical water quality assessment is based on the analysis of a list of a varying, but limited number of priority substances. Nowadays, the use of many of these compounds is restricted or banned, and concentrations of priority substances in surface waters are therefore decreasing. At the same time, industries have switched to a plethora of alternative compounds, which may enter the aquatic environment, seriously impacting water quality. Hence, priority substances lists are outdated, as the selected compounds are frequently absent, while many compounds with higher relevance are not listed as priority substances. Consequently, a large portion of toxic effects observed in surface waters cannot be attributed to compounds measured by water authorities, and toxic risks to freshwater ecosystems are thus caused by mixtures of a myriad of (un)known, unregulated compounds. Understanding of these risks requires a paradigm shift towards new monitoring methods that do not depend on chemical analysis of priority substances solely, but consider the biological effects of the entire micropollutant mixture first. Therefore, there is a need for effect-based monitoring strategies that employ bioassays to identify environmental risk. Responses in bioassays are caused by all bioavailable (un)known compounds and their metabolites, whether or not they are listed as priority substances.

Table 1. Example of the bioassay battery employed by the SIMONI approach of Van der Oost et al. (2017) that can be applied to assess surface water toxicity. Effect-based trigger values (EBT) were previously defined by Escher et al. (2018) (PAH, anti-AR and ER CALUX) and Van der Oost et al. (2017).

	Bioassay	Endpoint	Reference compound	EBT	Unit
in situ	Daphnia in situ	Mortality	n/a	20	% mortality
in vivo	Daphniatox	Mortality	n/a	0.05	TU
	Algatox	Algal growth inhibition	n/a	0.05	TU
	Microtox	Luminescence inhibition	n/a	0.05	TU
in vitro CALUX	cytotox	Cytotoxicity	n/a	0.05	TU
	DR	Dioxin (-like) activity	2,3,7,8-TCDD	50	pg TEQ/L
	PAH	PAH activity	benzo(a)pyrene	6.21	ng BapEQ/L
	PPARγ	Lipid metabolism inhibition	rosiglitazone	10	ng RosEQ/L
	Nrf2	Oxidative stress	curcumin	10	µg CurEQ/L
	PXR	Toxic compound metabolism	nicardipine	3	µg NicEQ/L
	p53 -S9	Genotoxicity	n/a	0.005	TU
	p53 +S9	Genotoxicity (after metabolism)	n/a	0.005	TU
	ER	Estrogenic activity	17ß-estradiol	0.1	ng EEQ/L
	anti-AR	Antiandrogenic activity	flutamide	14.4	µg FluEQ/L
	GR	Glucocorticoid activity	dexamethasone	100	ng DexEQ/L
izn vitro antibiotics	T	Bacterial growth inhibition (Tetracyclines)	oxytetracycline	250	ng OxyEQ/L
	Q	Bacterial growth inhibition (Quinolones)	flumequine	100	ng FlqEQ/L
	B+M	Bacterial growth inhibition (β-lactams and Macrolides)	penicillin G	50	ng PenEQ/L
	S	Bacterial growth inhibition (Sulfonamides)	sulfamethoxazole	100	ng SulEQ/L
	A	Bacterial growth inhibition (Aminoglycosides)	neomycin	500	ng NeoEQ/L

Bioassay Battery

The regular application of effect-based monitoring largely relies on the ease of use, endpoint specificity, costs and size of the used bioassays, as well as on the ability to interpret the measured responses. To ensure sensitivity to a wide range of potential stressors, while still providing specific endpoint sensitivity, a successful bioassay battery like the example given in Table 1 can include in situ whole organism assays (see section on Biomonitoring and in situ bioassays), and should include laboratory-based whole-organism in vivo (see section on In vivo bioassays) and mechanism-specific in vitro assays (see section on In vitro bioassays). Adverse effects in the whole-organism bioassays point to general toxic pressure and represent a high ecological relevance. In vitro or small-scale in vivo assays with specific drivers of adverse effects allow for focused identification and subsequent confirmation of (groups of) toxic compounds with specific modes of action. Bioassay selection can also be based on the Adverse Outcome Pathways (AOP) (see section on Adverse Outcome Pathways) concept that describes relationships between molecular initiating events and adverse outcomes. Combining different types of bioassays ranging from whole organism tests to in vitro assays targeting specific modes of action can thus greatly aid in narrowing down the number of candidate compound(s) that cause environmental risks. For example, if bioanalytical responses at a higher organisational level are observed (the orange and black pathways in Figure 1), responses in specific molecular pathways (blue, green, grey and red in Figure 1) can help to identify certain (groups of) compounds responsible for the observed effects.

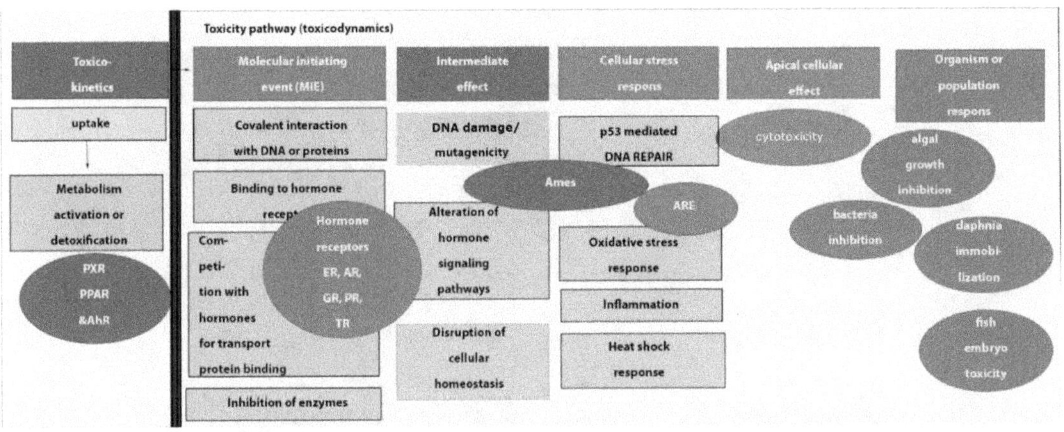

Figure 1. From toxicokinetics via molecular responses to population responses. Redrawn from Escher et al. (2018) by Wilma IJzerman.

Toxic and Bioanalytical Equivalent Concentrations

The severity of the adverse effect of an environmental sample in a bioassay is expressed as toxic equivalent (TEQ) concentrations for toxicity in in vivo assays or as bioanalytical equivalent (BEQ) concentrations for responses in in vitro bioassays. The toxic equivalent concentrations and bioanalytical equivalent concentrations represent the joint toxic potency of all unknown chemicals present in the sample that have the same mode of action (see section on Toxicodynamics and molecular interactions) as the reference compound and act concentration-additively (see section on Mixture toxicity). The toxic equivalent concentrations and bioanalytical equivalent concentrations are expressed as the concentration of a reference compound that causes an effect equal to the entire mixture of compounds present in an environmental sample. Figure 2 depicts a typical dose-response curve for a molecular in vitro assay that is indicative of the presence of compounds with a specific mode of action targeted by this in vitro assay. A specific water sample induced an effect of 38% in this assay, equivalent to the effect of approximately 0.02 nM bioanalytical equivalents.

Effect-based Trigger Values

The identification of ecological risks from bioassay battery responses follows from the comparison of bioanalytical signals to previously determined thresholds, defined as effect-based trigger values (EBT), that should differentiate between acceptable and poor water quality. Since bioassays potentially respond to the mixture of all compounds present in a sample, effect-based trigger values are expressed as toxic or bioanalytical equivalents of concentrations of model compounds for the respective bioassay (Table 1).

	Location					
Bioassay	**HN2**	**WN3**	**WN6**	**DO3**	**PM2**	**BD3**
Daphnia in situ	4.25	0.50	2.50	0.50	0.00	0.00
Microtox	0.00	0.00	0.04	0.02	0.00	0.00
Algatox	0.00	0.00	0.00	0.01	0.00	0.00
Daphniatox	0.00	0.00	0.00	0.02	0.04	0.00
cytotox CALUX polar	2.68	0.18	0.60	0.60	0.79	0.18
cytotox CALUX non polar	0.01	0.00	0.00	0.00	0.00	0.00
ER CALUX	1.43	5.71	0.63	0.71	0.32	0.00
a-AR CALUX	10.00	2.05	1.17	1.31	0.54	0.74
GR CALUX	0.00	0.00	0.00	0.49	0.00	0.00
Antibiotics	0.04	0.14	0.01	0.62	0.02	0.00
DR CALUX	0.65	0.83	0.41	0.26	0.12	0.15
PAH CALUX	0.24	0.48	0.26	0.15	0.07	0.06
PPAR CALUX	0.00	0.00	0.00	0.00	0.00	0.00
Nrf2 CALUX	1.52	0.56	0.51	0.29	0.53	0.30

Figure 2. Dose response relationship for a reference compound in an in vitro bioassay. The blue lines show that a specific water sample induced an effect of 38%, representing approximately 0.02 nM bioanalytical equivalents.

Ranking of Contaminated Sites based on Effect-based Risk Assessment

Once the toxic potency of a sample in a bioassay is expressed as toxic equivalent concentrations or bioanalytical equivalent concentrations, this response can be compared to the effect-based trigger value for that assay, thus determining whether or not there is a potential ecological risk from contaminants in the investigated water sample. The ecotoxicity profiles of the surface water samples generated by a bioassay battery allow for calculation and ranking of a cumulative ecological risk for the selected locations. In the example given in Figure 3, water samples of six locations were subjected to the SIMONI bioassay battery of Van der Oost et al. (2017), consisting of 17 *in situ*, *in vivo* and *in vitro* bioassays. Per site and per bioassay the response is compared to the corresponding effect-based trigger value and classified as 'no response' (green), 'response below the effect-based trigger value' (yellow) or 'response above the effect-based trigger value' (orange). Next, the cumulative ecological risk per location is calculated.

The resulting integrated ecological risk score allows ranking of the selected sites based on the presence of ecotoxicological risks rather than on the presence of a limited number of target compounds. This in turn permits water authorities to invest money where it matters most: identification of compounds causing adverse effects at locations with indicated ecotoxicological risks. Initially, the compounds causing the observed effect-based trigger value exceedance will not be known, however, this can subsequently be elucidated with targeted or non-target chemical analysis, which will only be necessary at locations with indicated ecological risks. A potential follow-up step could be to investigate the drivers of the observed effects by means of effect-directed analysis (see section on Effect-directed analysis).

	Location					
Bioassay	**HN2**	**WN3**	**WN6**	**DO3**	**PM2**	**BD3**
Daphnia in situ	4.25	0.50	2.50	0.50	0.00	0.00
Microtox	0.00	0.00	0.04	0.02	0.00	0.00
Algatox	0.00	0.00	0.00	0.01	0.00	0.00
Daphniatox	0.00	0.00	0.00	0.02	0.04	0.00
cytotox CALUX polar	2.68	0.18	0.60	0.60	0.79	0.18
cytotox CALUX non polar	0.01	0.00	0.00	0.00	0.00	0.00
ER CALUX	1.43	5.71	0.63	0.71	0.32	0.00
a-AR CALUX	10.00	2.05	1.17	1.31	0.54	0.74
GR CALUX	0.00	0.00	0.00	0.49	0.00	0.00
Antibiotics	0.04	0.14	0.01	0.62	0.02	0.00
DR CALUX	0.65	0.83	0.41	0.26	0.12	0.15
PAH CALUX	0.24	0.48	0.26	0.15	0.07	0.06
PPAR CALUX	0.00	0.00	0.00	0.00	0.00	0.00
Nrf2 CALUX	1.52	0.56	0.51	0.29	0.53	0.30
PRX CALUX	1.25	1.88	1.16	0.61	0.64	0.81
p53 CALUX -S9	0.06	0.03	0.03	0.01	0.02	0.00
p53 CALUX +S9	0.00	0.00	0.00	0.00	0.00	0.00
Integrated risk score	**2.63**	**1.29**	**0.98**	**0.62**	**0.31**	**0.22**

Figure 3. Heat map showing the response of 17 in situ, in vivo and in vitro bioassays to six surface water samples. The integrated risk score (SIMONI Risk Indication; Van der Oost et al., 2017) is classified as 'low risk' (green), 'potential risk' (orange) or 'risk' (red).

References

Escher, B. I., Aït-Aïssa, S., Behnisch, P. A., Brack, W., Brion, F., Brouwer, A., et al. (2018). Effect-based trigger values for in vitro and in vivo bioassays performed on surface water extracts supporting the environmental quality standards (EQS) of the European Water Framework Directive. Science of the Total Environment 628-629, 748-765.

Van der Oost, R., Sileno, G., Suarez-Munoz, M., Nguyen, M.T., Besselink, H., Brouwer, A. (2017). SIMONI (Smart Integrated Monitoring) as a novel bioanalytical strategy for water quality assessment: part I – Model design and effect-based trigger values. Environmental Toxicology and Chemistry 36, 2385-2399.

Additional Reading

Altenburger, R., Ait-Aissa, S., Antczak, P., Backhaus, T., Barceló, D., Seiler, T.-B., et al. (2015). Future water quality monitoring — Adapting tools to deal with mixtures of pollutants in water resource management. Science of the Total Environment 512-513, 540–551.

Escher, B.I., Leusch, F.D.L. (2012). Bioanalytical Tools in Water Quality Assessment. IWA publishing, London (UK).

Hamers, T., Legradi, J., Zwart, N., Smedes, F., De Weert, J., Van den Brandhof, E-J., Van de Meent, D., De Zwart, D. (2018). Time-Integrative Passive sampling combined with TOxicity Profiling (TIPTOP): an effect-based strategy for cost-effective chemical water quality assessment. Environmental Toxicology and Pharmacology 64, 48-59.

6.4.5. Biomonitoring: in situ bioassays and contaminant concentrations in organisms

Introduction

Several approaches and tools are available for diagnostic risk assessment. Tools specially developed for field assessments include the TRIAD approach (see section on TRIAD approach), in situ bioassays and biomonitoring. In ecotoxicology, biomonitoring is defined as the use of living organisms for the in situ assessment of environmental quality. Passive biomonitoring and active biomonitoring are distinguished. For passive biomonitoring, organisms are collected at the site of interest and their condition is assessed or the concentrations of specific target compounds in their tissues are analysed, or both. By comparing individuals from reference and contaminated sites an indication of the impact on local biota at the site of interest is obtained. For active biomonitoring, organisms are collected from reference sites and exposed in cages or artificial substrates at the study sites.

Ideally, reference organisms are simultaneously exposed at the site of origin to control for potential effects of the experimental set-up on the test organisms. As an alternative to field collected animals, laboratory cultured organisms may be employed. After exposure at the study sites for a certain period of time, the organisms are recollected and either their condition is analysed (in situ bioassay) or the concentrations of specific target compounds are measured in the organisms, or both.

The results of biomonitoring studies may be used for management decisions, e.g. when accumulation of contaminants has been demonstrated in the field and especially when the sources of the pollution have been identified. However, the use of biomonitoring studies in environmental management has not been captured in formal protocols or guidelines like those of the Water Framework Directive (WFD) or – to a lesser extent – the TRIAD approach and effect-based quality assessments. Biomonitoring studies are typically applied on an case-by-case basis and their application therefore strongly depends on the expertise and resources available for the assessment. The text below explains and discusses the most important aspects of biomonitoring techniques used in diagnostic risk assessment.

Selection of Biomonitoring Test Organisms

The selection of adequate organisms for biomonitoring partly follows the selection of test organisms for toxicity tests (see section on the Selection of test organisms). Suitable biomonitoring organisms:

- Are sedentary, since sedentary organisms may adapt more easily to the *in situ* experimental setup than more mobile organisms, for which caging may be an additional stress factor. Moreover, for sedentary organisms the relationship between the accumulated compounds and the environmental quality at the exposure site is straightforward, although this is more relevant to passive than to active biomonitoring.
- Are representative for the community of interest and native to the study sites, since this will ensure that the biomonitoring organisms tolerate local conditions other than contamination, preventing that stressors other than poor environmental conditions may affect their performance. Obviously, it is also undesirable to introduce exotic species into new environments.
- Are long living, at least substantially longer than the exposure duration and preferably large enough to obtain sufficient material for chemical analysis.
- Are easy to handle.

- Respond to a gradient of environmental quality, if the purpose of the biomonitoring study is to analyse the condition of the organisms after recollection (*in situ* bioassay).
- Accumulate contaminants without being killed, if the purpose of the biomonitoring study is to measure contaminant concentrations in the organisms after recollection.
- Are large enough to obtain sufficient biomass for the analysis of the target compounds above the limits of detection, if the purpose of the biomonitoring study is to measure contaminant concentrations in the organisms after recollection.

Based on the above listed criteria, in the marine environment mussels belonging to the genus Mytilus are predominantly selected. The genus Mytilus has the additional advantage of a global distribution, although represented by different species. This facilitates the comparison of contaminant concentrations in the organisms all around the globe. Lugworms have occasionally also been used for biomonitoring in marine systems. For freshwater, the cladoceran Daphnia magna is most frequently employed, although occasionally other species are selected, including mayflies, snails, worms, amphipods, isopods, caddisflies and fish. Given the positive experience with marine mussels, freshwater bivalves are also employed as biomonitoring organisms. Sometimes primary producers have been used, mainly periphyton. Due to the complexity of the sediment and soil compartments, few attempts have been made to expose organisms in situ, mainly restricted to chironomids on sediment.

In situ Exposure Devices

An obvious requirement of the *in situ* exposure devices is that the test organisms do not suffer from (sub)lethal effects of the experimental setup. If the organisms are large enough, cages may be used, like for freshwater and marine mussels. For daphnids, a simple glass jar with a permeable lid suffices. For riverine insects, the device should allow the natural flow of the stream to pass, but meanwhile prevent the organisms from escaping. In the device shown in Figure 1a, caddisfly larvae containing tubes are connected to floating tubes, maintaining the larvae at a constant depth of 65 cm. In the tubes, the caddisfly larvae are able to settle and build nets on artificial substrate, a plastic doormat with bristles standing out.

An elegant device for *in situ* colonization of periphyton was developed by Blanck (1985)(Figure 1b). Sand-blasted glass discs (1.5 cm2 surface) are used as artificial substratum for algal attachment. Substrata are placed vertically in the water, parallel to the current, by means of polyethylene racks, each rack supporting a total of 170 discs. After the colonization period, the periphyton containing glass discs can be harvested, offering the unique possibility to perform laboratory or field experiments with entire algal and microbial communities, replicated 170 times.

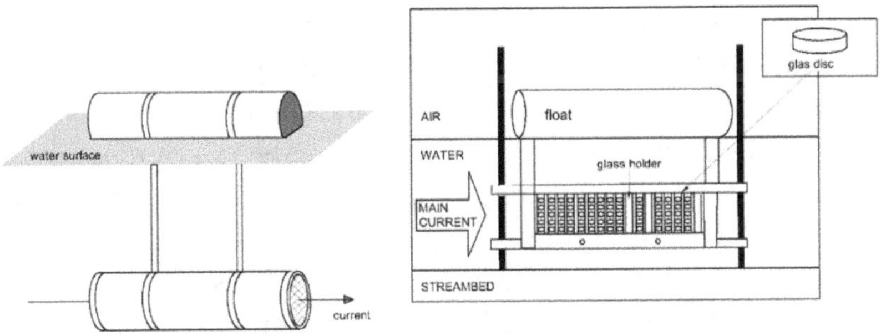

Figure 1. Left: Experimental set-up for in situ exposure of caddisfly larvae according to Stuijfzand et al. (1999, derived from Vuori (1995)). Right: Experimental set-up for in situ colonization of periphyton according to Ivorra et al. (1999), derived from Blanck (1985). Drawn by Wilma IJzerman.

In situ bioassays

After exposure at the study sites for a certain period of time, the organisms are recollected and their condition can be analysed (Figure 2). The endpoint is mostly survival, especially in routine monitoring programs. If the *in situ* exposure lasts long enough, also effects on species specific sublethal endpoints can be assessed. For daphnids and snails, this is reproduction and for isopods growth. For aquatic insects (mayflies, caddisflies, damselflies, chironomids), emergence has been assessed as a sensitive ecological relevant endpoint (Barmentlo et al., 2018).

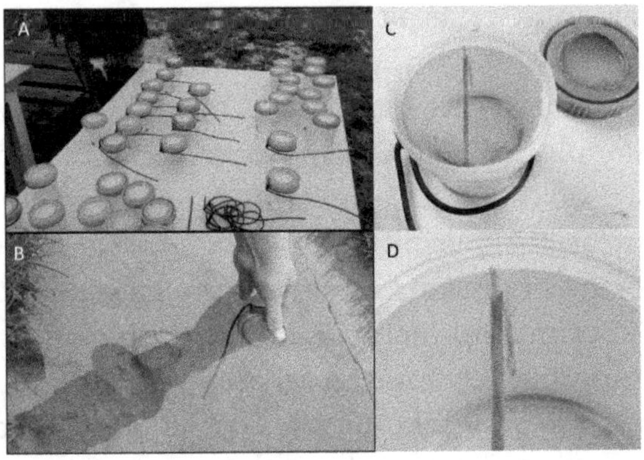

Figure 2. In situ exposure experiment. (A). Preparing damselfly containing jars. (B) Exposure of the in situ jars in ditches. (C). Retrieved jar containing a single damselfly larva. (D). Close up of the damselfly larva ready for inspection. Photos by Henrik Barmentlo.

In situ bioassays come closest to the actual field situation. Organisms are directly exposed at the site of interest and respond to all joint stressors present. Yet, this is also the limitation of the approach. If organisms do respond it remains unknown what causes the observed adverse effects. This could be (the combination of) any natural or anthropogenic physical or chemical stress factor. In situ bioassays can therefore be best combined with laboratory bioassays (see section on Bioassays) and the analysis of physico-chemical parameters, conform the TRIAD approach (see section on TRIAD approach). If the adverse effects are also observed in the bioassays under controlled laboratory conditions, then poor water quality is most likely the cause. The water sample may then be subjected to suspected target analysis, non-target analysis or effect directed analysis (EDA). If adverse effects are observed in situ but not in the laboratory, then the presence of hazardous compounds is most likely not the cause. Instead, the effects may be attributable to e.g. low pH, low oxygen concentrations, high temperatures etc, which may be verified by physico-chemical analysis in the field.

Online Biomonitoring

A specific application of in situ bioassays are the online systems for continuous water quality monitoring. In these systems, behaviour is generally the endpoint (see section on Endpoints). Organisms are exposed in a laboratory setting in situ (on shore or on a boat) in an experimental device to a continuous flow of surface water. If the water quality changes, the organisms respond by changing their behaviour. Above a certain threshold an alarm may go off and, for instance, the intake of surface water for drinking water preparation can be temporarily stopped.

Contaminant Concentrations in Organisms

As an addition or as an alternative to analysing the condition of the exposed biomonitoring organisms upon retrieval, contaminant concentrations in organisms can be analysed. This has several advantages over chemical analysis of environmental samples: biomonitoring organisms may be exposed for days to weeks at the site of interest, providing time integrated measurements of contaminant concentrations, in contrast to the chemical analysis of grab samples. This way, biomonitoring organisms actually serve as 'biological passive samplers' (see to section on Experimental methods of assessing available concentrations of organic chemicals). Another advantage of measuring contaminant concentrations in organisms is that they only take up the bioavailable (fraction of) substances, ecologically very relevant information, that remains unknown if chemical analysis is performed on water, sediment, soil or air samples. Yet, elevated concentrations in organisms do not necessarily imply toxic effects, and therefore these measurements are best complemented with determining their condition, as described above. Moreover, analysing contaminants in organisms may be more expensive than measurements of environmental samples, due to a more complex sample preparation. Weighing the advantages and disadvantages, the explicit strength of biomonitoring programs is that they provide insight into the spatial and temporal variation in bioavailable contaminant concentrations. In Figure 3 two examples are given. The left panel shows the concentrations of PCBs in zebra mussels at different sampling sites in Flanders, Belgium (Bervoets et al., 2004). The right panel shows the rapid (within 2 wk) Cd accumulation and depuration in biofilms translocated from a reference to a polluted site and from a polluted to a reference site, respectively (Ivorra et al., 1999).

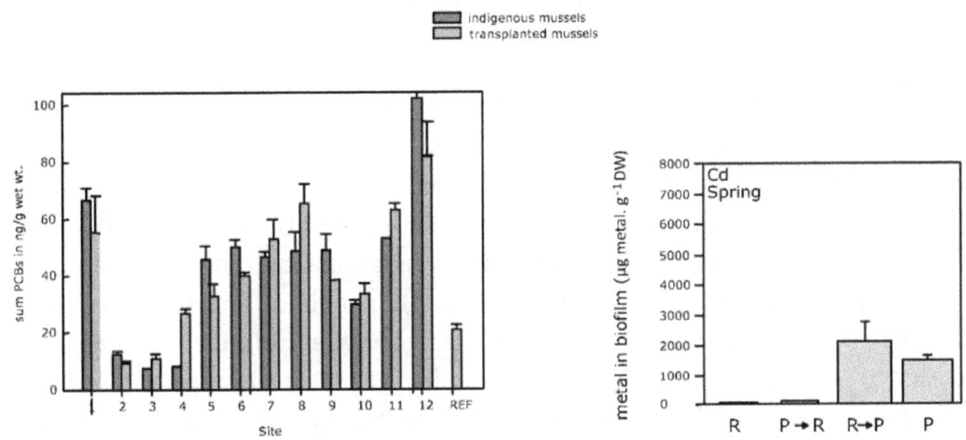

Figure 3. Left panel: Mean concentration of PCBs in 25 pooled zebra mussels at different sampling sites in Flanders, Belgium. Comparison between indigenous (black bars) and transplanted mussels (grey bars), from Bervoets et al. (2004). Right panel: Cd concentrations in local and translocated biofilms (R: Reference site; P: polluted site) from Ivorra et al. (1999). Drawn by Wilma IJzerman.

References

Barmentlo, S.H., Parmentier, E.M., De Snoo, G.R., Vijver, M.G. (2018). Thiacloprid-induced toxicity influenced by nutrients: evidence from in situ bioassays in experimental ditches. Environmental Toxicology and Chemistry 37, 1907-1915.

Bervoets, L., Voets, J., Chu, S.G., Covaci, A., Schepens, P., Blust, R. (2004). Comparison of accumulation of micropollutants between indigenous and transplanted zebra mussels (Dreissena polymorpha). Environmental Toxicology and Chemistry 23, 1973-1983.

Blanck, H. (1985). A simple, community level, ecotoxicological test systemusing samples of periphyton. Hydrobiologia 124, 251-261.

Ivorra, N., Hettelaar, J., Tubbing, G.M.J., Kraak, M.H.S., Sabater, S., Admiraal, W. (1999). Translocation of microbenthic algal assemblages used for in situ analysis of metal pollution in rivers. Archives of Environmental Contamination and Toxicology 37, 19-28.

Stuijfzand, S.C., Engels, S., Van Ammelrooy, E., Jonker, M. (1999). Caddisflies (Trichoptera: Hydropsychidae) used for evaluating water quality of large European rivers. Archives of Environmental Contamination and Toxicology 36, 186-192.

Vuori, K.M. (1995). Species- and population-specific responses of translocated hydropsychid larvae (Trichoptera, Hydropsychidae) to runoff from acid sulphate soils in the River Kyronjoki, western Finland. Freshwater Biology 33, 305-318.

6.4.6. TRIAD Approach for Site-specific Ecological Risk Assessment

Like the other diagnostic tools described in the previous sections (see sections on Effect-based monitoring In vivo bioassays and In vitro bioassays, Effect-directed analysis, and Effect-based water quality assessment and Biomonitoring), the TRIAD approach is a tool for site-specific ecological risk assessment of contaminated sites (Jensen et al., 2006; Rutgers and Jensen, 2011). Yet, it differs from the previous approaches by combining and integrating different techniques through a 'weight of evidence' approach. To this purpose, the TRIAD combines information on contaminant concentrations (environmental chemistry), the toxicity of the mixture of chemicals present at the site ((eco)toxicology), and observations of ecological effects (ecology) (Figure 1).

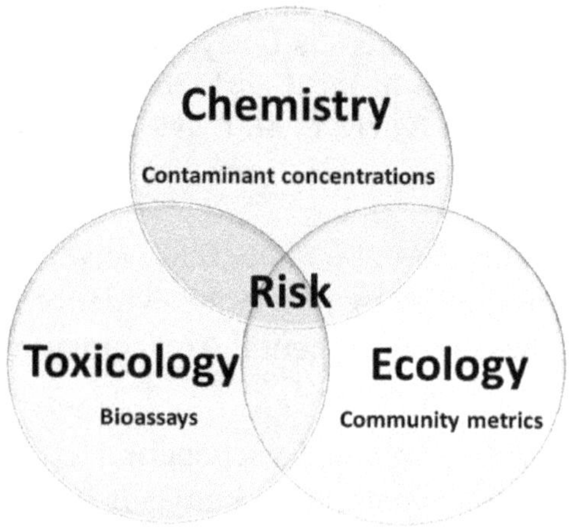

Figure 1: The TRIAD approach integrating information on contaminant concentrations (environmental chemistry), bioassays ((eco)toxicology) and ecological field inventories (ecology) into a weight of evidence for site-specific ecological risk assessment (adapted from Chapman, 1988).

The mere presence of contaminants is just an indication of potential ecological effects to occur. Additional data can help to better assess the ecological risks. For instance, information on actual toxicity of the contaminated site can be obtained from the exposure of test organisms to (extracts of) environmental samples (bioassays), while information on ecological effects can be obtained from an inventory of the community composition at the specific site. When these disciplines tend to converge to corresponding levels of ecological effects, a weight of evidence is established, making it possible to finalize the assessment and to support a decision for contaminated site management.

The TRIAD approach thus combines the information obtained from three lines of evidence (LoE):

- ***LoE Chemistry***: risk information obtained from the measured contaminant concentrations and information on their fate in the ecosystem and how they can evoke ecotoxicological effects. This can include exposure modelling and bioavailability considerations.
- ***LoE Toxicity***: risk information obtained from (eco)toxicity experiments exposing test organisms to (extracted) samples of the site. These bioassays can be performed on site or in the laboratory, under controlled conditions.
- ***LoE Ecology***: risk information obtained from the observation of actual effects in the field. This is deduced from data of ecological field surveys, most often at the community level. This information may include data on the composition of soil communities or other community metrics and on ecosystem functioning.

The three lines of evidence form a weight of evidence when they are converging, meaning that when the independent lines of evidence are indicating a comparable risk level, there is sufficient evidence for providing advice to decision makers about the ecological risk at a contaminated site. When there is no convergence in risk information obtained from the three lines of evidence, uncertainty is large. Further investigations are then required to provide a unambiguous advice.

Table 1. Basic data for site-specific environmental risk assessment (SS-ERA) sorted per line of evidence (LoE). Data and methods are described in Van der Waarde et al. (2001) and Rutgers et al. (2001).

Basic data		Unit	Site A	Site B	Ref
chemistry	pH		2.8	3.1	2.9
	SOM	%	2.4	2.1	3.2
	lutum	%	0.5	0.5	0.76
	lead conconcentration	mg/kg	290	880	11

	antimoon concentration	mmg/kg	3	3.1	3
	arsenicum concentration	mg/kg	2.5	2.9	2
	zinc concentration	mg/kg	7.6	7.2	5.9
	toxic pressure(Pb, Sb, As,Zn)	msPAF	0.49	0.71	0
Toxicology	Microtox assay	% inhibition	95	95	0
	Lettuce germination	% inhibition	55	10	0
	Lettuce growth	% inhibition	26	60	0
Ecology	nematodes abundance	n/100g soil	420	130	410
	nematodes Maturity Index 2-5 species composition		2.38	3.41	2.73
	Bait lamina	% activity	60	32	90

Tests and abbreviations used in the table:

- **Toxic Pressure metals** (sum TP metals): The toxic pressure of the mixture of metals in the sample, calculated as the potentially affected fraction in a Species Sensitivity Distribution with NOEC values (see Section on SSDs) and a simple rule for mixture toxicity (response addition; Section on Mixture toxicity).
- **Microtox:** A bioassay with the luminescent bacterium *Allivibrio fischeri*, formerly known as *Vibrio fischeri.* Luminescence is reduced when toxicity is high.
- **Lettuce Growth and Lettuce Germination:** A bioassay with the growth performance and the germination percentage of lettuce (seeds).
- **Bait Lamina:** The bait-lamina test consists of vertically inserting 16-hole-bearing plastic strips filled with a plant material preparation into the soil. This gives an indication of the feeding activity of soil animals.
- **Nematodes abundance and Nematodes Maturity Index 2-5:** The biomass and the Maturity Index (MI) of the nematode community in soil samples provide information about soil health (Van der Waarde et al. 2001).

The results of a site-specific ecological risk assessment (SS-ERA) applying the TRIAD approach are first organized basic tables for each sample and line of evidence separately. Table 1 shows an example. This table also collects supporting data, such as soil pH and organic matter content. Subsequently, these basic data are processed into ecological risk values by applying a risk scale running from zero (no effects) to one (maximum effect). An example of a metric used is the multi-substance Potentially Affected Fraction of species from the mixture of contaminants (see

Section on SSDs). These risk values are then collected in a TRIAD table (Table 2), for each endpoint separately, integrated per line of evidence individually, and finally integrated over the three lines of evidence. Also the level of agreement between the three lines of evidence is given a score. Weighting values are applied, e.g. equal weights for all ecological endpoints (depending on number of methods and endpoints), and equal weights for each line of evidence (33%). When differential weights are preferred, for instance when some data are judged as unreliable, or some endpoints are considered more important than others, the respective weight factors and the arguments to apply them must be provided in the same table and accompanying text.

Table 2. Soil Quality TRIAD table demonstrating scaled risk values for two contaminated sites (A, B) and a Reference site (based on real data, only for illustration purposes). Risk values are collected per endpoint, grouped according to respective Lines of Evidence (LoE), and finally integrated into a TRIAD value for risks. The deviation indicates a level of agreement between LoE (default threshold 0.4). For site B, a Weight of Evidence (WoE) is demonstrated (D<0.4) making decision support feasible. By default equal weights can be used throughout. Differential weights should be indicated in the table and described in the accompanying text.

Triad LoE	Parameter	Site A	Site B	Ref
Chemistry	Sum TP metals	0.49	0.71	0.00
	integration	0.49	0.77	0.00
Toxicology	Microtox	0.95	0.95	0.00
	Lettuce germination	0.55	0.09	0.00
	Lettuce growth	0.26	0.60	0.00
	integration	0.59	0.55	0.00
Ecology	Nematodes abundance	0.00	0.68	0.00
	Nematodes Maturity Index	0.00	0.30	0.00
	Bait lamina	0.30	0.64	0.00
	integration	0.10	0.54	0.00
	chemical risk	0.49	0.71	0.00
	ecotoxicity risk	0.59	0.55	0.00
	ecology risk	0.10	0.54	0.00
	final score for risk	*0.42*	*0.61*	*0.00*
	deviation	0.44	0.17	0.00

References

ISO (2017). ISO 19204: Soil quality -- Procedure for site-specific ecological risk assessment of soil contamination (soil quality TRIAD approach). International Standardization Organization, Geneva. https://www.iso.org/standard/63989.html.

Jensen, J., Mesman, M. (Eds.) (2006). LIBERATION, Ecological risk assessment of contaminated land, decision support for site specific investigations. ISBN 90-6960-138-9, Report 711701047, RIVM, Bilthoven, The Netherlands.

Rutgers, M., Bogte, J.J., Dirven-Van Breemen, E.M., Schouten, A.J. (2001) Locatiespecifieke ecologische risicobeoordeling – praktijkonderzoek met een Triade-benadering. RIVM-rapport 711701026, Bilthoven.

Rutgers, M., Jensen, J. (2011). Site-specific ecological risk assessment. Chapter 15, in: F.A. Swartjes (Ed.), Dealing with Contaminated Sites – from Theory towards Practical Application, Springer, Dordrecht. pp. 693-720.

Van der Waarde, J.J., Derksen, J.G.M, Peekel, A.F., Keidel, H., Bloem, J., Siepel, H. (2001) Risicobeoordeling van bodemverontreiniging met behulp van een triade benadering met chemische analyses, bioassays en biologische veldinventarisaties. Eindrapportage NOBIS 98-1-28, Gouda.

6.4.7. Eco-epidemiology

Introduction

Approaches for environmental protection, assessment and management differ between 'classical' stressors (such as excess nutrients and pH) and chemical pollution. For the 'classical' environmental stress factors, ecologists use monitoring data to develop concepts and methods to prevent and reduce impacts. Although there are some clear-cut examples of chemical pollution impacts [e.g., the decline in vulture populations in South East Asia due to diclofenac (Oaks et al. 2004), and the suit of examples in the book 'Silent Spring' (Carson 1962)], ecotoxicologists commonly have assessed the stress from chemical pollution by evaluating exposures vis a vis laboratory toxicity data. Current pollution often consists of complex mixtures of chemicals, with highly variable patterns in space and time. This poses problems when one wants to evaluate whether observed impacts in ecosystems can be attributed to chemicals or their mixtures. Eco-epidemiological methods have been established to discern such pollution stress. These methods provide the diagnostic tools to identify the impact magnitude and key chemicals that cause impacts in ecosystems. The use of these methods is further relevant for validating the laboratory-based risk assessment approaches developed by ecotoxicology.

The Origins of Eco-epidemiology

Risk assessments of chemicals provide insights in expected exposures and impacts, commonly for separate chemicals. These are predictive outcomes with a high relevance for decision making on environmental protection and management. The validation of those risk assessments is key to avoid wrong protection and management decisions, but it is complex. It consists of comparing predicted risk levels to observed effects. This begs the question on how to discern effects of chemical pollution in the field. This question can be answered based on the principles of ecological bio-assessments combined with those of human epidemiology. A bio-assessment is a study of stressors and ecosystem attributes, made to delineate causes of impacts via (often) statistical associations between biotic responses and particular stressors. Epidemiology is defined as the study of the distribution and causation of health and disease conditions in specified populations. Applied epidemiology serves as a scientific basis to help counteracting the spreading of human health problems. Dr. John Snow is often referred to as the 'father of epidemiology'. Based on observations on the incidence, locations and timings of the 1854 cholera outbreak in London, he attributed the disease to contaminated water taken from the Broad Street pump well, counteracting the prevailing idea that the disease was caused by transmission via air. His proposals to control the disease were effective. Likewise, eco-epidemiology – in its ecotoxicological context – has been defined as the study of the distribution and causation of impacts of multiple stressor exposures in ecosystems. In its applied form, it supports the reduction of ecological impacts of chemical pollution. Human-health eco-epidemiology is concerned with environment-mediated disease.

The first literature mention of eco-epidemiological analyses on chemical pollution stems from 1984 (Bro-Rasmussen and Løkke 1984). Those authors described eco-epidemiology as a discipline necessary to validate the risk assessment models and approaches of ecotoxicology. In its initial years, progress in eco-epidemiological research was slow due to practical constraints such as a lack of monitoring data, computational capacity and epidemiological techniques.

Current Eco-epidemiology

Current eco-epidemiological studies in ecotoxicology aim to diagnose the impacts of chemical pollution in ecosystems, and utilize a combination of approaches in order to diagnose the role of chemical mixtures in causing ecological impacts in the field. The combination of approaches consists of:

- Collection of monitoring data on abiotic characteristics and the occurrence and/or abundance of biotic species, for the environmental compartment under study;

- If needed: data optimization, usually to align abiotic and biotic monitoring data, including the chemicals;
- Statistical analysis of the data set using eco-epidemiological techniques to delineate impacts and probable causes, according to the approaches followed in 'classical' ecological bio-assessments;
- Interpretation and use of the outcomes for either validation of ecotoxicological models and approaches, or for control of the impacts sensu Dr. Snow.

Key Examples of Chemical Effects in Nature

Although impacts of chemicals in the environment were known before 1962, Rachel Carson's book Silent Spring (see Section on the history of Environmental toxicology) can be seen as early and comprehensive eco-epidemiological study that synthesized the available information of impacts of chemicals in ecosystems. She considered effects of chemicals a novel force in natural selection when she wrote: "If Darwin were alive today the insect world would delight and astound him with its impressive verification of his theories of survival of the fittest. Under the stress of intensive chemical spraying the weaker members of the insect populations are being weeded out."

Clear examples of chemical impacts on species are still reported. Amongst the best-known examples is a study on vultures. The population of Indian vultures declined more than 95% due to diclofenac exposure which was used intensively as a veterinary drug (Oaks et al. 2004). The analysis of chemical impacts in nature becomes however more complex over time. The diversity of chemicals produced and used has vastly increased, and environmental samples contain thousands of chemicals at often low concentrations. Hence, contemporary eco-epidemiology is complex. Nonetheless, various studies demonstrated that contemporary mixture exposures affect species assemblages. Starting from large-scale monitoring data and following the four steps mentioned above, De Zwart et al. (2006) were able to show that effects on fish species assemblages could be attributed to both habitat characteristics and chemical mixtures. Kapo and Burton Jr (2006) showed the impacts of multiple stressors and chemical mixtures in aquatic species assemblages with similar types of data, but slightly different techniques. Eco-epidemiological studies of the effects of chemicals and their mixtures currently represent different geographies, species groups, stressors and chemicals/mixtures that are considered. The potential utility eco-epidemiological studies was reviewed by Posthuma et al. (2016). The review showed that mixture impacts occur, and that they can be separated from natural variability and multiple-stressor impacts. That means that water managers can develop management plans to counteract stressor impacts. Thereby, the study outcomes are used to prioritize management to sites that are most affected, and to chemicals that contribute most to those effects. Based on sophisticated statistical analyses, Berger et al. (2016) suggested chemicals can induce effects in the environment at concentrations much lower than expected based on laboratory experiments. Schäfer et al. (2016) argued that eco-epidemiological studies that cover both mixtures and other stressors are essential for environmental quality assessment and management. In practice, however, the analysis of the potential impacts of chemical mixtures is often still separate from the analysis of impacts of other stressors.

Steps in Eco-epidemiological Analysis

Various regulations require collection of monitoring data, followed by bio-assessment, such as the EU Water Framework Directive (see section on the Water Framework Directive). Therefore, monitoring data sets are increasingly available. The data set is subsequently curated and/or optimized for the analyses. Data curation and management steps imply amongst others that taxonomic names of species are harmonized, and that metrics for abiotic and biotic variables represent the conditions for the same place and time as much as possible. Next, the data set is expanded with novel variables, e.g. a metric for the toxic pressure exerted by chemical mixtures. An example of such a metric is the multi-substance Potentially Affected Fraction of species (msPAF). This metric transfers measured or predicted concentrations into the Potentially Affected Fraction of species (PAF), the values of which are then aggregated for a total mixture (De Zwart and Posthuma 2005). This is crucial, as adding each chemical of interest as a separate variable implies an increasingly expanding number of required sampling sites to maintain statistical power to diagnose impacts and probable causation.

The interpretation of the outcomes of the statistical analyses of the data set is the final step. Here, it must be acknowledged that statistical association is not equal to causation, and that care must be taken to explain the findings as indicative for mixture effects. Depending on the context of the study, this may then trigger a refined assessment, or alignment with other methods to collect evidence, or a direct use in an environmental management program.

Eco-epidemiological Methods

A very basic eco-epidemiological method is quantile regression. Whereas common regression methods explore the magnitude of the change of the mean of the response variable (e.g., biodiversity) in relation to a predictor variable (e.g., pollutant stress), the quantile regression looks at the tails of the distributions of the response variable. How this principle operates is illustrated in Figure 1. When a monitoring data set contains one stressor variable at different levels (i.e., a gradient of data), the observations typically take the shape of a common stressor-response relationship (see section on Concentration-effect relationships). If the monitoring sites are affected by an extra stressor, the maximum-performance under the first stressor cannot be reached, so that the area under the curve contains the XY-points for this situation. Further addition of stressor variables and levels fills this space under the curve. When the raw data plotted as XY show an 'empty area' lacking XY-points, e.g. in the upper right corner, it is likely that the stressor variable can be identified as a stressor that limits the response variable, for example: chemicals limit biodiversity. The quantile regression calculates an upper percentiles (e.g., the 95th percentile) of the Y-values in assigned subgroups of X-values ("bins"). Such a procedure yields a picture such as Figure 1.

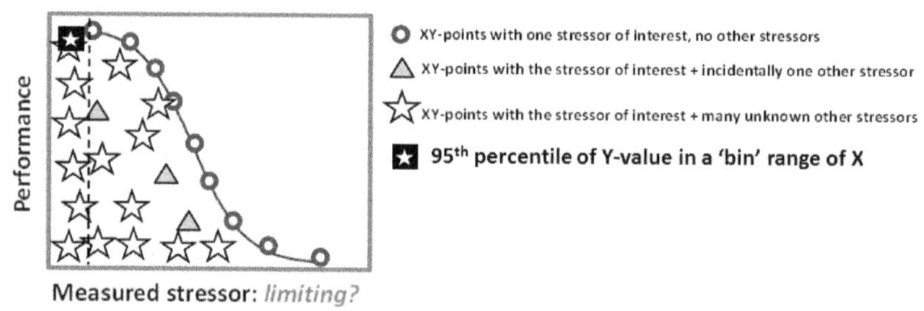

Figure 1. The principle of quantile regression in identification of a predictor variable (= stressor) that acts as a limiting factor to a response variable (= performance). It is common to derive e.g. the 95th percentile of the Y values in a 'bin' of X values to derive a stressor-impact curve. As illustration, the 95th percentile is marked only for the first bin of X values, with the blackened star.

More complex methods for analysis of (bio)monitoring data have been developed and applied. The methods are closely associated to those developed for, and utilized in, applied ecology. Well-known examples are 'species distribution models' (SDM), which are used to describe the abundance or presence of species as a function of multiple environmental variables. A well-known SDM is the bell-shaped curve relating species abundances to water pH: numbers of individuals of a species are commonly low at low and high pH, and the SDM is characterized as an optimum model for species abundance (Y) versus pH (X). Statistical models can also describe species abundance, presence or biodiversity, as a function of multiple stressors, for example via Generalized Linear Models. These have the general shape of:

$$\text{Log(Abundance)} = (a.\,pH + a'\,pH^2) + (b.\,OM + b'\,OM^2) + \ldots\ldots + e,$$

With a, a', b and b' being estimated from fitting the model to the data, whilst pH and OM are the abiotic stressor variables (acidity and Organic Matter, respectively);

the quadratic terms are added to allow for optimum and minimum shaped relationships. When SSD models (see Section on Species Sensitivity Distribution) are used to predict the multi-substance Potentially Affected Fraction of species, the resulting mixture stress proxy can be analysed together with the other stressor variables. Data analyses from monitoring data from the United States and the Netherlands have, for example, shown that the abundance of >60% of the taxa is co-affected by mixtures of chemicals. An example study is provided by Posthuma et al. (2016).

Prospective Mixture Impact Assessments

In addition to the retrospective analysis of monitoring data, in search of chemical impacts, recent studies also show examples of prospective studies of effects of mixtures. Different land uses imply different chemical use patterns, summarized as 'signatures'. That is, agricultural land use will yield intermittent emissions of crop-specific plant protection products, aligning with the growing season. Emissions from populated areas will show continuous emission of household chemicals and discontinuous emissions of chemicals in street run-off associated to heavy rain events. The application of emission, fate and ecotoxicity models showed that aquatic ecosystems are subject to the 'signatures', with associated predicted impact magnitudes (Holmes et al. 2018; Posthuma et al. 2018). Although such prospective assessments did not yet prove ecological impacts, they can assist in avoiding impacts by preventing the emission 'signatures' that are identified as potentially most hazardous.

The use of Eco-epidemiological Output

Eco-epidemiological analysis outputs serve two purposes, closely related to prospective and retrospective risk assessment of chemical pollution:

- Validation of ecotoxicological models and approaches;
- Derivation of control measures, to reduce impacts of diagnosed probable causes of impacts.

If needed, multiple lines of evidence can be combined, such as in the Triad approach (see section on TRIAD) or approaches that consider more than three lines of evidence (Chapman and Hollert, 2006). The higher the importance of a good diagnosis, the better the user may rely on multiple lines of evidence.

First, the validation of ecotoxicological models and approaches is crucial, to avoid that important environmental protection, assessment and management activities rely on approaches that have limited relationship to field effects. Eco-epidemiological analyses have, for example, been used to validate the protective benchmarks used in the chemical-oriented environmental policies.

Second, the outcomes of an eco-epidemiological analysis can be used to control causes of impacts to ecosystems. Some studies have, for example, identified a statistical association between observed impacts (species expected but absent) and pollution of surface waters with mixtures of metals. Though local experts first doubted this association due to lack of industrial activities with metals in the area, they later found the association relevant given the presence of old spoil heaps from past mining activities. Metals appeared to leach into the surface waters at low rates, but the leached mixtures appeared to co-vary with species missing (De Zwart et al. 2006).

References

Berger, E., Haase, P., Oetken, M., Sundermann, A. (2016). Field data reveal low critical chemical concentrations for river benthic invertebrates. Science of The Total Environment 544, 864-873.

Bro-Rasmussen, F., Løkke, H. (1984). Ecoepidemiology - a casuistic discipline describing ecological disturbances and damages in relation to their specific causes; exemplified by chlorinated phenols and chlorophenoxy acids. Regulatory Toxicology and Pharmacology 4, 391-399.

Carson, R. (1962). Silent spring. Boston, Houghton Mifflin.

Chapman, P.M., Hollert, H. (2006). Should the sediment quality triad become a tetrad, a pentad, or possibly even a hexad? Journal of Soils and Sediments 6, 4-8.

De Zwart, D., Dyer, S.D., Posthuma, L., Hawkins, C.P. (2006). Predictive models attribute effects on fish assemblages to toxicity and habitat alteration. Ecological Applications 16, 1295-1310.

De Zwart, D., Dyer, S.D., Posthuma, L., Hawkins, C.P. (2006). Use of predictive models to attribute potential effects of mixture toxicity and habitat alteration on the biological condition of fish assemblages. Ecological Applications 16, 1295-1310.

De Zwart, D., Posthuma, L. (2005). Complex mixture toxicity for single and multiple species: Proposed methodologies. Environmental Toxicology and Chemistry 24,: 2665-2676.

Holmes, C.M., Brown, C.D., Hamer, M., Jones, R., Maltby, L., Posthuma, L., Silberhorn, E., Teeter, J.S., Warne, M.S.J., Weltje, L. (2018). Prospective aquatic risk assessment for chemical mixtures in agricultural landscapes. Environmental Toxicology and Chemistry 37, 674-689.

Kapo, K.E., Burton Jr, G.A. (2006). A geographic information systems-based, weights-of-evidence approach for diagnosing aquatic ecosystem impairment. Environmental Toxicology and Chemistry 25, 2237-2249.

Oaks, J.L., Gilbert, M., Virani, M.Z., Watson, R.T., Meteyer, C.U., Rideout, B.A., Shivaprasad, H.L., Ahmed, S., Chaudhry, M.J., Arshad, M., Mahmood, S., Ali, A., Khan, A.A. (2004). Diclofenac residues as the cause of vulture population decline in Pakistan. Nature 427(6975), 630-633.

Posthuma, L., Brown, C.D., de Zwart, D., Diamond, J., Dyer, S.D., Holmes, C.M., Marshall, S., Burton, G.A. (2018). Prospective mixture risk assessment and management prioritizations for river catchments with diverse land uses. Environmental Toxicology and Chemistry 37, 715-728.

Posthuma, L., De Zwart, D., Keijzers, R., Postma, J. (2016). Water systems analysis with the ecological key factor 'toxicity'. Part 2. Calibration. Toxic pressure and ecological effects on macrofauna in the Netherlands. Amersfoort, the Netherlands, STOWA.

Posthuma, L., Dyer, S.D., de Zwart, D., Kapo, K., Holmes, C.M., Burton Jr, G.A. (2016). Eco-epidemiology of aquatic ecosystems: Separating chemicals from multiple stressors. Science of The Total Environment 573, 1303-1319.

Posthuma, L., Suter, II, G.W., Traas, T.P. (Eds.) (2002). Species Sensitivity Distributions in Ecotoxicology. Boca Raton, FL, USA, Lewis Publishers.

Schäfer, R.B., Kühn, B., Malaj, E., König, A., Gergs, R. (2016). Contribution of organic toxicants to multiple stress in river ecosystems. Freshwater Biology 61, 2116–2128

6.5. REGULATORY FRAMEWORKS

Introduction

There is no single, overarching global regulatory framework to manage the risks of all chemicals. Instead, different regulations or directives have been developed for different categories of chemicals. These categories are typically related to the usage of the chemicals. Important categories are industrial chemicals (solvents, plasticizers, etc.), plant protection products, biocides and human and veterinary drugs. Some chemicals may belong to more than one category. Zinc, for example, is used in the building industry, but it also has biocidal applications (antifouling agent) and zinc oxide is used as a veterinary drug. In the European Union, each chemical category is subject to specific regulations or directives providing the legal conditions and requirements to guarantee a safe production and use of chemicals. A key element of all legal frameworks is the requirement that sufficient data on a chemical should be made available. Valid data on production and identity (e.g. chemical structure), use volumes, emissions, environmental fate properties and the (eco)toxicity of a chemical are the essential building blocks for a sound assessment and management of environmental risks. Rules for the minimum data set that should be provided by the actors involved (e.g. producers or importers) are laid down in various regulatory frameworks. With this data, both hazard and risk assessments can be carried out according to specified technical guidelines. The outcome of the assessment is then used for risk management, which is focused on minimizing any risk by taking measures, ranging from requests for additional data to restrictions on the particular use or a full-scale ban of a chemical.

REACH

REACH is a regulation of the European Union, adopted to improve the protection of human health and the environment from the risks that can be posed by chemicals, while enhancing the competitiveness of the EU chemicals industry. REACH stands for Registration, Evaluation, Authorisation and Restriction of Chemicals. The REACH regulation entered into force on 1st June 2007 to streamline and improve the former legislative frameworks on new and existing chemical substances. It replaced approximately forty community regulations and directives by one single regulation. REACH establishes procedures for collecting and assessing information on the properties, hazards and risks of substances. REACH applies to a very broad spectrum of chemicals: from industrial to household applications, and very much more. It requires that EU manufacturers and importers register their chemical substances if produced or imported in annual amounts of > 1 tonne, unless the substance is exempted from registration under REACH. At quantities of > 10 tonnes, the manufacturers, importers, and down-stream users are responsible to show that their substances do not adversely affect human health or the environment.

The amount of standard information required to show safe use depends on the quantity of the substance that is manufactured or imported. Before testing on vertebrate animals like fish and mammals, the use of alternative methods must be considered. The European Chemical Agency (ECHA) coordinates and facilitates the REACH program. For production volumes above 10 tonnes per year, industry has to prepare a risk assessment, taking into account all risk management measures envisaged, and document this in a chemical safety assessment (CSA). A CSA should include an exposure assessment, hazard or dose-response assessment, and a risk characterization showing risks ratios below 1.0, i.e. safe use (see sections on REACH Human and REACH Eco).

Classification, Labelling and Packaging (CLP)

The EU CLP regulation requires manufacturers, importers or downstream users of substances or mixtures to classify, label and package their hazardous chemicals appropriately before placing them on the market. When relevant information (e.g. ecotoxicity data) on a substance or mixture meets the classification criteria in the CLP regulation, the hazards of a substance or mixture are identified by assigning a certain hazard class and category. An important CLP hazard class is 'Hazardous to the aquatic environment', which is divided into categories based on the criteria, for example, Category Acute 1, representing the most (acute) toxic chemicals (LC_{50}/EC_{50} ≤ 1 mg/L). CLP also sets detailed criteria for the labelling elements, such as the well-known pictograms (Figure 1).

Figure 1: Pictogram used to indicate hazardousness to the environment. (source: https://www.pictogrammenwinkel.nl/index.php?main_page=product_info&products_id=5988, with permission)

Plant Protection Products Regulation

Plant protection products (PPPs) are pesticides that are mainly used to keep crops healthy and prevent them from being damaged by disease and infestation. They include among others herbicides, fungicides, insecticides, acaricides, plant growth regulators and repellents (see section on Crop Protection Products). PPPs fall under the EU Regulation (EC) No 1107/2009 which determines that PPPs cannot be placed on the market or used without prior authorization. The European Food and Safety Authority (EFSA) coordinates the EU regulation on PPPs.

Biocides Regulation

The distinction between biocides and PPP is not always straightforward, but as a general rule of thumb the PPP regulation applies to substances used by farmers for crop protection while the biocides regulation covers all other pesticide applications. Different applications of the same active ingredient, one as a PPP and the other as a biocide, may thus fall under different regulations. Biocides are used to protect humans, animals, materials or articles against harmful organisms like pests or bacteria, by the action of the active substances contained in the biocidal product. Examples of biocides are antifouling agents, preservatives and disinfectants.

According to the EU Biocidal Products Regulation (BPR), all biocidal products require an authorization before they can be placed on the market, and the active substances contained in the biocidal product must be previously approved. The European Chemical Agency (ECHA) coordinates and facilitates the BPR. More or less similar to other legislations, for biocides the environmental risk assessment is mainly performed by comparing compartmental concentrations (PEC) with the concentration below which unacceptable effects on organisms will most likely not occur (PNEC).

Veterinary and Human Pharmaceuticals Regulation

Since 2006, EU law requires an environmental risk assessment (ERA) for all new applications for a marketing authorization of human and veterinary pharmaceuticals. For both products, guidance documents have been developed for conducting an ERA based on two phases. The first phase estimates the exposure of the environment to the drug substance. Based on an action limit the assessment may be terminated. In the second phase, information about the fate and effects in the environment is obtained and assessed. For conducting an ERA a base set, including ecotoxicity data, is required. For veterinary medicines, the ERA is part of a risk-benefit analysis, in which the positive therapeutic effects are weighed against any environment risks, whereas for human medicines the environmental concerns are excluded from the risk-benefit analysis. The European Medicines Agency (EMA) is responsible for the scientific evaluation, supervision and safety monitoring of medicines in the EU.

Harmonization of Testing

Testing chemicals is an important aspect of risk assessment, e.g. testing for toxicity, for degradation or for a physicochemical property like the K_{ow} (see Chapter 3). The outcome of a test may vary depending on the conditions, e.g. temperature, test medium or light conditions. For this reason there is an incentive to standardize the test conditions and to harmonize the testing procedures between agencies and countries. This would also avoid duplication of testing, and leading to a more efficient and effective testing system.

The Organization for Economic Co-operation and Development (OECD) assists its member governments in developing and implementing high-quality chemical management policies and instruments. One of the key activities to achieve this goal is the development of harmonized guidelines to test and assess the risks of chemicals leading to a system of mutual acceptance of chemical safety data among OECD countries. The OECD also developed Principles of Good Laboratory Practice (GLP) to ensure that studies are of sufficient quality and rigor and are verifiable. The OECD also facilitates the development of new tools to obtain more safety information and maintain quality while reducing costs, time and animal testing, such as the OECD QSAR toolbox.

6.5.1. REACH Human

Human Risk Assessment Under REACH

The REACH Regulation aims to ensure a high level of protection of human health and the environment, including the promotion of alternative methods for assessment of hazards of substances, as well as the free circulation of substances on the internal market while enhancing competitiveness and innovation. Risk assessment under REACH aims to realize such a level of protection for humans that the likelihood of adverse effects occurring is low, taking into account the nature of the potentially exposed population (including sensitive groups) and the severity of the effect(s). Industry therefore has to prepare a risk assessment (in REACH terminology: chemical safety assessment, CSA) for all relevant stages in the life cycle of the chemical, taking into account all risk management measures envisaged, and document this in the chemical safety report (CSR). Risk characterization in the context of a CSA is the estimation of the likelihood that adverse effect levels occur due to actual or predicted exposure to a chemical. The human populations considered, or protection goals, are workers, consumers and humans exposed via the environment. In risk characterization, exposure levels are compared to reference levels to yield "risk characterization ratios" (RCRs) for each protection goal. RCRs are derived for all endpoints (e.g. skin and eye irritation, sensitization, repeated dose toxicity) and time scales. It should be noted that these RCRs have to be derived for all stages in the life-cycle of a compound.

Environmental Exposure Assessment for Humans

Humans can be exposed through the environment directly via inhalation of indoor and ambient air, soil ingestion and dermal contact, and indirectly via food products and drinking water (Figure 1). REACH does not consider direct exposure via soil.

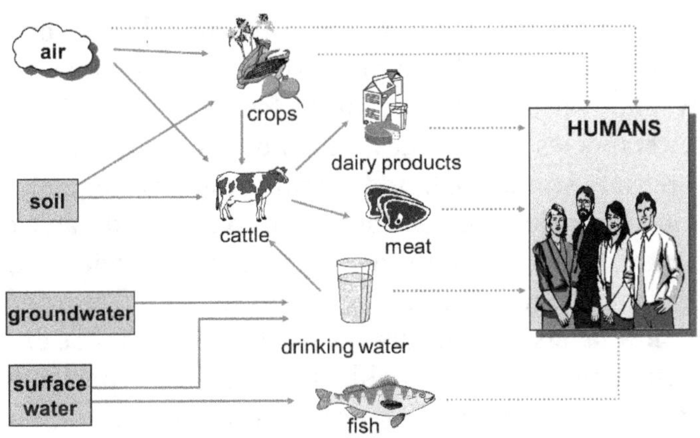

Figure 1. Main exposure routes considered in REACH for environmental exposure of humans.

In the REACH exposure scenario, assessment of human exposure through the environment can be divided into three steps:

- Determination of the concentrations in intake media (air, soil, food, drinking water);
- Determination of the total daily intake of these media;
- Combining concentrations in the media with total daily intake (and, if necessary, using a factor for bioavailability through the route of uptake concerned).

A fourth step may be the consideration of aggregated exposure taking into account exposure to the same substance in consumer products and at the workplace. Moreover, there may be similar substances, acting via the same mechanism of action, that may have to be considered in the exposure assessment, for instance, as a worst case, by applying the concept of dose or concentration addition. The section on Environmental realistic scenarios (PECs) – Human explains the concept of exposure scenarios and how concentrations in environmental compartments are derived.

Hazard Identification and Dose-response Assessment

The aim of hazard identification is to classify chemicals and to select key data for the dose-response assessment to derive a safe reference level, which in REACH terminology is called the DNEL (Derived No Effect Level) or DMEL (Derived Minimal Effect Level). For human end-points, a distinction is made between substances considered to have a threshold for toxicity and those without a threshold. For threshold substances, a No-Observed-Adverse Effect Level (NOAEL) or Lowest-Observed-Adverse-Effect Level (LOAEL) is derived, typically from toxicity studies with laboratory animals such as rats and mice. Alternatively a Benchmark Dose (BMD) can be derived by fitting a dose-response model to all observations. These toxicity values are then extrapolated to a DNEL using assessment factors to correct for uncertainty and variability. The most frequently used assessment factors are those for interspecies differences and those for intraspecies variability (see section on Setting safe standards). Additionally, factors can be applied to account for remaining uncertainties such as those due to a poor database. For substances considered to exert their effect by a non-threshold mode of action, especially mutagenicity and carcinogenicity, it is generally assumed, as a default assumption, that even at very low levels of exposure residual risks cannot be excluded. That said, recent progress has been made on establishing scientific, 'health-based' thresholds for some genotoxic carcinogens. For non-threshold genotoxic carcinogens it is recommended to derive a DMEL, if the available data allow. A DMEL is a cancer risk value considered to be of very low concern, e.g. a 1 in a million tumour risk after lifetime exposure to the chemical and using a conservative linear dose-response model. There is as yet no EU-wide consensus on acceptable levels of cancer risk.

Risk Characterization

Safe use of substances is demonstrated when:

- RCRs are below one, both at local and regional level. For threshold substances, the RCR is the ratio of the estimated exposure (concentration or dose) and the DNEL; for non-threshold substances the DMEL is used.
- The likelihood and severity of an event such as an explosion occurring due to the physicochemical properties of the substance as determined in the hazard assessment is negligible.

A risk characterization needs to be carried out for each exposure scenario (see Section on Environmental realistic scenarios (PECs) – Human) and human population. The assessment consists of a comparison of the exposure of each human population known to be or likely to be exposed with the appropriate DNELs or DMELs and an assessment of the likelihood and severity of an event occurring due to the physicochemical properties of the substance.

Example of a Deterministic Assessment (Vermeire et al., 2001)

Exposure Assessment

Based on an emission estimation for processing of dibutylphthalate (DBP) as a softener in plastics, the concentrations in environmental compartments were estimated. Based on modelling as schematically presented in Figure 1, the total human dose was determined to be 93 ug.kg bw-1.

PEC-air	2.4	μg.m-3
PEC-surface water	2.8	μg.l-1
PEC- grassland soil	0.15	mg.kg-1
PEC-porewater agric. soil	3.2	μg.l-1
PEC-porewater grassl. soil	1.4	μg.l-1
PEC-groundwater	3.2	μg.l-1
Total Human Dose	93	μg.kgbw-1.d-1

Effects Assessment

The total dose should be compared to a DNEL for humans. DBP is not considered a genotoxic carcinogen but is toxic to reproduction and therefore the risk assessment is based on endpoints assumed to have a threshold for toxicity. The lowest NOAEL of DBP was observed in a two-generation reproduction test in rats and at the lowest dose-level in the diet (52 mg.kgbw-1.d-1 for males and 80 mg.kgbw-1.d-1 for females) a reduced number of live pups per litter and decreased pup weights were seen in the absence of maternal toxicity. The lowest dose level of 52 mg.kgbw-1.d-1 was chosen as the NOAEL. The DNEL was derived by the application of an overall assessment factor of 1000, accounting for interspecies differences, human variability and uncertainties due to a non-chronic exposure period.

Risk Characterisation

The deterministic estimate of the RCR would be based on the deterministic exposure estimate of 0.093 mg.kgbw-1.d-1 and the deterministic DNEL of 0.052 mg.kgbw-1.d-1. The deterministic RCR would then be 1.8, based on the NOAEL. Since this is higher than one, this assessment indicates a concern, requiring a refinement of the assessment or risk management measures.

Additional Reading

Van Leeuwen C.J., Vermeire T.G. (Eds.) (2007) Risk assessment of chemicals: an introduction. Springer, Dordrecht, The Netherlands, ISBN 978-1-4020-6102-8 (e-book), https://doi.org/10.1007/978-1-4020-6102-8.

Vermeire, T., Jager, T., Janssen, G., Bos, P., Pieters, M. (2001) A probabilistic human health risk assessment for environmental exposure to dibutylphthalate. Journal of Human and Ecological Risk Assessment 7, 1663-1679.

6.5.2. REACH Environment

Introduction

REACH establishes procedures for collecting and assessing information on the properties, hazards and risks of substances. At quantities of > 10 tonnes, the manufacturers, importers, and down-stream users must show that their substances do not adversely affect human health or the environment for the uses and operational conditions registered. The amount of standard information required to show safe use depends on the quantity of the substance that is manufactured or imported. This section explains how risks to the environment are assessed in REACH.

Data Requirements

As a minimum requirement, all substances manufactured or imported in quantities of 1 tonne or more need to be tested in acute toxicity tests on Daphnia and algae, while also information should be provided on biodegradability (Table 1). Physical-chemical properties relevant for environmental fate assessment that have to be provided at this tonnage level are water solubility, vapour pressure and octanol-water partition coefficient. At 10 tonnes or more, this should be supplemented with an acute toxicity test on fish and an activated sludge respiration inhibition test. At this tonnage level, also an adsorption/desorption screening and a hydrolysis test should be performed. If the chemical safety assessment, performed at 100 tonnes or more in case a substance is classified based on hazard information, indicates the need to investigate further the effects on aquatic organisms, the chronic toxicity on these aquatic species should be determined. If the substance has a high potential for bioaccumulation (for instance a log $_{Kow}$ > 3), also the bioaccumulation in aquatic species should be determined. The registrant should also determine the acute toxicity to terrestrial species or, in the absence of these data, consider the equilibrium partitioning method (EPM) to assess the hazard to soil organisms. To further investigate the fate of the substance in surface water, sediment and soil, simulation tests on its degradation should be conducted and when needed further information on the adsorption/desorption should be provided. At 1000 tonnes or more, chronic tests on terrestrial and sediment-living species should be conducted if further refinement of the safety assessment is needed. Before testing vertebrate animals like fish and mammals, the use of alternative methods and all other options must be considered to comply with the regulations regarding (the reduction of) animal testing.

Table 1 Required ecotoxicological and environmental fate information as defined in REACH

1-10 t/y	Acute Aquatic toxicity (invertebrates, algae) Ready biodegradability
10-100 t/y	Acute Aquatic toxicity (fish) Activated sludge respiration, inhibition Hydrolysis as a function of pH Adsorption/ desorption screening test
100-1000 t/y	Chronic Aquatic toxicity (invertebrates, fish) Bioaccumulation Surface water, soil and sediment simulation (degradation) test Acute terrestrial toxicity Further information on adsorption/desorption
≥ 1000 t/y	Further fate and behaviour in the environment of the substance and/or degradation products Chronic terrestrial toxicity Sediment toxicity Avian toxicity

Safety Assessment

For substances that are classified based on hazard information the registrant should assess the environmental safety of a substance, by comparing the predicted environmental concentration (PEC) with the Predicted No Effect Concentration (PNEC), resulting in a Risk Characterisation Ratio (RCR=PEC/PNEC). The use of the substance is considered to be safe when the RCR <1.

Chapter 16 of the ECHA guidance offers methods to estimate the PEC based on the tonnage, use and operational conditions, standardised through a set of use descriptors, particularly the Environmental Release Categories (ERCs). These ERCs are linked to conservative default release factors to be used as a starting point for a first tier environmental exposure assessment. When substances are emitted via waste water, the physical-chemical and fate properties of the chemical substance are then used to predict its behaviour in the Wastewater Treatment Plant (WWTP). Subsequently, the release of treated wastewater is used to estimate the concentration in fresh and marine surface water. The concentration in sediment is estimated from the PEC in water and experimental or estimated sediment-water partitioning coefficient (Kpsed). Soil concentrations are estimated from deposition from air and the application of sludge from an WWTP. The guidance offers default values for all relevant parameters, thus a generic local PEC can be calculated and considered applicable to all local emissions in Europe, although the default values can be adapted to specific conditions if justified. The local risk for wide-dispersive uses (e.g. from consumers or small, non- industrial companies) is estimated for a default WWTP serving 10,000 inhabitants.

In addition, a regional assessment is conducted for a standard area, a region represented by a typical densely populated EU area located in Western Europe (i.e. about 20 million inhabitants, distributed in a 200 x 200 km2 area). For calculating the regional PECs, a multi-media fate-modelling approach is used (e.g. the SimpleBox model; see Section on Multicompartment fate modelling). All releases to each environmental compartment for each use, assumed to constitute a constant and continuous flux, are summed and averaged over the year and steady-state concentrations in the environmental compartments are calculated. The regional concentrations are used as background concentrations in the calculation of the local concentrations.

The PNEC is calculated using the lowest toxicity value and an assessment factor (AF) related to the amount of information (see section Setting safe standards or chapter 10 of the REACH guidance . If only the minimum set of aquatic acute toxicity data is available, i.e. LC50s or EC50s for algae, daphnia and fish, a default value of 1000 is used. When one, two or three or more long-term tests are available, a default AF of 100, 50 and 10 is applied to No Observed Effect Concentrations (NOECs), respectively. The idea behind lowering the AF when more data become available is that the amount of uncertainty around the PNEC is being reduced.

In the absence of ecotoxicological data for soil and/or sediment-dwelling organisms, the PNECsoil and/or PNECsed may be provisionally calculated using the EPM. This method uses the PNECwater for aquatic organisms and the suspended matter/water partitioning coefficient as inputs. For substances with a log Kow >5 (or with a corresponding log Kp value), the PEC/PNEC ratio resulting from the EPM is increased by a factor of 10 to take into account possible uptake through the ingestion of sediment. If the PEC/PNEC is greater than 1 a sediment test must be conducted. If one, two or three long-term No Observed Effect Concentrations (NOECs) from sediment invertebrate species representing different living and feeding conditions are available, the PNEC can be derived using default AFs of 100, 50 and 10, respectively.

For data rich chemicals, the PNEC can be derived using Species Sensitivity Distributions (SSD) or other higher-tier approaches.

6.5.3. Pesticides (EFSA)

under review

6.5.4. Environmental Risk Assessment of Pharmaceuticals in Europe

Introduction

Pharmaceuticals are a crucial element of modern medicine and confer significant benefits to society. About 4,000 active pharmaceutical ingredients are being administered worldwide in prescription medicines, over-the-counter medicines, and veterinary medicines. They are designed to be efficacious and stable, as they need to pass different barriers i.e. skin, the gastrointestinal system (GIT), or even the blood-brain barrier before reaching the target cells. Each target system has a different pH and different lipophilicity and the GIT is in addition colonised with specific bacteria, specialized to digest, dissolve and disintegrate organic molecules. As a consequence of this stability, most of the pharmaceutical ingredients are stable in the environment as well and could cause effects on non-target organisms.

The active ingredients comprise a variety of synthetic chemicals produced by pharmaceutical companies in both the industrialized and the developing world at a rate of 100,000 tons per year.

While pharmaceuticals are stringently regulated in terms of efficacy and safety for patients, as well as for target animal safety, user and consumer safety, the potential effects on non-target organisms and environmental effects are regulated comparably weakly.

The authorisation procedure requires an environmental risk assessment (ERA) to be submitted by the applicants for each new human and veterinary medicinal product. The assessment encompasses the fate and behaviour of the active ingredient in the environment and its ecotoxicity based on a catalogue of standardised test guidelines.

In the case of veterinary pharmaceuticals, constraints to reduce risk and thus ensure safe usage can be stipulated in most cases. In the case of human pharmaceuticals, it is far more difficult to ensure risk reduction through restriction of the drug's use due to practical and ethical reasons. Because of their unique benefits, a restriction is not reasonable. This is reflected in the legal framework, as a potential effect on the environment is not included in the final benefit risk assessment for a marketing authorisation.

Exposure Pathways

Human Pharmaceuticals

Human pharmaceuticals enter the environment mainly via surface waters through sewage systems and sewage treatment plants. The main exposure pathways are excretion and non-appropriate disposal. Typically only a fraction of the medicinal product taken is metabolised by the patients, meaning that the main share of the active ingredient is excreted unchanged into the wastewater system. Furthermore, sometimes the metabolites themselves are pharmacologically active. Yet, no wastewater treatment plant is able to degrade all active ingredients. So medicinal products are commonly found in surface water, to some extent in ground water, and sometimes even in drinking water. However, the concentrations in drinking water are orders of magnitude lower than therapeutic concentrations. An additional exposure pathway for human pharmaceuticals is the spreading of sewage sludge on soil, if the sludge is used as fertilizer on farmland. See for more details the Link "The Drugs We Wash Away: Pharmaceuticals, Drinking Water and the Environment".

Veterinary Pharmaceuticals

Veterinary pharmaceuticals on the other hand enter the environment mainly via soil, either indirectly, if the slurry and manure from mass livestock production is spread onto agricultural land as fertiliser, or directly from pasture animals. Moreover, pasture animals might additionally excrete directly into surface water. Pharmaceuticals can also enter the environment via the detour of manure used in biogas plants.

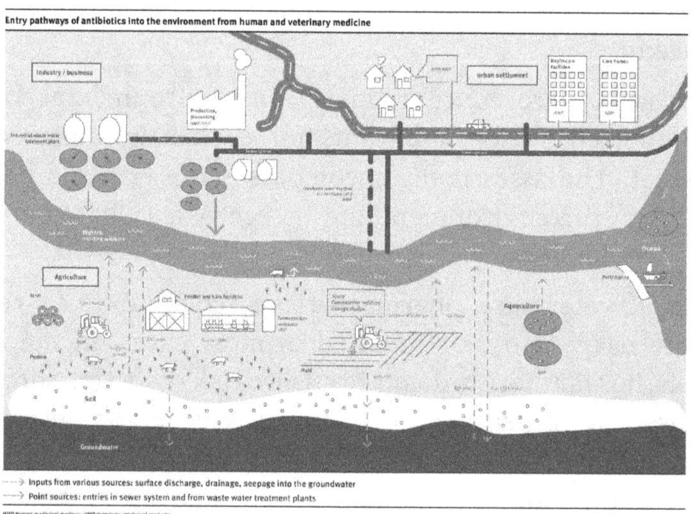

Figure 1: Entry path paths of human and veterinary medicinal products. See text for more details (reproduced with permission from the German Environmental Agency).

Assessment Schemes

Despite the differences mentioned above, the general scheme of the environmental risk assessment of human and veterinary pharmaceuticals is similar. Both assessments start with an exposure assessment. Only if specific trigger values are reached an in-depth assessment of fate, behaviour and effects of the active ingredient is necessary.

Environmental Risk Assessment of Human Pharmaceuticals

In Europe, an ERA for human pharmaceuticals has to be conducted according to the Guideline on Environmental Risk Assessment of Medicinal Products for Human Use (EMA 2006). This ERA consists of two phases. Phase I is a pre-screening for estimating the exposure in surface water, and if this Predicted Environmental Exposure Concentration (PEC) does not reach the action limit of 0.01 μg/L, in most cases, the ERA can stop. In case this action limit is reached or exceeded, a base set of aquatic toxicology, and fate and behaviour data need to be supplied in phase II Tier A. A risk assessment, comparing the PEC with the Predicted No Effect Concentration (PNEC), needs to be conducted. If in this step a risk is still identified for a specific compartment, a substance and compartment-specific refinement and risk assessment in Phase II Tier B needs to be conducted (Figure 2).

Phase I: Estimation of Exposure

In Phase I, the PEC calculation is restricted to the aquatic compartment. The estimation should be based on the drug substance only, irrespective of its route of administration, pharmaceutical form, metabolism and excretion. The initial calculation of the PEC in surface water assumes:

- The predicted amount used per capita per year is evenly distributed over the year and throughout the geographic area (Doseai);
- A fraction of the overall market penetration (Fpen), in other words 'how many people will take the medicinal product? Normally a default value of 1% is used;
- The sewage system is the drug's main route of entry into surface water.

The following formula is used to estimate the PEC in surface water:

$$PEC_{surfacewater} = \frac{(DOSE_{ai} * F_{pen})}{(WASTE_{inhab} * DILUTION)}$$

$PEC_{surfacewater}$ = mg/l

$DOSE_{ai}$ = Maximum daily dose consumed per capita [$mg.inh^{-1}.d^{-1}$]

F_{pen} = Fraction of market penetration (= 1% by default)

$WASTE_{inhab}$ = Amount of wastewater per inhabitant per day (= 200 l by default)

DILUTION = Dilution Factor (= 10 by default)

Three factors of this formula, i.e. F_{pen}, $Waste_{inhab}$ and the Dilution Factor, are default values, meaning that the PECsurfacewater in Phase I entirely depends on the dose of the active ingredient. The Fpen can be refined by providing reasonably justified market penetration data, e.g. based on published epidemiological data.

If the PECsurfacewater value is equal to or above 0.01 μg/l (mean dose ≥ 2 mg cap^{-1} d^{-1}), a Phase II environmental fate and effect analysis should be performed. Otherwise, the ERA can stop. However, in some cases, the action limit may not be applicable. For instance, medicinal substances with a $\log_{Kow} > 4.5$ are potential PBT candidates and should be screened for persistence (P), bioaccumulation potential (B), and toxicity (T) independently of the PEC value. Furthermore, some substances may affect vertebrates or lower animals at concentrations lower than 0.01 μg/L. These substances should always enter Phase II and a tailored risk assessment strategy should be followed which addresses the specific mechanism of action of the substance. This is often true for e.g. hormone active substances (see section on Endocrine disruption). The required tests in a Phase II assessment (see below) need to cover the most sensitive life stage, and the most sensitive endpoint needs to be assessed. This means for instance that for substances affecting reproduction, the organism needs to be exposed to the substance during gonad development and the reproductive output needs to be assessed.

Phase II: Environmental Fate and Effects Analysis

A Phase II assessment is conducted by evaluating the PEC/PNEC ratio based on a base set of data and the predicted environmental concentration from Tier A. If a potential environmental impact is indicated, further testing might be needed to refine PEC and PNEC values in Tier B.

Under certain circumstances, effects on sediment-dwelling organisms and terrestrial environmental fate and effects analysis are also required. Experimental studies should follow standard test protocols, e.g. OECD guidelines. It is not acceptable to use QSAR estimation, modelling and extrapolation from e.g. a substance with a similar mode of action and molecular structure (read across). This is in clear contrast to other regulations like e.g. REACH.

Human pharmaceuticals are used all year round without any major fluctuations and peaks. The only exemption are substances used against cold and influenza. These substances have a clear peak in the consumption in autumn and winter times. In developed countries in Europe and North America, antibiotics display a similar peak as they are prescribed to support the substances used against viral infections. The guideline reflects this exposure scenario and asks explicitly for long-term effect tests for all three trophic levels: algae, aquatic invertebrates and vertebrates (i.e., fish).

In order to assess the physio chemical fate, amongst other tests the sorption behaviour and fate in a water/sediment system should be determined.

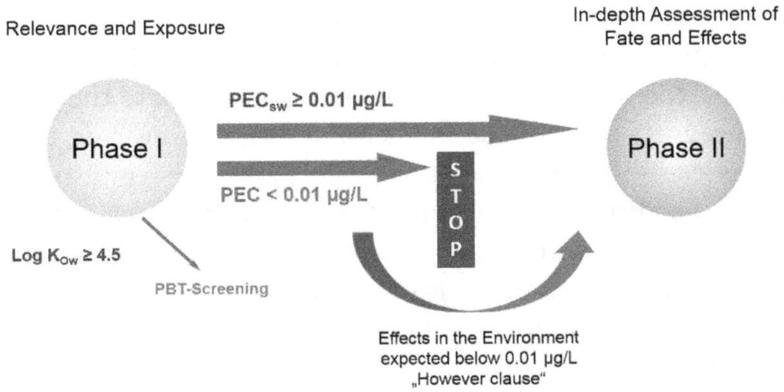

Figure 2: Scheme of conducting an ERA for Human Medicinal Products according to the EMA guideline

If, after refinement, the possibility of environmental risks cannot be excluded, precautionary and safety measures may consist of:

- An indication of potential risks presented by the medicinal product for the environment.
- Product labelling, Summary Product Characteristics (SPC), Package Leaflet (PL) for patient use, product storage and disposal.

Labelling should generally aim at minimising the quantity discharged into the environment by appropriate mitigation measures

Environmental Risk Assessment of Veterinary Pharmaceuticals

In the EU, the Environmental Risk Assessment (ERA) is conducted for all veterinary medicinal products. The structure of an ERA for Veterinary Medicinal Products (VMPs) is quite similar to the ERA for Human Medicinal Products. It is also tier based and starts with an exposure assessment in Phase I. Here, the potential for environmental exposure is assessed based on the intended use of the product. It is assumed that products with limited environmental exposure will have negligible environmental effects and thus can stop in Phase I. Some VMPs that might otherwise stop in Phase I as a result of their low environmental exposure, may require additional hazard information to address particular concerns associated with their intrinsic properties and use. This approach is comparable to the assessment of Human Pharmaceutical Products, see above.

Phase I: Estimation of Environmental Exposure

For the exposure assessment, a decision tree was developed (Figure 3). The decision tree consists of a number of questions, and the answers of the individual questions will conclude in the extent of the environmental exposure of the product. The goal is to determine if environmental exposure is sufficiently significant to consider if data on hazard properties are needed for characterizing a risk. Products with a low environmental exposure are considered not to pose a risk to the environment and hence these products do not need further assessment. However, if the outcome of Phase I assessment is that the use of the product leads to significant environmental exposure, then additional environmental fate and effect data are required. Examples for products with a low environmental exposure are, among others are products for companion animals only and products that result in a Predicted Environmental Concentration in soil (PECsoil) of less than 100 µg/kg, based on a worst-case estimation.

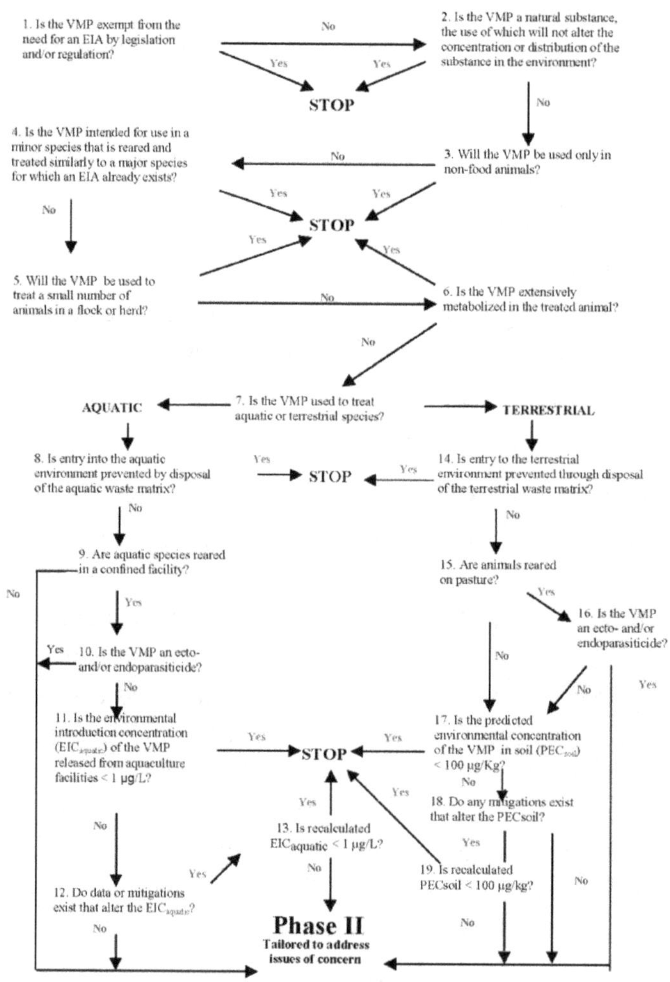

Figure 3: Phase I Decision Tree for Veterinary Medicinal Products (VMPs); (VICH 2000)

Phase II: Environmental Fate and Effects Analysis

A Phase II assessment is necessary if either the trigger of 100 μg/kg in the terrestrial branch or the trigger of 1 μg/L in the aquatic branch is reached. It is also necessary, if the substance is a parasiticide for food producing animals. A Phase II is also required for substances that would in principle stop in Phase I, but there are indications that an environmental risk at very low concentrations is likely due to their hazardous profile (e.g., endocrine active medicinal products). This is comparable to the assessment for Human Pharmaceutical Products.

For Veterinary Pharmaceutical Products also the Phase II assessment is subdivided into several Tiers, see Figure 4. For Tier A, a base set of studies assessing the physical-chemical properties, the environmental fate, and effects of the active ingredient is necessary. For Tier A, acute effect tests are suggested, assuming a more peak like exposure scenario due to e.g. applying manure and dung on fields and meadows, in contrast to the permanent exposure of human pharmaceuticals. If for a specific trophic level, e.g. dung fauna or algae, a risk is identified (PEC/PNEC ≥1) (see Introduction to Chapter 6), long-term tests for this level have to be conducted in Tier B. For the trophic levels, without an identified risk, the assessment can stop. If the risk still applies with these long-term studies, a further refinement with field studies in Tier C can be conducted. Here a co-operation with a competent authority is strongly recommended, as these tests are tailored, reflected by the individual design of these field studies. In addition, and independent of this, risk mitigation measures can be imposed to reduce the exposure concentration (PEC). These can be, beside others, that animals must remain stabled for a certain amount of time after the treatment, to ensure that the concentration of active ingredient in excreta is low enough to avoid adverse effects on dung fauna and their predators. Alternatively, the treated animals are denied access to water as the active ingredient has harmful effects on aquatic organisms.

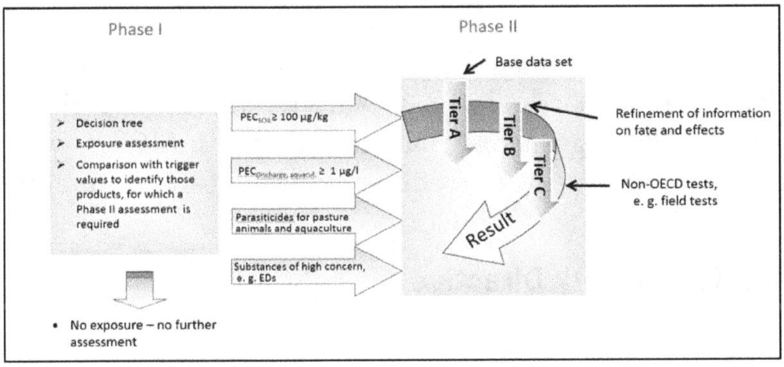

Figure 4: Scheme for conducting an ERA for Veterinary Medicinal Products (VMPs) according to the EMA guidelines (VICH 2000; VICH 2004).

Conclusion

The Environmental Risk Assessment of Human and Veterinary Medicinal Products is a straightforward, tiered-based process with the possibility to exit at several steps in the assessment procedure. Depending on the dose, the physico-chemical properties, and the anticipated use, this can be quite early in the procedure. On the other hand, for very potent substances with specific modes of action the guidelines are flexible enough to allow specific assessments covering these modes of action.

The ERA guideline for human medicinal products entered into application 2006 and many data gaps exist for products approved prior to 2006. Although there is a legal requirement for an ERA dossier for all marketing authorisation applications, new applications for pharmaceuticals on the market before 2006 are only required to submit ERA data under certain circumstances (e.g. significant increase in usage). Even for some of the blockbusters, like Ibuprofen, Diclofenac, and Metformin, full information on fate, behaviour and effects on non-target organisms is currently lacking. Furthermore, systematic post-authorisation monitoring and evaluation of potential unintended ecotoxicological effects does not exist. The market authorisation for pharmaceuticals does not expire, in contrast to e.g. an authorisation of pesticides, which needs to be renewed every 10 years.

For Veterinary Medicinal Products, an in-depth ERA is necessary for food producing animals only. An ERA for non-food animals can stop with question 3 in Phase I (Figure 3) as it is considered that the use of products for companion animals leads to negligible environmental concentrations, which might not be necessarily the case. Here, the guideline does not reflect the state of the art of scientific and regulatory knowledge. For example, the market authorisation, as a pesticide or biocide, has been withdrawn or strongly restricted for some potent insecticides like imidacloprid and fipronil which both are authorised for use in companion animals.

Further Reading

Pharmaceuticals in the Environment:https://www.umweltbundesamt.de/en/publikationen/pharmaceuticals-in-the-environment-the-global Recommendations for reducing micro-pollutants in waters: https://www.umweltbundesamt.de/publikationen/recommendations-for-reducing-micropollutants-in

6.5.5. European Water Framework Directive

Introduction

Early water legislation on the European level only began in the seventies with standards for rivers and lakes used for drinking water abstraction and aiming to control the discharge of particular harmful substances. In the early eighties, quality targets were set for drinking water, fishing waters, shellfish waters, bathing waters and groundwater. The main emission control instrument was the Dangerous Substances Directive. Within a decade, the Urban Waste Water Treatment Directive (1991), the Nitrates Directive (1991), the Drinking Water Directive (1998) and the Directive for Integrated Pollution Prevention and Control (1996) followed. Finally, on 23 October 2000, the "Directive 2000/60/EC of the European Parliament and of the Council establishing a framework for the Community action in the field of water policy" or, in short, the EU Water Framework Directive (WFD) was adopted (European Commission, 2000). The key aim of this directive is to achieve good ecological and good chemical status for all waters by 2027. This is captured in the following objectives:

- expanding the scope of water protection to all waters (not only waters intended for particular uses), surface waters and groundwater;
- achieving "good status" for all waters by a set deadline;
- water management based on river basins;
- combined approach of emission limit values and water quality standards;
- ensuring that the user bears the true costs of providing and using water;
- getting the citizen involved more closely;
- streamlining legislation.

Instead of administrative or political boundaries, the natural geographical and hydrological unit (the river basin) was set as the unit of water management. For each river basin, independent from national frontiers, a river basin management plan needs to be established and updated every six years. Herein, the general protection of ecological water quality, specific protection of unique and valuable habitats, protection of drinking water resources, and protection of bathing water are integrated, assessed and, where necessary, translated into an action plan. Basically, the key requirement of the Directive is that the environment as an entity is protected to a high level, in other words the protection of the ecological integrity applies to all waters. Within five months after the WFD came into force, the Common Implementation Strategy (CIS) was established. The CIS includes, for instance, guidance documents on technical aspects, key events and additional resource documents related to different aspects of the implementation. The links at the end of this chapter provide access to these additional key documents.

WFD Methodology

The WFD integrated approach to manage water bodies consists of three key components of aquatic ecosystems: water quality, water quantity, and physical structure. Furthermore, it implies that the ecological status of water bodies must be determined with respect to the near-natural reference conditions, which represent a 'high ecological status'. The determination of the 'good ecological status' (Figure 1) is based on the quality of the biological community, the hydrological characteristics and the chemical characteristics that may slightly deviate from these reference conditions. To describe reference conditions, a typology of water bodies is needed. In the WFD, water bodies are categorized as rivers, lakes, transitional or coastal waters. Within each of these categories, the type of water body must be differentiated, based on an ecoregion approach in combination with at least three additional factors: altitude, catchment area and geology (and depth for lakes). The objective of the typology is to ensure that type-specific biological reference conditions can be determined. Furthermore, it may be necessary or is allowed to use additional descriptors (called optional descriptors) to achieve sufficient differentiation. Waters in each category can be classified as natural, heavily modified or artificial dependent on their origin and human-induced changes.

The WFD requires to include ecological status and chemical status classification schemes for surface water bodies that differ for the four major water types, i.e. rivers, lakes, transitional waters and coastal waters. Rivers and lakes are assessed in relation to their ecological and chemical reference status, and heavily modified and artificial water bodies in relation to their ecological potential and chemical status. The classification schemes of ecological status and potential make use of several quality elements (QEs; Annex V):

- Biological quality elements; composition and abundance of algae, macroalgae, higher water plants, benthic invertebrates, and fish;
- Hydro-morphological quality elements (e.g. water flow, substrate, river morphology);
- General physicochemical quality elements (e.g. nutrients, chloride, oxygen condition, temperature, transparency, salinity, river basin-specific pollutants);
- Environmental Quality Standards (EQSs) for synthetic and non-synthetic pollutants.

For the ecological status and ecological potential classification schemes, the Directive provides normative definitions of the degree of human disturbance to each relevant quality element that is consistent with each of the classes for (potential) ecological status (Figure 1). These definitions have been expanded and used in the development of classification tools (assessment systems) and appropriate numeric class boundaries for each quality element. The results of applying these classification tools or assessment systems are used to determine the status (quality class) of each water body or group of water bodies.

Once reference conditions are established, the departure from these can be measured. Boundaries have been defined for the degree of deviation from the reference conditions for each of the WFD ecological status classes. Annex V 1.4.1 of the Directive states: "the results of the (classification) system shall be expressed as ecological quality ratios (EQRs) for the purposes of classification of ecological status. These ratios shall represent the relationship between the values of the biological parameters observed for a given body of surface water and the values for these parameters in the reference conditions applicable to that body. The ratio shall be expressed as a numerical value between zero and one, with high ecological status represented by values close to one and bad ecological status by values close to zero." (Figure 1). Boundaries are thus the values that separate the 5 classes.

The reference conditions form the anchor point for the whole ecological assessment. The outcome or score of all WFD Quality Elements is combined to inform the overall quality classification of a water body. Hereby, the one-out-all-out principle applies, meaning that the lowest score for an individual Quality Element decides the final score.

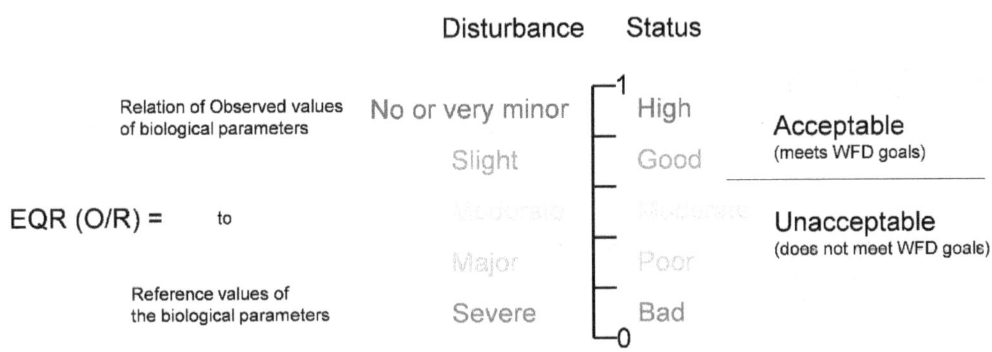

Figure 1. Ecological Quality Ratio (altered after Vincent et al., 2002).

Priority Substances

The values of the environmental quality standards (values for specific pollutants) are set to ensure that the water body is not exposed to acute and chronic toxicity for aquatic organisms, no accumulation in the ecosystem and losses of habitats and biodiversity occurs, as well as there is no threat to human health. Substances that were identified to present significant risks to or via the aquatic environment are listed as priority substances. According to the Directive on Environmental Quality Standards (Directive 2008/105/EC) good chemical status is reached for a water body when it complies with the Environmental Quality Standard (EQS) for all priority substances and eight other pollutants that are not in the priority substances list. The EQSs define a limit on the concentrations of 33 priority substances, 13 of them designated as priority hazardous substances in surface waters. These concentration limits are derived following the methodologies explained in Section 6.3.4.

Furthermore, the Directive on EQSs offers the possibility of applying EQSs for sediment and biota, instead of those for water. It opens the possibility of designating mixing zones adjacent to discharge points where concentrations of the priority substances might be expected to exceed their EQS. Furthermore, authorities can add basin or catchment specific EQSs.

Environmental Quality Standards (EQSs) can be expressed as a maximum allowable concentration (EQS-MAC) or as an annual average value (EQS-AA). For all priority substances, Member States need to establish an inventory of emissions, discharges and losses. To improve the legislation, the EU also 1) introduced biota standards for several substances, 2) provided improvements on the efficiency of monitoring and the clarity of reporting with regard to certain substances behaving as ubiquitous persistent, bioaccumulative and toxic (PBT) substances, and 3) added a watch-list mechanism designed to allow targeted EU-wide monitoring of substances of possible concern to support the prioritization process in future reviews of the priority substances list.

Status Classification

Together the classification of surface water bodies follows the scheme provided in Figure 2.

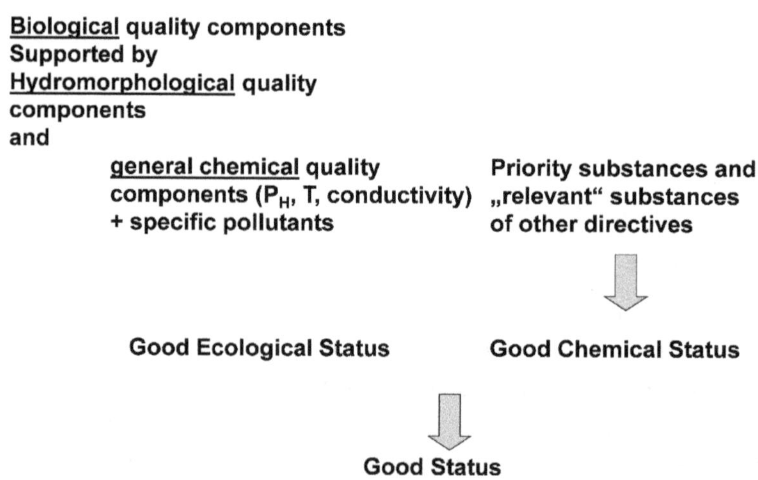

Figure 2. Elements of 'good status' of surface waters.

In summary, under the WFD, the ecological quality status assessment of surface water bodies is primarily based on biological quality elements phytoplankton, fish, and benthic flora and fauna. In the Netherlands, the worst Biological Quality Element score is taken as the overall final score (one-out-all-out principle). Furthermore, adequate assessment of stream and river hydro-morphology requires the consideration of any modifications to flow regime, sediment transport, river morphology, lateral channel mobility (or channel migration), and river continuity. For (groups of) substances, the WFD requires assessment of their relevance. A substance is relevant, when it exceeds its Environmental Quality Standard, meaning the boundary between good and moderate status is exceeded and a de-classification takes place when. The overall assessment follows the scheme given in Figure 3.

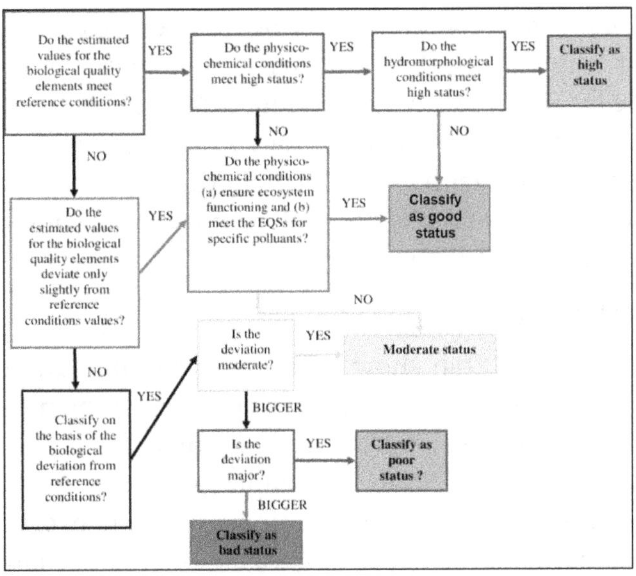

Figure 3. Decision tree for determining the ecological status of surface water bodies based on biological, hydromorphological and physicochemical quality elements according the normative definitions in Annex V: 1.2. (WFD).

References

European Commission (2000). Directive 2000/60/EC. Establishing a framework for community action in the field of water policy. European Commission PE-CONS 3639/1/100 Rev 1, Luxembourg.

Vincent, C., Heinrich, H., Edwards, A., Nygaard, K., Haythornthwaite, J. (2002). Guidance on typology, reference conditions and classification systems for transitional and coastal waters. CIS working group, 2, 119.

6.5.6. Policy on Soil and Groundwater Regulation

History

As a bomb hit, soil contamination came onto the political agenda in the United States and in Europe through a number of disasters in the late 1970s and early 1980s. Starting point was the 1978 Love Canal disaster in upper New York State, USA, in which a school and a number of residences had been built on a former landfill for chemical waste disposal with thousands of tonnes of dangerous chemical wastes, and became a national media event. In Europe in 1979, the residential site of Lekkerkerk in the Netherlands became an infamous national event. Again, a residential area had been built on a former waste dump, which included chemical waste from the painting industry, and with channels and ditches that had been filled in with chemical waste-containing materials.

Since these events, soil contamination-related policies emerged one after the other in different countries in the world. Crucial elements of these policies were a benchmark date for a ban on bringing pollutants in or on the soil ('prevention'), including a strict policy, e.g. duty of care, for contaminations that are caused after the benchmark date, financial liability for polluting activities, tools for assessing the quality of soil and groundwater, and management solutions (remediation technologies and facilities for disposal).

Evolution in Soil Policies

Objectives in soil policies often show evolution over time and changes go along with developing new concepts and approaches for implementing policies. In general, soil policies often develop from a maximum risk control until a functional approach. The corresponding tools for implementation usually develop from a set of screening values towards a systemic use of frameworks, enabling sound environmental protection while improving the cost-benefit-balance. Consequently, soil policy implementation usually goes through different stages. In general terms, four different stages can be distinguished, i.e., related to maximum risk control, to the use of screening values, to the use of frameworks and based on a functional approach. Maximum risk control follows the precaution principle and is a stringent way of assessing and managing contamination by trying to avoid any risk. Procedures based on screening values allow for a distinction in polluted and non-polluted sites for which the former, the polluted sites, require some kind of intervention. The scientific underpinning of the earliest generations of screening values was limited and expert judgement played an important role. Later, more sophisticated screening values emerged, based on risk assessment. This resulted in screening values for individual contaminants within the contaminant groups metals and metalloids, other inorganic contaminants (e.g., cyanides), polycyclic aromatic hydrocarbons (PAHs), monocyclic aromatic hydrocarbons (including BETX (benzene, toluene, xylene)), persistent organic pollutants (including PCBs and dioxins), volatile organic contaminants (including trichloroethylene, tetrachloroethylene, 1,1,1-trichloroethane, and vinyl chloride), petroleum hydrocarbons and, in a few countries only, asbestos. For some contaminants such as PAHs, sum-screening values for groups were derived in several countries, based on toxicity equivalents. In a procedure based on *frameworks*, often the same screening values generally act as a trigger for further, more detailed site-specific investigations in one or two additional assessment steps. In the *functional approach*, soil and groundwater must be suited for the land use it relates to (e.g., agricultural or residential land) and the functions (e.g., drinking water abstraction, irrigation water) it performs. Some countries skip the maximum risk control and sometimes also the screening values stages and adopt a framework and/or a functional approach.

European Collaboration and Legislation

In Europe, collaboration was strengthened by concerted actions such as CARACAS (concerted action on risk assessment for contaminated sites in the European Union; 1996 - 1998) and CLARINET (Contaminated Land Rehabilitation Network for Environmental Technologies; 1998 - 2001). These concerted actions were followed up by fruitful international networks that are still are active today. These are the Common Forum, which is a network of contaminated land policy makers, regulators and technical advisors from Environment Authorities in European Union member states and European Free Trade Association countries, and NICOLE (Network for Industrially Co-ordinated Sustainable Land Management in Europe), which is a leading forum on industrially co-ordinated sustainable land management in Europe. NICOLE is promoting co-operation between industry, academia and service providers on the development and application of sustainable technologies.

In 2000, the EU Water Framework Directive (WFD; Directive 2000/60/EC) was adopted by the European Commission, followed by the Groundwater Directive (Directive 2006/118/EC) in 2006 (European parliament and the council of the European Union, 2019b). The environmental objectives are defined by the WFD. Moreover, ‘good chemical status’ and the ‘no deterioration clause’ account for groundwater bodies. ‘Prevent and limit’ as an objective aims to control direct or indirect contaminant inputs to groundwater, and distinguishes for ‘preventing hazardous substances’ to enter groundwater as well as ‘limiting other non-hazardous substances’.

Moreover, the European Commission adopted a Soil Thematic Strategy, with soil contamination being one out of the seven identified threats. A proposal for a Soil Framework Directive, launched in 2006, with the objective to protect soils across the EU, was formally withdrawn in 2014 because of a lack of support from some countries.

Policies in the World

Today, most countries in Europe and North America, Australia and New Zealand, and several countries in Asia and Middle and South America, have regulations on soil and groundwater contamination. The policies, however, differ substantially in stage, extent and format. Some policies only cover prevention, e.g., blocking or controlling the inputs of chemicals onto the soil surface and in groundwater bodies.

Other policies cover prevention, risk based quality assessment and risk management procedures and include elaborated technical tools, which enable a decent and uniform approach. In particular the larger countries such as the USA, Germany and Spain, policies differ between states or provinces within the country. And even in countries with a policy on the federal level, the responsibilities for different steps in the soil contamination chain are very different for the different layers of authorities (at the national, regional and municipal level).

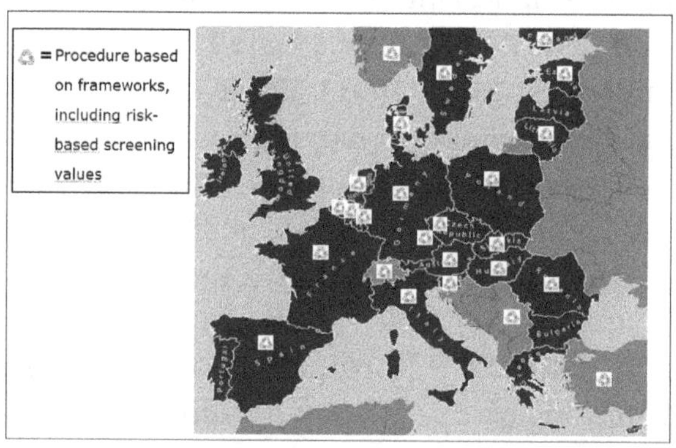

Figure 1. European countries that have a soil policy procedure based on frameworks (see text), including risk-based screening values. Figure prepared by Frank Swartjes.

In Figure 1 the European countries are shown that have a procedure based on frameworks (as described above), including risk-based screening values. It is difficult, if not impossible, to summarise all policies on soil and groundwater protection worldwide. Alternatively, some general aspects of these policies are given here. A fair first basic element in nearly all soil and groundwater policies, relating to prevention of contamination, is the declaration of a formal point in time after which polluting soil and groundwater is considered an illegal act. For soil and groundwater quality assessment and management, most policies follow the risk-based land management approach as the ultimate form of the functional approach described above. Central in this approach are the risks for specific targets that need to be protected up to a specified level.

Different protection targets are considered. Not surprisingly, 'human health' is the primary protection target that is adopted in nearly all countries with soil and groundwater regulations. Moreover, the ecosystem is an important protection target for soil, while for groundwater the ecosystem as a protection target is under discussion. Another interesting general characteristic of mature soil and groundwater policies is the function-specific approach. The basic principle of this approach is that land must be suited for its purpose. As a consequence, the appraisal of a contaminated site in a residential area, for instance, follows a much more stringent concept than that of an industrial site.

Risk assessment tools

Risk assessment tools often form the technical backbone of policies. Since the late 1980s risk assessment procedures for soil and groundwater quality appraisal were developed. In the late 1980s the exposure model CalTOX was developed by the Californian Department of Toxic Substances Control in the USA, a few years later the CSOIL model in the Netherlands (Van den Berg, 1991/ 1994/ 1995). In Figure 2, the flow chart of the Dutch CSOIL exposure model is given as an example. Three elements are recognized in CSOIL, like in most exposure models: (1) contaminant distribution over the soil compartments; (2) contaminant transfer from (the different compartments of) the soil into contact media; and (3) direct and indirect exposure to humans. The major exposure pathways are exposure through soil ingestion, crop consumption and inhalation of indoor vapours (Elert et al., 2011). Today, several exposure models exist (see Figure 3 for some 'national' European exposure models). However, these exposure models may give quite different exposure estimates for the same exposure scenario (Swartjes, 2007).

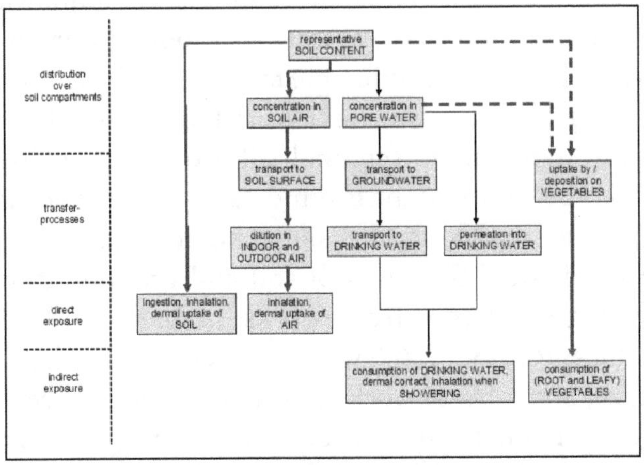

Figure 2. Flow chart of the Dutch CSOIL exposure model.

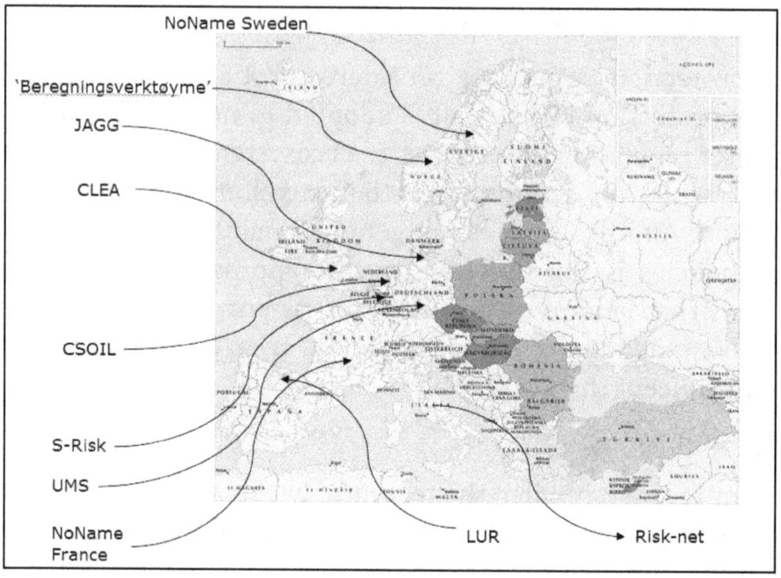

Figure 3. Some 'national' European soil exposure models, projected on the country in which they are used. Figure prepared by Frank Swartjes.

Moreover, procedures were developed for ecological risk assessment, including the Species Sensitivity Distributions (see section on SSDs), based on empirical relations between concentration in soil or groundwater and the percentage of species or ecological processes that experience adverse effects (PAF: potentially Affected Fraction). For site specific risk assessment, the TRIAD approach was developed, based on three lines of evidence, i.e., chemically-based, toxicity-based and using data from ecological field surveys (see section on the TRIAD approach).

In the framework of the HERACLES network, another attempt was made to summarizing different EU policies on polluted soil and groundwater. A strong plea was made for harmonisation of risk assessment tools (Swartjes et al., 2009). The authors also described a procedure for harmonization based on the development of a toolbox with standardized and flexible risk assessment tools. Flexible tools are meant to cover national or regional differences in cultural, climatic and geological (e.g., soil type, depth of the groundwater table) conditions. It is generally acknowledged, however, that policy decisions should be taken on the national level. In 2007, an analysis of the differences of soil and groundwater screening values and of the underlying regulatory frameworks, human health and ecological risk assessment procedures (Carlon, 2007) was launched. Although screening values are difficult to compare, since frameworks and objectives of screening values differ significantly, a general conclusion can be drawn for e.g. the screening values at the potentially unacceptable risk level (often used as *'action' values*, i.e. values that trigger further research or intervention when exceeded). For the 20 metals, most soil screening values (from 13 countries or regions) show between a factor of 10 and 100 difference between the lowest and highest values. For the 23 organic pollutants considered, most soil screening values (from 15 countries or regions) differ by a factor of between 100 and 1000, but for some organic pollutants these screening values differ by more than four orders of magnitude. These conclusions are merely relevant from a policy viewpoint. Technically, these conclusions are less relevant, since, the screenings values are derived from a combination of different protection targets and tools and based on different policy decisions. Differences in screening values are explained by differences in geographical and biological and socio-cultural factors in different countries and regions, different national regulatory and policy decisions and variability in scientific/ technical tools.

References

Carlon, C. (Ed.) (2007). Derivation methods of soil screening values in Europe. A review and evaluation of national procedures towards harmonisation, JRC Scientific and Technical report EUR 22805 EN.

Elert, M., Bonnard, R., Jones, C., Schoof, R.A., Swartjes, F.A. (2011). Human Exposure Pathways. Chapter 11 in: Swartjes, F.A. (Ed.), Dealing with Contaminated Sites. From theory towards practical application. Springer Publishers, Dordrecht.

Swartjes, F.A. (2007). Insight into the variation in calculated human exposure to soil contaminants using seven different European models. Integrated Environmental Assessment and Management 3, 322–332.

Swartjes, F.A., D'Allesandro, M., Cornelis, Ch., Wcislo, E., Müller, D., Hazebrouck, B., Jones, C., Nathanail, C.P. (2009). Towards consistency in risk assessment tools for contaminated sites management in the EU. The HERACLES strategy from the end of 2009 onwards. National Institute for Public Health and the Environment (RIVM), Bilthoven, The Netherlands, RIVM Report 711701091.

Van den Berg, R. (1991/1994/1995). Exposure of humans to soil contamination. A quantitative and qualitative analyses towards proposals for human toxicological C-quality standards (revised version of the 1991/ 1994 reports). National Institute for Public Health and the Environment (RIVM), Bilthoven, The Netherlands, RIVM-report no. 725201011.

Further Reading

Swartjes, F.A. (Ed.) (2011). Dealing with Contaminated Sites. From theory towards practical application. Springer Publishers, Dordrecht.

Rodríguez-Eugenio, N., McLaughlin, M., Pennock, D. (2018). Soil Pollution: a hidden reality. Rome, FAO.

6.5.7. Drinking Water

In preparation

6.6. RISK MANAGEMENT AND RISK COMMUNICATION

In preparation

6.7. RISK PERCEPTION

Introduction

If risk perception had a first law like toxicology has with Paracelsus' "Sola dosis facit venenum" (see section on History of Environmental Toxicology) it would be:

"People fear things that do not make them sick and get sick from things they do not fear."

People can, for instance, worry extremely over a newly discovered soil pollution site in their neighborhood, which they hear about at a public meeting they have come to with their diesel car, and, when returning home, light an extra cigarette without thinking to relieve the stress.

Sources: *Left: File photo: Kalamazoo City Commissioner Don Cooney and Kalamazoo residents march to protest capping the Allied Paper Landfill, May 2013. https://www.wmuk.org/post/will-bioremediation-work-allied-microbial-ecologist-skeptical. Right: https://desmotivaciones.es/352654/Obama*

The explanation for this first law is quite easy. The annual risk of getting sick, being injured or to die has only limited influence on the perception of a risk. Other factors are more important. Figure 1 is a model of risk perception in its most basic form.

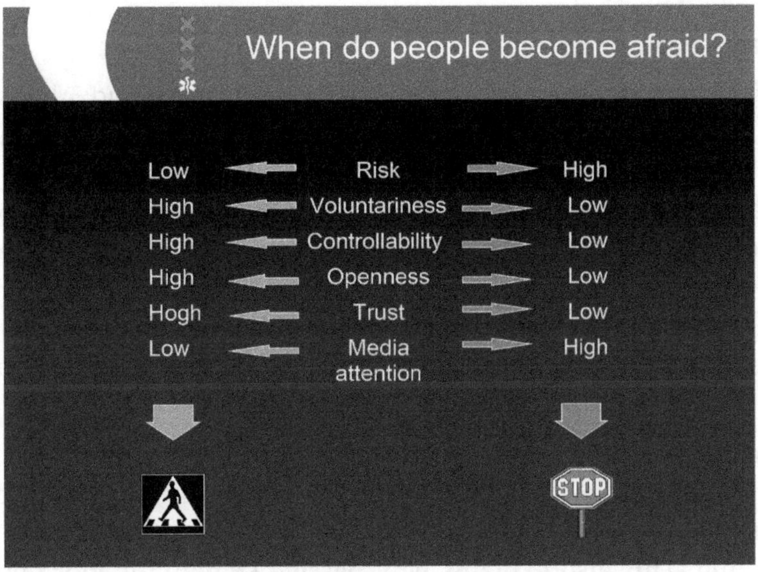

Figure 1. Simplified model of risk perception

In the middle of this figure, there is a list with several factors which determine risk perception to a large extent. In any given situation, they can each end up at the left, safe side or on the right, dangerous side. The model is a simplification. Research since the late sixties of the previous century has led to a collection of many more factors that often are connected to each other (for lectures of some well-known researchers see examples 1, 2, 3, 4 and 5).

Why do People Fear Soil Pollution?

An example can show this interconnection and the discrepancy between the annual health risks (at the top of Figure 1) and other factors. The risk of harmful health effects for people living on polluted soil is often very small. The factor 'risk' thus ends at the left, safe side. Most of the other factors end up at the right. People do not voluntary choose to have a soil pollution in their garden. They have absolutely no control over the situation and an eventual sanitation. For this, they depend on authorities and companies. Nowadays, trust in authorities and companies is very small. Many people will suspect that these authorities care more about their money than about the health and well-being of their citizens and neighbours. A newly discovered soil pollution will get local media attention and this will certainly be the case if there is controversy. If the untrusted authorities share their conclusion that the risks are low, many people will suspect that they withhold information and are not completely open. Especially saying that there is 'no cause for alarm' will only make people worry more (see a funny example). People will not believe the conclusion of authorities that the risk is low, so effectively all factors end up at the dangerous side.

Why Smokers are not Afraid

For smoking a cigarette, the evaluation is the other way around. Almost everybody knows that smoking is dangerous, but people light their cigarette themselves. Most people at least think they have control over their smoking habit as they can decide to stop at any moment (but being addicted, they probably highly overestimate their level of control). For their information or taking measures, people do not depend on others and no information is withheld about smoking. Some smokers suffer from what is called optimistic bias, the idea that misery only happens to other. They always have the example of their grandfather who started smoking at 12 and still ran the marathon at 85.

People can be upset if they learn that cigarette companies purposely make cigarettes more addictive. It makes them feel the company takes over control which people greatly resent. This, and not the health effects, can make people decide to quit smoking. This also explains why passive smoking is more effective than active smoking in influencing people's perceptions. Although the risk of passive smoking is 100 times smaller than the risk of active smoking, most factors end up at the right, dangerous side, making passive smoking maybe 100 times more objectionable and worrisome than active smoking.

Experts React like Lay People at Home

Many people are surprised to find out that the calculated or estimated health risks influences risk perception so little. But we experience it in our own daily lives, especially when we add another factor in the model: advantages.

All of us perform risky activities because they are necessary, come with advantages or sometimes out of sheer fun. Most of us take part in daily traffic with an annual risk of dying far higher than 1 in a million. Once, twice or even more times a year we go on a holiday with a multitude of risks: transport, microbes, robbery, divorce. The thrill seekers of us go diving, mountain climbing, parachute jumping without even knowing the annual fatality rates. If the stakes are high, people can knowingly risk their life in order to improve it, as the thousands of immigrants trying to cross the Mediterranean illustrate, or even to give their life for a higher cause, like soldiers at war (Winston Churchill in 1940: "I have nothing to offer but blood, toil, tears and sweat").

An example at the other side can show it maybe even clearer. No matter how small the risk, it can be totally unacceptable and nonsensical. Suppose the government starts a new lottery with an extremely small chance of winning, say one in a billion. Every citizen must play and tickets are free. So far nothing strange, but there is a twitch. The main and only price of the lottery is a public execution broadcasted live on national TV. The government will probably not make itself very popular with this absurd lottery. When government, as still is done, tells people they have to accept a small risk because they accept larger risks of activities they choose themselves, it makes people feel they have been given a ticket in the above mentioned lottery. This is how people can feel if the government tells them the risk of the polluted soil they live on is extremely small and that it would be wiser for them to quit smoking.

All Risks have a Context

A main lesson which can be learned from the study of risk perception is that risks always occur in a context. A risk is always part of a situation or activity which has many more characteristics than only the chance of getting sick, being injured or to die. We do not judge risks, we judge situations and activities of which the risk is often only a small part. Risk perception occurs in a rich environment. After 50 years of research a lot has been discovered, but predicting how angry of afraid people will be in a new, unknown situation is still a daunting task.

INDEX

D

E

F

G

H

www.ingramcontent.com/pod-product-compliance
Lightning Source LLC
LaVergne TN
LVHW062115200926
848484LV00007B/1678
* 9 7 8 1 7 8 7 1 5 2 0 4 5 *